AF616691

Food Allergies in Modern Life

Food Allergies in Modern Life

Editors

Sara Manti
Gian Luigi Marseglia
Salvatore Leonardi

MDPI • Basel • Beijing • Wuhan • Barcelona • Belgrade • Manchester • Tokyo • Cluj • Tianjin

Editors
Sara Manti
University of Catania
Italy

Gian Luigi Marseglia
University of Pavia
Italy

Salvatore Leonardi
University of Catania
Italy

Editorial Office
MDPI
St. Alban-Anlage 66
4052 Basel, Switzerland

This is a reprint of articles from the Special Issue published online in the open access journal *Nutrients* (ISSN 2072-6643) (available at: https://www.mdpi.com/journal/nutrients/special_issues/food_allergies_in_modern_life).

For citation purposes, cite each article independently as indicated on the article page online and as indicated below:

LastName, A.A.; LastName, B.B.; LastName, C.C. Article Title. *Journal Name* **Year**, *Volume Number*, Page Range.

ISBN 978-3-0365-5167-8 (Hbk)
ISBN 978-3-0365-5168-5 (PDF)

Contents

About the Editors

Sara Manti

MD, Pediatrician, PhD, University Resercher at Pediatric Unit, Dept. of Human and Pediatric Pathology "Gaetano Barresi", University of Messina, Messina, Italy. Her research experience is focused on Paediatrics, Allergology, Pulmonology. She is on the Editorial Board of national and international Journals. She has published book chapters and articles in peer-reviewed journals. She has reviewed several papers for several peer-review journals. She has given several national lectures. She has received several national Awards and Research Grants. She serves as Executive Member of the Italian Society of Pediatric Allergology and Immunology (SIAIP).

Gian Luigi Marseglia

Full Professor of Pediatrics, Chief of Pediatric Clinic, Chief of Department of Clinical, Diagnostic and Pediatric Surgical Sciences, Chief of School of Specialization in Pediatrics, Fondazione IRCCS Policlinico San Matteo, University of Pavia. Field of interest: Paediatrics, Allergology, Immunology. He is on the Editorial Board of national and international Journals. He has published book chapters and articles in peer-reviewed journals. He has given several national lectures. He has received several national and international Awards and Research Grants. He serves as a President of the Italian Society of Pediatric Allergology and Immunology (SIAIP).

Salvatore Leonardi

Associate Professor of Pediatrics, PhD. Chief of Pediatric and Respiratory Unit, Department of Clinical and Experimental Medicine, University of Catania. He was graduated as pediatric specialist and pulmonary fellowship training at the Pediatric Department of University of Catania, Italy. He was graduated as specialist in Gastroenterology and Digestive Endoscopy in the Gastroenterology School of Specialization of the University of Catania in Italy. His research experience is focused on cystic fibrosis, pneumology and allergology in childhood. He is on the Editorial Board of Allergy Asthma Proceedings Journal, World Journal of Gastroenterology, Hepatitis Monthly, and Frontiers. He has published book chapters and 222 articles in peer-reviewed journals. He has reviewed several papers for several peer-review journals. He has given several national lectures. He has received several national Awards and Research Grants. He serves as Executive Member of the Italian Society for Pediatric Respiratory Diseases (SIMRI).

Preface to "Food Allergies in Modern Life"

Food allergy is a complex and multifactorial disease whose causes, mechanisms, and effects are not yet fully understood. Food allergy is an increasing public health problem, affecting up to 10% of children and causing a significant burden on affected patients due to dietary restrictions, fear of accidental ingestions, reduced quality of life, and risk of severe reactions. Since there is no specific food allergy treatment, the only available management for food allergy is limited to strict dietary avoidance, prompt recognition of adverse symptoms, and emergency treatment of adverse reactions.

This book aims to provide an updated overview of the causes and current management of food allergy, also reporting original data to improve gaps in knowledge, encourage the implementation of food allergy management, delineate a roadmap to safety for patients at risk for adverse reactions, and provide an update on changes in the treatment landscape for food allergy. This book also aims to provide a practical, readable reference for clinicians, pediatricians, allergists, nutritionists, scientists, and students to diagnose and manage food allergies accurately in the hospital and private practice setting. Each of the chapters can stand alone, but when placed together, all chapters show a complete picture of current food allergy research.

Sara Manti, Gian Luigi Marseglia, and Salvatore Leonardi
Editors

 nutrients

Review

Malnutrition in Eosinophilic Gastrointestinal Disorders

Martina Votto [1], Maria De Filippo [1], Francesca Olivero [1], Alessandro Raffaele [2], Emanuele Cereda [3], Mara De Amici [1,4], Giorgia Testa [1], Gian Luigi Marseglia [1,†] and Amelia Licari [1,*,†]

1 Pediatric Clinic, Department of Pediatrics, Fondazione IRCSS-Policlinico San Matteo, University of Pavia, 27100 Pavia, Italy; martinavotto@gmail.com (M.V.); maria_defilippo@hotmail.it (M.D.F.); francescaolivero31@gmail.com (F.O.); m.deamici@smatteo.pv.it (M.D.A.); giorgia_testa@hotmail.it (G.T.); gl.marseglia@smatteo.pv.it (G.L.M.)
2 Pediatric Surgery Unit, Department of Maternal and Child Health, Fondazione IRCCS Policlinico San Matteo, 27100 Pavia, Italy; a.raffaele@smatteo.pv.it
3 Clinical Nutrition and Dietetics Unit, Fondazione IRCCS Policlinico San Matteo, 27100 Pavia, Italy; e.cereda@smatteo.pv.it
4 Immuno-Allergology Laboratory of the Clinical Chemistry Unit, Fondazione IRCCS Policlinico San Matteo, 27100 Pavia, Italy
* Correspondence: amelia.licari@unipv.it
† These authors contributed equally to this work.

Abstract: Primary eosinophilic gastrointestinal disorders (EGIDs) are emerging chronic/remittent inflammatory diseases of unknown etiology, which may involve any part of the gastrointestinal (GI) tract, in the absence of secondary causes of GI eosinophilia. Eosinophilic esophagitis is the prototype of eosinophilic gastrointestinal disorders and is clinically characterized by symptoms related to esophageal inflammation and dysfunction. A few studies have assessed the nutritional status of patients with eosinophilic gastrointestinal disorders, showing conflicting results. This review summarizes the current evidence on the nutritional status of patients with EGIDs, focusing on the pediatric point of view and also speculating potential etiological mechanisms.

Keywords: children; adolescents; eosinophilic esophagitis; eosinophilic gastrointestinal disorders; growth; failure to thrive; malnutrition; undernutrition; obesity; vitamin

Citation: Votto, M.; De Filippo, M.; Olivero, F.; Raffaele, A.; Cereda, E.; De Amici, M.; Testa, G.; Marseglia, G.L.; Licari, A. Malnutrition in Eosinophilic Gastrointestinal Disorders. *Nutrients* **2021**, *13*, 128. https://doi.org/10.3390/nu13010128

Received: 30 November 2020
Accepted: 27 December 2020
Published: 31 December 2020

1. Introduction

Primary eosinophilic gastrointestinal disorders (EGIDs) are emerging chronic/remittent inflammatory diseases of unknown etiology, which may involve any part of the gastrointestinal (GI) tract, leading to eosinophilic mucosal infiltration in the absence of secondary causes of intestinal eosinophilia [1–3]. While eosinophilic esophagitis (EoE) is a well-characterized disease with established guidelines [4,5], nonesophageal EGIDs, including eosinophilic gastritis, gastroenteritis, and colitis, remain a clinical enigma [1]. Although their pathogenic mechanisms are still unknown, EGIDs seems to be commonly associated with atopy and, to a lesser extent, autoimmunity [1,2]. EoE pathogenesis has been more extensively studied, and advances concerning the genetic and environmental contributors and cellular and molecular etiology have been achieved [6]. EGIDs seem to be multifactorial diseases resulting from genetic predisposition, environmental risk factors, and intestinal dysbiosis, leading to the activation of T-helper type 2 (Th2) inflammation and impaired epithelial barrier [1,7]. To date, no studies have extensively assessed malnutrition in patients with EGIDs.

In all its forms, malnutrition includes undernutrition, inadequate intake of vitamins and/or minerals, overweight, and obesity [8]. Undernutrition is a common complication of several chronic inflammatory GI diseases, mainly coeliac disease (CD) and Crohn's disease, often associated with weight loss, failure to thrive, malabsorption, and vitamin deficiency. However, obesity and overweight are the main comorbidities of gastroesophageal reflux disease (GERD) and functional GI disorders, and are well-known risk factors of hepatic steatosis [9,10].

This review aims to summarize the current evidence on the nutritional status and malnutrition in patients with EGIDs, mainly focusing on the pediatric patients' population and highlining the lack of nutritional management algorithms.

A review of articles was performed via the online database PubMed (Table 1), following PRISMA guidelines [11]. The literature review was performed in December 2020, including all publication years. All studies that met the following criteria were included: (1) case reports, case series, and cross-sectional and cohort studies published in English in peer-reviewed journals; (2) participants were children and adult patients diagnosed with EGIDs. Potentially eligible publications were manually screened and reviewed, and nonrelevant publications were excluded (Figure 1).

Table 1. Search strategy.

PubMed: "Eosinophilic gastrointestinal disorders" AND "malnutrition." Publication date: all years.
PubMed: "Eosinophilic gastrointestinal disorders" AND "obesity." Publication date: all years.
PubMed: "Eosinophilic gastrointestinal disorders" AND "vitamin." Publication date: all years.

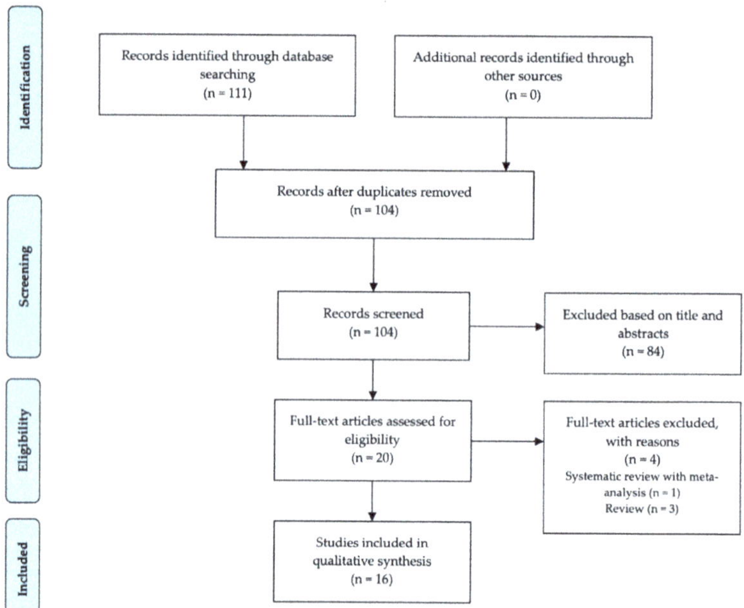

Figure 1. Process of literature screening.

2. Obese and Overweight EGID Patients

Obesity is a global public health problem associated with many chronic diseases, including type 2 diabetes, arterial hypertension, cardiovascular diseases, and asthma [12]. Growing evidence supports the association between obesity and immune disorders, such as cancer, autoimmunity, and atopy [13]. Some studies have suggested that pediatric obesity epidemy and obesity-related inflammation might at least in part be responsible for the significantly raised prevalence of allergic diseases [13]. The relationship between asthma and obesity in children is widely demonstrated, and several observational studies have reported that obese children are more frequently affected by a severe phenotype of asthma, refractory to conventional therapies [14–17]. Additionally, data from the National Health and Nutrition Examination Study III (NHANES III) have described a positive association between body mass index (BMI) and atopy rates [17]. However, a real link between obesity and other allergic disorders, such as allergic rhinitis, atopic dermatitis, as well as EGIDs, has not yet been extensively established [18]. A few studies have assessed the role of body weight and BMI in children and adolescents with EoE, and no articles were published on EGIDs distal to the esophagus (Table 2). There is evidence that most adults with EoE mainly have a good nutritional status and expected BMI values [19–27]. Despite feeding or swallowing issues, EoE children did not generally report nutritional deficiency or impaired growth [23]. Rezende et al. found that 82.8% of the enrolled EoE children had a good nutritional state, 11.4% were overweight, whereas 5.7% were underweight [27]. Moreover, Jensen et al., 2019 reported that EoE children might present a slight impairment of height at diagnosis and achieve their expected growth, regardless of treatment modality [21]. Finally, children with GERD and EoE had a weight-for-length (WFL) Z score at the 18th–13th percentiles; thus, they did not meet the criteria for failure to thrive (FTT) [24].

Table 2. Studies reporting a normal or high BMI of children and adult patients with EoE. No study has been published on non-esophageal eosinophilic gastrointestinal disorders (EGIDs).

Author, Year	Country	Study Design	Sample Size	Population	Outcomes
Zdanowicz et al., 2020 [19]	Poland	Single-center retrospective study	36 EoE patients	Children	No difference was observed in the prevalence of failure to thrive between children with EoE and controls (30.6% vs. 19.14%).
Alexander et al., 2020 [20]	U.S.A.	Retrospective cohort study	223 EoE patients	Adults	PPI non-responding EoE patients were younger ($p = 0.001$), had a lower BMI (27.3 vs. 28.6 kg/m^2, $p = 0.04$), and higher peripheral eosinophil count ($p = 0.006$) than responders, suggesting that these variables might be risk factors for PPI non-response in EoE.
Jensen et al., 2019 [21]	U.S.A.	Retrospective multicenter study	409 EoE patients	Children (<18 years)	Children with EoE had a slight impairment of height at diagnosis; thus, they were not malnourished. Additionally, they generally maintained their expected growth regardless of treatment modality. Subtle changes were noted for patients treated with elemental diets in combination with other therapeutical approaches.
Kovačić et al., 2019 [22]	Croatia	Cross-sectional study	32 EoE patients	Children (<18 years)	Most of the enrolled patients were well-nourished, and a normal BMI Z score was found in 75% of the patients. There was no difference in BMI Z score between baseline and 12 months follow-up (median −0.3 vs. −0.3 SD, $p = 0.862$).
Tanaka et al., 2019 [23]	Japan	Cross-sectional study	27 EoE patients	Adults	Subjects with EoE had higher BMI values than those without EoE (23.4 kg/m^2 vs. 22.3 kg/m^2, $p = 0.005$). Additionally, they had a higher proportion of bronchial asthma and hiatal hernia compared to controls (25.9% vs. 5.2%; $p < 0.00129.6$% vs. 14.7%; $p = 0.049$).
Mehta et al., 2018 [24]	U.S.A.	Prospective study	91 patients (GERD = 38, EoE = 53)	Children (0–7 years)	Children with GERD and EoE had greater eating issues than healthy controls and did not report nutritional deficiency or impaired growth. Additionally, children with GERD and EoE had a WFL Z score at the 18th and 13th percentiles; thus, they did not meet FTT criteria.

Table 2. *Cont.*

Author, Year	Country	Study Design	Sample Size	Population	Outcomes
Wolf et al., 2017 [25]	U.S.A.	Prospective case-control study	417 patients (EoE = 120, healthy controls = 297)	Adults	BMI was lower in EoE cases than controls (25 kg/m^2 vs. 28 kg/m^2, p = 0.002), but it was not in the underweight range. Additionally, BMI was lower in EoE patients with esophageal narrowing, suggesting that a low weight in a patient suspected of having EoE should raise concern for esophageal remodeling.
Lee et al., 2015 [26]	U.S.A.	Cross-sectional study	57 EoE patients	Adults	The median BMI was 25.5 kg/m^2, defined as overweight. There was no significant difference between the mean ages at diagnosis and different BMI categories (<25, 25–30, and >30 kg/m^2). Rural and urban adult groups did not differ in BMI categories (24 kg/m^2 ± 8.2 vs. 27 kg/m^2 ± 11.7, p = 0.271).
Rezende et al., 2014 [27]	Brazil	Cross-sectional study	35 EoE patients	Children (<18 years)	A good nutritional state was observed in 82.8% of the enrolled children. In particular, 11.4% of enrolled children were overweight, whereas 5.7% were underweight.

BMI, body mass index; EoE, eosinophilic esophagitis; GERD, gastroesophageal reflux disease; PPI, proton pump inhibitor; WFL, weight-for-length.

To date, no research has investigated the possible pathogenetic role of obesity in EGID development. Putative explanations could probably be found in environmental and genetic risk factors and EGID-related comorbidities. The overall prevalence of EGIDs seems to higher in developed Western countries, where childhood obesity and atopic diseases were significantly increased through time [7,28]. Indeed, obesity and the Western lifestyle, mainly characterized by high calorie/fat consumption and reduced physical activity, might be directly related to the increased risk of developing allergic diseases, such as EGIDs [13]. In a study in mice, Silva et al. demonstrated that obesity aggravated the immune histopathological characteristics of the EoE experimental model, reducing the regulatory cytokines profile (low expression of forkhead box P3, FOXP3, and interleukin 10, IL-10), increasing the inflammatory mediators (IL-5 and thymic stromal lymphopoietin, TSLP), and promoting tissue remodeling [29]. These fascinating data might provide new insights about obesity as a possible EoE risk factor that might impair esophageal inflammation and symptoms.

Another possible pathogenetic mechanism might be the relationship between EoE and GERD. Diagnosis of GERD has also increased, especially in developed countries [7]. In half of the infants with refractory vomiting and regurgitation, GERD was also expressed in the underlying cow's milk allergy, and improved with a hydrolyzed formula [30]. Several studies reported that GERD might play a possible pathogenetic role in esophageal eosinophilia, more relevant in PPI-responsive patients [31]. Indeed, EoE and GERD are not mutually exclusive and might coexist [4]. Although there are no exact data, four mechanisms have been proposed to explain this association: (1) GERD only causes esophageal eosinophilia; (2) GERD and EoE coexist but are independent phenomena; (3) EoE induces GERD; (4) GERD contributes to or induces EoE [7,31]. Acid reflux alters the esophageal epithelial barrier, leading to high intestinal permeability, with a subsequent passage of food allergens and release of inflammatory and eosinophil chemoattractant molecules might trigger EoE in susceptible subjects [32].

On the other hand, the esophageal eosinophilic inflammation is also associated with the production of different proinflammatory cytokines that might impair peristalsis and the esophageal acid clearance [7,33]. The subepithelial fibrosis, a delayed complication of EoE, might also promote esophageal dysmotility and GERD-related symptoms [31]. It is well described that being overweight and obese contribute to the development and worsening of GERD frequency and symptoms [34,35]. Obesity is notoriously involved in the pathogenesis of GERD [23]. Visceral fat might mechanically induce reflux events, increasing the intra-abdominal pressure [36]. Additionally, abdominal fat is metabolically active, activating macrophages, increasing and releasing proinflammatory cytokines and adipokines such as leptin [23,36].

Genes, obesity, and atopic diseases are linked. This association is well described in asthma patients, whereas no studies have been reported on EGID subjects. The β2-adrenergic (ADRB2) and glucocorticoid (NR3C1) receptor genes have been involved in the development of asthma and obesity [13]. Similarly, polymorphisms of the fractalkine receptor gene (CX3CR1) have been associated with asthma, atopy, and obesity [16]. However, no studies have described a genetic correlation between obesity/overweight and EGIDs.

Finally, EoE is characterized by chronic inflammation, specifically affecting the esophagus and generally sparing other GI tracts. This feature could clarify why EoE is not related to intestinal malabsorption and does not affect the bodyweight of adult patients.

The relationship between EGIDs, overweight, and obesity is still speculative, and further studies are required to confirm these clinical findings.

3. Undernutrition and Failure to Thrive in EGIDs Patients

Although poorly investigated, EGIDs may also be complicated by undernutrition and FTT for pathogenetic mechanisms similar to those reported in inflammatory bowel disease (IBD) patients [37]. FTT is one of the most commonly described clinical complications in children with EoE [3,38], although the exact prevalence has never been documented. Retrospective studies have reported that the prevalence of FTT ranges from 10.5% to 24% of EoE patients with different age-related rates (Table 3) [39–44]. In a large retrospective study, Spergel et al. demonstrated that FTT mainly characterized young children (2.8 ± 3.2 years) [44]. Moreover, Alhmoud et al. reported FTT and weight loss only in children with EoGE, and 15% of these had severe mucosal involvement leading to malabsorption [41].

Table 3. Studies reporting underweight and failure to thrive in children and adult patients with EGIDs.

Author, Year	Country	Study Design	Sample Size	Population	Outcomes
Hoofien et al., 2019 [39]	Europe	Multicentric retrospective study	410 EoE patients	Children	The most frequent indications for endoscopy were dysphagia (38%), gastroesophageal reflux (31.2%), food impaction (24.4%), and FTT (10.5%).
Chehade et al., 2018 [40]	U.S.A.	Multicentric study	705 EoE patients	Children and adults	FTT was present in 21.3% of enrolled subjects and was significantly common in children. Common pediatric comorbidities were neurological/developmental disorders, gastric tube placement, prematurity, atopic dermatitis, and food allergy.
Alhmoud et al., 2016 [41]	U.S.A.	Retrospective study	13 EoGE patients	Children and adults	FTT and weight loss were observed only in children. Two children (15%) had severe mucosal involvement leading to malabsorption, FTT, and weight loss.
Paquet et al., 2016 [42]	Canada	Retrospective study	62 EoE patients	Children	Sixty-two children were enrolled. Of these, 15 (24%) met at least one criterion for FTT.
Colson et al., 2014 [43]	France	Retrospective study	59 EoE patients	Children	Most children had negative WFH z scores, and 10% had nutritional indices compatible with moderate malnutrition. Nutrition therapy (elemental and six food elimination diets) did not impair nutritional status.
Spergel et al., 2009 [44]	U.S.A.	Retrospective study	620 EoE patients	Children	FTT/feeding issues and GERD-like symptoms were the most common presentations in the youngest children. (118 patients).

EoE, eosinophilic esophagitis; EoGE, eosinophilic gastroenteritis; FTT, failure to thrive; GERD, gastroesophageal reflux disease; WFL, weight-for-length.

Several factors may negatively impact the nutritional status of EGIDs patients (Table 4), mostly children. Firstly, children with EoE more likely present feeding disorders, recurrent vomiting, or regurgitation due to the esophageal inflammation and dysfunction, which can severely impair the adequate intake of foods and nutrients [2,3]. EGIDs are emerging GI disorders, therefore the diagnostic delay was often reported in adolescents and adults, who can consequently develop esophageal strictures due to the chronic inflammation and fibrous tissue deposition, prolonging clinical symptoms and patient feeding discomfort [45].

Table 4. Potential factors that may negatively influence the nutritional status of patients with EGIDs.

Chronic esophageal inflammation leading to typical GI symptoms: recurrent vomiting and regurgitation, loss of appetite, food impaction, GERD-like symptoms
Diagnostic delay may increase the risk of esophageal stricture and prolong GI discomforting symptoms
The low compliance to therapies may sustain esophageal inflammation, also allowing a low grade of antigen exposure
Swallowing disorders and fear of food impaction may compromise feeding behavior, allowing the development of food avoidance, anorexia, and anxiety
Restrictive food-elimination diets may reduce adequate food oral intake and lead to low levels of vitamins
Atopic (IgE mediated food allergy, atopic dermatitis) and non-atopic comorbidities (CD, IBD, type 1 diabetes mellitus, ASDs, CF) may be associated with FTT, low growth, reduced food oral intake, vitamins deficiency, and undernutrition
Multisite GI eosinophilic inflammation with subsequent abnormal permeability may be a possible reason for nutrients loss and higher caloric and protein requirements in patients with EGIDs distal to the esophagus

ASDs, autism spectrum disorders; CD, coeliac disease; CF, cystic fibrosis; FTT, failure to thrive; GERD, gastroesophageal reflux disease; GI, gastrointestinal; IBD, inflammatory bowel disease.

Secondly, low compliance to treatment is one of the main reasons for therapeutic failure and persistent active EoE, especially in adolescents and adults [46]. Chronic GI symptoms and impaired oral food intake, due to the sustained esophageal inflammation and continued low-grade antigen exposure, through limited dietary compliance are other possible explanations for undernutrition.

Thirdly, children, adolescents, and adults with previous food impaction episodes may have a high risk of developing anxiety and eating disorders, such as nervous anorexia and food avoidance, leading to an inadequate nutrient intake [46,47]. In a case-control study, Wu et al. found that most children with EGIDs had feeding behavioral problems compared to healthy controls [48]. Another study showed that 16.5% of EGID children had feeding issues, such as food refusal, low volume, and variety of intake, grazing, and spitting food out [49]. Moreover, 21% of these children were also complicated by FTT, suggesting that feeding issues may impair the regular childhood oral intake contributing to undernutrition and growth failure [49].

Additionally, a retrospective multicentric U.S. study of Consortium of Eosinophilic Gastrointestinal Disease Researchers (CEGIR) reported that 41% of children and adolescents with nonesophageal EGIDs might have a multisite GI inflammation [50]. This finding suggests that the persistent GI inflammation and subsequent abnormal intestinal permeability may be possible reasons for nutrients loss and higher caloric and protein requirements in patients with EGIDs distal to the esophagus [24].

Moreover, the association between EoE and other allergic conditions is well established and might be a potential further reason for FTT and undernutrition in EGIDs children. Children with EGIDs have an excessive prevalence of atopic dermatitis, IgE-mediated food allergy, asthma, and allergic rhinitis, potentially affecting the expected growth [51]. Moreover, several reports have suggested that EGIDs may also be frequently associated with chronic non-allergic comorbidities that might compromise adequate child growth, feeding behavior, and quality of life [46]. In a cross-sectional study, Capuccilli et al. demonstrated that children with EoE also had higher rates of coexisting non-atopic diseases, including IBD (0.7%) and CD (5.6%), as well as a higher prevalence of autism spectrum disorders (ASDs) (7.5%), type 1 diabetes mellitus (1.2%) and cystic fibrosis (0.9%) [52].

Finally, an important unanswered question is whether therapies can influence FTT. Paquet et al. have reported that EoE-related FTT resolved in 62% of affected children, suggesting that medical interventions might be helpful not only for disease-remission but also for clinical complications [42]. However, these results cannot be generalized because this study was retrospective and based on a small number of patients (15 patients with EoE + FTT). On the other hand, it was widely described that impaired growth and inadequate intake of macro- and micronutrients are possible complications of restrictive food elimination diets, which are pivotal therapeutical approaches of several pediatric illnesses, including EGIDs [1]. Several clinical factors might induce protein–calorie malnutrition and impaired food intake with weight loss, FTT, and delayed puberty. These findings underly the importance of assessing potential risk factors that may bring dietary limitations and normal growth of children with EGIDs.

4. Vitamin D Deficiency in EGIDs

Low serum vitamin D levels have been proposed to explain the increased prevalence of atopic and autoimmune diseases in Western countries [53]. Several efforts have focused on the role of vitamin D in the contribution of chronic dysregulated inflammation and its modulation [53]. Prevalence of EoE is higher in Western countries and cold climate zones, suggesting a possible association with low serum vitamin D levels [7]. Increasingly, significant evidence has shown a consistent link between vitamin D deficiency—due to the quality of diet, lack of exposure to sunlight—and the risk of atopy, as already described for asthma, allergic rhinitis, food allergy, and atopic dermatitis [7].

A systematic review has reported that low vitamin D prevalence varied widely in enrolled studies (0–52%) and did not improve with therapy [24,54] (Table 5). Low levels of vitamin D were described in 42% of adults and 50% of children with EoE, prevailing in patients with symptoms of food impaction [54,55]. In a case-control study of 69 children, Waterhouse et al. reported that patients with EoE and GERD had low vitamin D levels compared to normal controls, but without a significant difference [56]. To date, no study assessed other vitamins in EGIDs and serum vitamin D in patients with EGIDs beyond the esophagus.

Table 5. Studies reporting levels of vitamin D in children and adult patients with EoE.

Author, Year	Country	Study Design	Sample Size	Population	Outcomes
Mehta et al., 2018 [24].	U.S.A.	Prospective study	91 patients (GERD = 38, EoE = 53)	Children (0–7 years)	Enrolled children had adequate nutrient intakes, except for vitamin D levels that were low in both groups.
Slack et al., 2015 [54].	U.S.A.	Cross-sectional study	69 EoE patients	Children and adults	The median vitamin D level was 28.9 ng/mL. Patients with low vitamin D levels were older (25.5 years) and had a higher body mass index (25.2 kg/m^2). Vitamin D insufficiency was not associated with IgE and surrogate markers of severity (dilation in adults or hospitalization or emergency visits in children).

EoE, eosinophilic esophagitis; GERD, gastroesophageal reflux disease.

Although there is emerging evidence of vitamin D in the development of the immune system and pathogenesis of allergic diseases, such as asthma, atopic dermatitis, and food allergy, no studies have evaluated its possible role in EGIDs development and remission [53].

Furthermore, based on the design of available studies (cross-sectional data analysis) no cause–effect relationship can be inferred. It is reasonable to argue that toddlers and young children with EoE could present with feeding difficulty and refusal, with subsequent nutrient deficiencies, thus malnutrition. Besides, food elimination diets, mostly milk-free diets, could increase the risk of vitamin D deficiency in EoE patients, as reported in children with cow's milk allergy [57,58].

5. Management of EGIDs Patients: From Traditional Tools and Treatments to Future Insights

Diagnoses of EGIDs are not always straightforward and require chronic GI symptoms, coupled with suggestive endoscopic findings, prevalent eosinophilic inflammation (≥15 eosinophils/high-power field (HPF) for EoE) in biopsy specimens, and the exclusion of other causes of GI eosinophilia [1,4,5]. Symptoms of EGID are generally heterogeneous and often overlap with other conditions and may occur concomitantly. In EoE, the eosinophilic inflammation leads to progressive esophageal dysfunction, mainly characterized by feeding refusal and vomiting in children, and dysphagia, heartburn, and food bolus impaction in adolescents and adult patients [3]. Patients do not always appear to have feeding or eating disorders; only 24% of younger patients showed a failure to thrive. As reported in this review, most patients were normal weight or even obese. A meticulous evaluation of the patient's symptoms should be recommended, and the clinician should ask the right questions to detect suspicious eating habits (Table 6) [59].

Table 6. Useful questions to ask patients with EoE (Adapted from Muir et al., 2019) [59].

Does the patient take longer than others to eat?
Does the patient have to be reminded to chew a lot?
Does the patient need to cut food, especially steak, into small pieces?
Does the patient always need to drink during the meals?
Does the patient eat steak or crusty bread?

Although several research efforts have produced fascinating progress in the diagnosis and management of EGIDs, especially EoE, the only currently available tool to confirm the clinical suspicion is GI endoscopy with a biopsy [4,5]. Nevertheless, surrogate measures for EoE activity and response to therapy, such as the esophageal String test, transnasal esophagoscopy, and Cytosponge, have emerged as effective, less invasive tools for obtaining esophageal tissue samples [60,61].

Since EoE was initially identified in the mid-1990s, multiple EoE treatment strategies have been developed. Dietary treatment represented the first-line therapeutical approach for EGIDs [1,4,5]. Elemental (exclusive amino acid-based formulas) and six-food (milk, wheat, egg, soy, fish and shellfish, nuts) elimination diet (SFED) are the two main nutritional methods for EGID management with high rates of remission [1,4,5]. Trials have reported that a significant proportion of EoE patients achieved histologic remission on less restrictive (two/four food elimination) diets. Thus, personalized dietary strategies might offer the greatest success, improving the nutritional status and quality of life of affected subjects [60]. Successful targeted removal of specific foods based on allergy tests have been reported as case reports. However, targeted food removal might not be effective and is not recommended, because response to therapy did not seem to correspond to food allergies identified by skin prick testing or measuring serum food-specific IgE concentrations [62].

Swallowed steroids are alternative EGID treatments to diet-based interventions. The two most common approaches include swallowed fluticasone and viscous budesonide [4,5]. Comparisons between elimination diets and swallowed steroids are difficult, due to the heterogeneity of available studies. Meta-regression analyses showed that both therapeutical approaches are generally equivalent at inducing histologic remission in EGIDs patients [63].

Unfortunately, a significant population of patients with EGIDs has persistent active disease. Therefore, several ongoing efforts identify promising biological therapies beyond diet or steroid strategies [60,64]. Future efforts should be targeted to particular EGID endotypes using traditional and biologic therapies to achieve a new and high disease control degree.

How to Manage Malnutrition in Children with EGIDs?

This study suggests that a multidisciplinary approach (allergist, gastroenterologist, nutritionist, psychologist) is a key winner of EGIDs management (Figure 2), especially in children with allergic and non-allergic phenotypes. Moreover, the nutritional status assessment may help recognize patients with an inadequate nutrient intake, especially if they require restrictive food elimination diets (Figure 3).

Figure 2. The multidisciplinary approach of children and adolescents with eosinophilic gastrointestinal disorders.

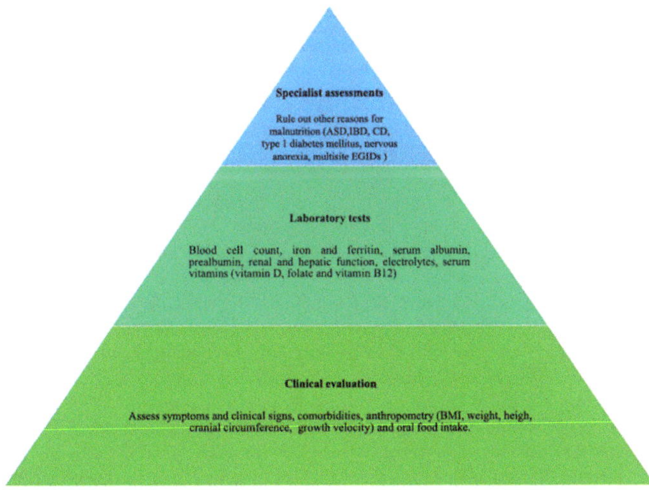

Figure 3. Nutritional status assessment of children and adolescents with eosinophilic gastrointestinal disorders.

This review summarized evidence on pediatric EGIDs malnutrition and underly conflicting findings. While some studies have reported normal or high BMI, especially in adults with coexisting GERD, FTT might mostly afflict young children. As reported for allergic diseases, EGIDs may also show vitamin D deficiency. However, no study has assessed how intestinal inflammation or EGIDs therapies may impact serum vitamin D and bone metabolism. Despite an inadequate investigation, EGID malnutrition is a relevant clinical field that requires further efforts to strengthen the efficacy of therapies and improve the patients' quality of life.

Author Contributions: Conceptualization, M.V. and A.L.; methodology, M.V., F.O. and M.D.F.; revision of the literature, M.V. and M.D.F.; writing—original draft preparation, M.V.; writing—review and editing, A.R., E.C., M.D.A., G.T.; visualization, M.D.A., G.T. supervision, G.L.M. and A.L. All authors have read and agreed to the published version of the manuscript.

Funding: This research received no external funding.

Conflicts of Interest: The authors declare no conflict of interest.

References

1. Licari, A.; Votto, M.; D'Auria, E.; Castagnoli, R.; Caimmi, S.M.E.; Marseglia, G.L. Eosinophilic gastrointestinal diseases in children: A practical review. *Curr. Pediatr. Rev.* **2020**, *16*, 106–114. [CrossRef] [PubMed]
2. Cianferoni, A.; Spergel, J.M. Eosinophilic esophagitis and gastroenteritis. *Curr. Allergy Asthma Rep.* **2015**, *15*, 58. [CrossRef] [PubMed]
3. Furuta, G.T.; Katzka, D.A. Eosinophilic Esophagitis. *N. Engl. J. Med.* **2015**, *373*, 1640–1648. [CrossRef] [PubMed]
4. Dellon, E.S.; Liacouras, C.A.; Molina-Infante, J.; Furuta, G.T.; Spergel, J.M.; Zevit, N.; Spechler, S.J.; Attwood, S.E.; Straumann, A.; Aceves, S.S.; et al. Updated International Consensus Diagnostic Criteria for Eosinophilic Esophagitis: Proceedings of the AGREE Conference. *Gastroenterology* **2018**, *155*, 1022–1033. [CrossRef] [PubMed]
5. Spergel, J.M.; Dellon, E.S.; Liacouras, C.A.; Hirano, I.; Molina-Infante, J.; Bredenoord, A.J.; Furuta, G.T. Participants of AGREE. Summary of the updated international consensus diagnostic criteria for eosinophilic esophagitis: AGREE conference. *Ann. Allergy Asthma Immunol.* **2018**, *121*, 281–284. [CrossRef] [PubMed]
6. O'Shea, K.M.; Aceves, S.S.; Dellon, E.S.; Gupta, S.K.; Spergel, J.M.; Furuta, G.T.; Rothenberg, M.E. Pathophysiology of Eosinophilic Esophagitis. *Gastroenterology* **2018**, *154*, 333–345. [CrossRef]
7. Votto, M.; Marseglia, G.L.; De Filippo, M.; Brambilla, I.; Caimmi, S.M.E.; Licari, A. Early life risk factors in pediatric EoE: Could we prevent this modern disease? *Front. Pediatr.* **2020**, *8*, 263. [CrossRef]
8. World Health Organization (WHO). Available online: https://www.who.int/news-room/q-a-detail/malnutrition (accessed on 28 November 2020).
9. Tambucci, R.; Quitadamo, P.; Ambrosi, M.; De Angelis, P.; Angelino, G.; Stagi, S.; Verrotti, A.; Staiano, A.; Farello, G. Association between obesity/overweight and functional gastrointestinal disorders in children. *J. Pediatr. Gastroenterol. Nutr.* **2019**, *68*, 517–520. [CrossRef]
10. Fifi, A.C.; Velasco-Benitez, C.; Saps, M. Functional abdominal pain and nutritional status of children. A school-based study. *Nutrients* **2020**, *12*, 2559. [CrossRef]
11. Moher, D.; Liberati, A.; Tetzlaff, J.; Altman, D.G.; PRISMA Group. Preferred reporting items for systematic reviews and meta-analyses: The PRISMA statement. *BMJ* **2009**, *339*, e1000097. [CrossRef]
12. Lee, E.Y.; Yoon, K.H. Epidemic obesity in children and adolescents: Risk factors and prevention. *Front Med.* **2018**, *12*, 658–666. [CrossRef] [PubMed]
13. Umano, G.R.; Pistone, C.; Tondina, E.; Moiraghi, A.; Lauretta, D.; Miraglia Del Giudice, E.; Brambilla, I. Pediatric obesity and the immune system. *Front. Pediatr.* **2019**, *7*, 487. [CrossRef] [PubMed]
14. Figueroa-Munoz, J.I.; Chinn, S.; Rona, R.J. Association between obesity and asthma in 4-11-year-old children in the UK. *Thorax* **2001**, *56*, 133–137. [CrossRef] [PubMed]
15. Del Giudice, M.M.; Marseglia, G.L.; Leonardi, S.; Tosca, M.A.; Marseglia, A.; Perrone, L.; Ciprandi, G. Fractional exhaled nitric oxide measurements in rhinitis and asthma in children. *Int. J. Immunopathol. Pharmacol.* **2011**, *24*, 29–32. [CrossRef] [PubMed]
16. Chen, Y.C.; Dong, G.H.; Lin, K.C.; Lee, Y.L. Gender difference of childhood overweight and obesity in predicting the risk of incident asthma: A systematic review and meta-analysis. *Obes. Rev.* **2013**, *14*, 222–231. [CrossRef] [PubMed]
17. von Mutius, E.; Schwartz, J.; Neas, L.M.; Dockery, D.; Weiss, S.T. Relation of body mass index to asthma and atopy in children: The National Health and Nutrition Examination Study III. *Thorax* **2001**, *56*, 835–838. [CrossRef] [PubMed]
18. Boulet, L.P. Obesity and atopy. *Clin. Exp. Allergy* **2015**, *45*, 75–86. [CrossRef]
19. Zdanowicz, K.; Kucharska, M.; Sobaniec-Lotowska, M.E.; Lebensztejn, D.M.; Daniluk, U. Eosinophilic Esophagitis in Children in North-Eastern Poland. *J. Clin. Med.* **2020**, *9*, 3869. [CrossRef]

20. Alexander, R.; Alexander, J.A.; Akambase, J.; Harmsen, W.S.; Geno, D.; Tholen, C.; Katzka, D.A.; Ravi, K. Proton pump inhibitor therapy in eosinophilic esophagitis: Predictors of nonresponse. *Dig. Dis. Sci.* 2020. [CrossRef]
21. Jensen, E.T.; Huang, K.Z.; Chen, H.X.; Landes, L.E.; McConnell, K.A.; Almond, M.A.; Safta, A.M.; Johnston, D.T.; Durban, R.; Jobe, L.; et al. Longitudinal growth outcomes following first-line treatment for pediatric patients with eosinophilic esophagitis. *J. Pediatr. Gastroenterol. Nutr.* **2019**, *68*, 50–55. [CrossRef]
22. Kovačić, M.; Unić, J.; Mišak, Z.; Jadrešin, O.; Konjik, V.; Kolaček, S.; Hojsak, I. One-year outcomes in children with eosinophilic esophagitis. *Esophagus* **2019**, *16*, 162–167. [CrossRef] [PubMed]
23. Tanaka, F.; Fukumoto, S.; Morisaki, T.; Otani, K.; Hosomi, S.; Nagami, Y.; Kamata, N.; Taira, K.; Nakano, A.; Kimura, T.; et al. Obesity and hiatal hernia may be non-allergic risk factors for esophageal eosinophilia in Japanese adults. *Esophagus* **2019**, *16*, 309–315. [CrossRef] [PubMed]
24. Mehta, P.; Furuta, G.T.; Brennan, T.; Henry, M.L.; Maune, N.C.; Sundaram, S.S.; Menard-Katcher, C.; Atkins, D.; Takurukura, F.; Giffen, S.; et al. Nutritional state and feeding behaviors of children with eosinophilic esophagitis and gastroesophageal reflux disease. *J. Pediatr Gastroenterol. Nutr.* **2018**, *66*, 603–608. [CrossRef] [PubMed]
25. Wolf, W.A.; Piazza, N.A.; Gebhart, J.H.; Rusin, S.; Covey, S.; Higgins, L.L.; Beitia, R.; Speck, O.; Woodward, K.; Cotton, C.C.; et al. Association between body mass index and clinical and endoscopic features of eosinophilic esophagitis. *Dig. Dis. Sci.* **2017**, *62*, 143–149. [CrossRef]
26. Lee, Y.J.; Redd, M.; Bayman, L.; Frederickson, N.; Valestin, J.; Schey, R. Comparison of clinical features in patients with eosinophilic esophagitis living in an urban and rural environment. *Dis. Esophagus* **2015**, *28*, 19–24. [CrossRef]
27. Rezende, E.R.; Barros, C.P.; Ynoue, L.H.; Santos, A.T.; Pinto, R.M.; Segundo, G.R. Clinical characteristics and sensitivity to food and inhalants among children with eosinophilic esophagitis. *BMC Res. Notes* **2014**, *7*, 47.
28. Licari, A.; Votto, M.; Scudeller, L.; De Silvestri, A.; Rebuffi, C.; Cianferoni, A.; Marseglia, G.L. Epidemiology of nonesophageal eosinophilic gastrointestinal diseases in symptomatic patients: A systematic review and meta-analysis. *J. Allergy Clin. Immunol. Pract.* **2020**, *8*, 1994–2003. [CrossRef]
29. Silva, F.M.C.E.; Oliveira, E.E.; Ambrósio, M.G.E.; Ayupe, M.C.; Souza, V.P.; Gameiro, J.; Reis, D.R.L.; Machado, M.A.; Macedo, G.C.; Mattes, J.; et al. High-fat diet-induced obesity worsens TH2 immune response and immunopathologic characteristics in murine model of eosinophilic oesophagitis. *Clin. Exp. Allergy* **2020**, *50*, 244–255. [CrossRef]
30. Pensabene, L.; Salvatore, S.; D'Auria, E.; Parisi, F.; Concolino, D.; Borrelli, O.; Thapar, N.; Staiano, A.; Vandenplas, Y.; Saps, M. Cow's milk protein allergy in infancy: A risk factor for functional gastrointestinal disorders in children? *Nutrients* **2019**, *10*, 1716. [CrossRef]
31. Cheng, E.; Souza, R.F.; Spechler, S.J. Eosinophilic esophagitis: Interactions with gastroesophageal reflux disease. *Gastroenterol. Clin. N. Am.* **2014**, *43*, 243–256. [CrossRef]
32. Tobey, N.A.; Hosseini, S.S.; Argote, C.M.; Dobrucali, A.M.; Awayda, M.S.; Orlando, R.C. Dilated intercellular spaces and shunt permeability in nonerosive acid-damaged esophageal epithelium. *Am. J. Gastroenterol.* **2004**, *99*, 13–22. [CrossRef] [PubMed]
33. Cao, W.; Cheng, L.; Behar, J.; Fiocchi, C.; Biancani, P.; Harnett, K.M. Proinflammatory cytokines alter/reduce esophageal circular muscle contraction in experimental cat esophagitis. *Am. J. Physiol. Gastrointest. Liver Physiol.* **2004**, *287*, 1131–1139. [CrossRef] [PubMed]
34. El-Serag, H.B.; Graham, D.Y.; Satia, J.A.; Rabeneck, L. Obesity is an independent risk factor for GERD symptoms and erosive esophagitis. *Am. J. Gastroenterol.* **2005**, *100*, 1243–1250. [CrossRef] [PubMed]
35. Vaishnav, B.; Bamanikar, A.; Maske, P.; Reddy, A.; Dasgupta, S.; Dasgupta, S. Gastroesophageal reflux disease and its association with body mass index: Clinical and endoscopic study. *J. Clin. Diagn. Res.* **2017**, *11*, OC01–OC04. [CrossRef] [PubMed]
36. Nam, S.Y.; Choi, I.J.; Ryu, K.H.; Park, B.J.; Kim, Y.-W.; Kim, H.B.; Kim, J. The effect of abdominal visceral fat, circulating inflammatory cytokines, and leptin levels on reflux esophagitis. *J. Neurogastroenterol. Motil.* **2015**, *21*, 247–254. [CrossRef] [PubMed]
37. Lim, H.-S.; Kim, S.; Hong, S.J. Food Elimination Diet and Nutritional Deficiency in Patients with Inflammatory Bowel Disease. *Clin. Nutr. Res.* **2018**, *7*, 48–55. [CrossRef]
38. Liacouras, C.A.; Spergel, J.; Gober, L.M. Eosinophilic esophagitis: Clinical presentation in children. *Gastroenterol. Clin. N. Am.* **2014**, *43*, 219–229. [CrossRef]
39. Hoofien, A.; Dias, J.A.; Malamisura, M.; Rea, F.; Chong, S.; Oudshoorn, J.; Nijenhuis-Hendriks, D.; Otte, S.; Papadopoulou, A.; Romano, C.; et al. Pediatric Eosinophilic Esophagitis: Results of the European Retrospective Pediatric Eosinophilic Esophagitis Registry (RetroPEER). *J. Pediatr. Gastroenterol. Nutr.* **2019**, *68*, 552–558. [CrossRef]
40. Chehade, M.; Jones, S.M.; Pesek, R.D.; Burks, A.W.; Vickery, B.P.; Wood, R.A.; Leung, D.Y.; Furuta, G.T.; Fleischer, D.M.; Henning, A.K.; et al. Phenotypic Characterization of Eosinophilic Esophagitis in a Large Multicenter Patient Population from the Consortium for Food Allergy Research. *J. Allergy Clin. Immunol. Pract.* **2018**, *6*, 1534–1544. [CrossRef]
41. Alhmoud, T.; Hanson, J.A.; Parasher, G. Eosinophilic Gastroenteritis: An Underdiagnosed Condition. *Dig. Dis. Sci.* **2016**, *61*, 2585–2592. [CrossRef]
42. Paquet, B.; Bégin, P.; Paradis, L.; Drouin, E.; Roches, A.D. High rate of failure to thrive in a pediatric cohort with eosinophilic esophagitis. *Ann. Allergy Asthma Immunol.* **2016**, *116*, 73–74. [CrossRef] [PubMed]

43. Colson, D.; Kalach, N.; Soulaines, P.; Vannerom, Y.; Campeotto, F.; Talbotec, C.; Chatenoud, L.; Hankard, R.; Dupont, C. The impact of dietary therapy on clinical and biologic parameters of pediatric patients with eosinophilic esophagitis. *J. Allergy Clin. Immunol. Pract.* **2014**, *2*, 587–593. [CrossRef] [PubMed]
44. Spergel, J.M.; Brown-Whitehorn, T.F.; Beausoleil, J.L.; Franciosi, J.; Shuker, M.; Verma, R.; A Liacouras, C. 14 years of eosinophilic esophagitis: Clinical features and prognosis. *J. Pediatr. Gastroenterol. Nutr.* **2009**, *48*, 30–36. [CrossRef] [PubMed]
45. Schoepfer, A.M.; Safroneeva, E.; Bussmann, C.; Kuchen, T.; Portmann, S.; Simon, H.U.; Straumann, A. Delay in diagnosis of eosinophilic esophagitis increases risk for stricture formation in a time-dependent manner. *Gastroenterology* **2013**, *145*, 1230–1236. [CrossRef]
46. Votto, M.; Castagnoli, R.; De Filippo, M.; Brambilla, I.; Cuppari, C.; Marseglia, G.L.; Licari, A. Behavioral issues and quality of life in children with eosinophilic esophagitis. *Minerva Pediatr.* **2020**, *72*, 424–432. [CrossRef]
47. Mehta, H.; Groetch, M.; Wang, J. Growth and nutritional concerns in children with food allergy. *Curr. Opin. Allergy Clin. Immunol.* **2013**, *13*, 275–279. [CrossRef]
48. Wu, Y.P.; Franciosi, J.P.; Rothenberg, M.E.; Hommel, K.A. Behavioral feeding problems and parenting stress in eosinophilic gastrointestinal disorders in children. *Pediatr. Allergy Immunol.* **2012**, *23*, 730–735. [CrossRef]
49. Mukkada, V.A.; Haas, A.; Maune, N.C.; Capocelli, K.E.; Henry, M.; Gilman, N.; Petersburg, S.; Moore, W.; Lovell, M.A.; Fleischer, D.M.; et al. Feeding dysfunction in children with eosinophilic gastrointestinal diseases. *Pediatrics* **2010**, *126*, e672–e677. [CrossRef]
50. Pesek, R.D.; Reed, C.C.; Muir, A.B.; Fulkerson, P.C.; Menard-Katcher, C.; Falk, G.W.; Kuhl, J.; Martin, E.K.; Magier, A.Z.; Consortium of Eosinophilic Gastrointestinal Disease Researchers (CEGIR); et al. Increasing rates of diagnosis, substantial co-occurrence, and variable treatment patterns of eosinophilic gastritis, gastroenteritis, and colitis based on 10-year data across a multicenter consortium. *Am. J. Gastroenterol.* **2019**, *114*, 984–994. [CrossRef]
51. Capucilli, P.; Hill, D.A. Allergic Comorbidity in Eosinophilic Esophagitis: Mechanistic Relevance and Clinical Implications. *Clin. Rev. Allergy Immunol.* **2019**, *57*, 111–127. [CrossRef]
52. Capucilli, P.; Cianferoni, A.; Grundmeier, R.W.; Spergel, J.M. Comparison of comorbid diagnoses in children with and without eosinophilic esophagitis in a large population. *Ann. Allergy Asthma Immunol.* **2018**, *121*, 711–716. [CrossRef]
53. Mailhot, G.; White, J.H. Vitamin D and Immunity in Infants and Children. *Nutrients* **2020**, *12*, 1233. [CrossRef]
54. Slack, M.A.; Ogbogu, P.U.; Phillips, G.; Platts-Mills, T.A.; Erwin, E.A. Serum vitamin D levels in a cohort of adult and pediatric patients with eosinophilic esophagitis. *Ann. Allergy Asthma Immunol.* **2015**, *115*, 45–50. [CrossRef]
55. Linglart, A.; Rothenbuhler, A.; Adamsbaum, C.; Colson, D.; Soulaines, P.; Dupont, C. The impact of pediatric eosinophilic esophagitis on bone metabolism. *J. Allergy Clin. Immun.* **2015**, *135*, AB46. [CrossRef]
56. Waterhouse, K.; Katta, A.; Teckman, J.; Foy, T.; Derdoy, J.; Jain, A.; Knutsen, A.; Becker, B. Vitamin D levels in children with eosinophilic esophagitis and gastroesophageal reflux. *J. Allergy Clin. Immun.* **2011**, *127*, AB107. [CrossRef]
57. Fissinger, A.; Mages, K.C.; Solomon, A.B. Vitamin deficiencies in pediatric eosinophilic esophagitis: A systematic review. *Pediatr. Allergy Immunol.* **2020**. [CrossRef]
58. Noimark, L.; Cox, H.E. Nutritional problems related to food allergy in childhood. *Pediatr. Allergy Immunol.* **2008**, *19*, 188–195. [CrossRef]
59. Muir, A.B.; Brown-Whitehorn, T.; Godwin, B.; Cianferoni, A. Eosinophilic esophagitis: Early diagnosis is the key. *Clin. Exp. Gastroenterol.* **2019**, *12*, 391–399. [CrossRef]
60. Slack, I.F.; Schwartz, J.T.; Mukkada, V.A.; Hottinger, S.; Abonia, J.P. Eosinophilic Esophagitis: Existing and Upcoming Therapies in an Age of Emerging Molecular and Personalized Medicine. *Curr. Allergy Asthma Rep.* **2020**, *20*, 30. [CrossRef]
61. Venkatesh, R.D.; Dellon, E.S. This String's Attached: The Esophageal String Test for Detecting Disease Activity in Eosinophilic Esophagitis. *Gastroenterology* **2020**. [CrossRef]
62. Walker, M.M.; Potter, M.; Talley, N.J. Eosinophilic gastroenteritis and other eosinophilic gut diseases distal to the oesophagus. *Lancet Gastroenterol. Hepatol.* **2018**, *4*, 271–280. [CrossRef]
63. Cotton, C.C.; Eluri, S.; Wolf, W.A.; Dellon, E.S. Six-food elimination diet and topical steroids are effective for eosinophilic esophagitis: A meta-regression. *Dig. Dis. Sci.* **2017**, *62*, 2408–2420. [CrossRef]
64. Licari, A.; Castagnoli, R.; Marseglia, A.; Olivero, F.; Votto, M.; Ciprandi, G.; Marseglia, G.L. Dupilumab to Treat Type 2 Inflammatory Diseases in Children and Adolescents. *Paediatr. Drugs* **2020**, *22*, 295–310. [CrossRef]

MDPI

Review

Non–IgE- or Mixed IgE/Non–IgE-Mediated Gastrointestinal Food Allergies in the First Years of Life: Old and New Tools for Diagnosis

Mauro Calvani [1,*], Caterina Anania [2], Barbara Cuomo [3], Enza D'Auria [4], Fabio Decimo [5], Giovanni Cosimo Indirli [6], Gianluigi Marseglia [7], Violetta Mastrorilli [8], Marco Ugo Andrea Sartorio [4], Angelica Santoro [9] and Elisabetta Veronelli [10]

1 Operative Unit of Pediatrics, S. Camillo-Forlanini Hospital, 00152 Rome, Italy
2 Immunology and Allergology Unit, Department of Mother-Child, Urological Science, Sapienza University of Rome, 00161 Rome, Italy; caterina.anania@uniroma1.it
3 Operative Complex Unit of Pediatrics, Belcolle Hospital, 00100 Viterbo, Italy; cuomoba@gmail.com
4 Department of Pediatrics, Vittore Buzzi Children's Hospital, University of Milan, 20154 Milan, Italy; enza.dauria@unimi.it (E.D.); marco.sartorio@unimi.it (M.U.A.S.)
5 Department of Woman, Child and of General and Specialized Surgery, University of Campania "Luigi Vanvitelli", 80100 Naples, Italy; fabio.decimo@unicampania.it
6 Pediatric Allergology and Immunology (SIAIP) for Regions Puglia and Basilicata, 73100 Lecce, Italy; gindirli@libero.it
7 Pediatric Clinic, Pediatrics Department, Policlinico San Matteo, University of Pavia, 27100 Pavia, Italy; giamar04@unipv.it
8 Operative Complex Unit of Pediatrics and Emergency, Giovanni XXIII Hospital, 70056 Bari, Italy; violetta.mastrorilli@libero.it
9 Pediatric Clinic, Mother-Child Department, University of Parma, 43121 Parma, Italy; angelica.santoro204@gmail.com
10 Food Allergy Committee of the Italian Society of Pediatric Allergy and Immunology (SIAIP), Pediatric Department, Garbagnate Milanese Hospital, ASST Rhodense, 70056 Garbagnate Milanese, Italy; bettyvero@libero.it
* Correspondence: mi5660@mclink.it; Tel.: +39-333-2000943

Citation: Calvani, M.; Anania, C.; Cuomo, B.; D'Auria, E.; Decimo, F.; Indirli, G.C.; Marseglia, G.; Mastrorilli, V.; Sartorio, M.U.A.; Santoro, A.; et al. Non–IgE- or Mixed IgE/Non–IgE-Mediated Gastrointestinal Food Allergies in the First Years of Life: Old and New Tools for Diagnosis. *Nutrients* **2021**, *13*, 226. https://doi.org/10.3390/nu13010226

Received: 25 November 2020
Accepted: 9 January 2021
Published: 14 January 2021

Abstract: non-IgE and mixed gastrointestinal food allergies present various specific, well-characterized clinical pictures such as food protein-induced allergic proctocolitis, food protein-induced enterocolitis and food protein-induced enteropathy syndrome as well as eosinophilic gastrointestinal disorders such as eosinophilic esophagitis, allergic eosinophilic gastroenteritis and eosinophilic colitis. The aim of this article is to provide an updated review of their different clinical presentations, to suggest a correct approach to their diagnosis and to discuss the usefulness of both old and new diagnostic tools, including fecal biomarkers, atopy patch tests, endoscopy, specific IgG and IgG4 testing, allergen-specific lymphocyte stimulation test (ALST) and clinical score (CoMiss).

Keywords: non-IgE gastrointestinal food allergy; eosinophilic gastrointestinal disorders; fecal biomarkers; IgG and IgG4; allergen-specific lymphocyte stimulation test; oral food challenge; atopy patch test; clinical score; endoscopy

1. Introduction

Food allergy (FA) is defined as an adverse health effect arising from a specific immune response that occurs reproducibly on exposure to a given food [1]. Based on the immunological mechanism involved, FA may be further classified as (a) IgE-mediated, the most well-understood form, which is caused by immunoglobulin E (IgE) antibodies against food antigens; (b) non-IgE mediated, in which the immune response is thought to act mainly through cell-mediated mechanisms; (c) or mixed, in which both IgE-mediated and cell-mediated immunological mechanisms are involved in the reaction.

IgE-mediated FA are the most common. They are easily characterized by the presence of specific serum IgE (sIgE) or a positive skin prick test (SPT). They occur most frequently in the first years of life, giving rise to urticaria/angioedema, oral allergic syndrome, rhinitis, or acute asthma and anaphylaxis [1].

Non-IgE FA are characterized by cutaneous reactions (such as atopic dermatitis, contact dermatitis and herpetiform dermatitis), respiratory reactions (such as Heiner's syndrome) or gastrointestinal reactions, which we will discuss in more detail below [2].

Non-IgE and mixed FA are less understood, despite their frequency: a Swedish population study showed that 36% of 118 children diagnosed with cow's milk allergy (CMA) by an oral challenge test were negative to specific IgE and SPT for cow's milk [3].

The diagnosis of non-IgE and mixed FA is mainly clinical and is not always easy. In contrast with IgE FA, the onset of symptoms is delayed and they may have a chronic presentation, making their association with the allergen less evident [4]. Furthermore, there is a lack of laboratory tests to assist in diagnosis. In most cases non-IgE FA are diagnosed on the basis of compatible symptoms and the demonstration that symptoms disappear once the suspected food has been eliminated and reappear when it is reintroduced [5].

Oral food challenge (OFC) is considered the gold standard for the diagnosis of IgE and non-IgE FA [6]. This complex test involves the oral administration of the suspected allergen in a controlled and standardized setting, thus requiring considerable healthcare resources (physician, nurse, hospital facilities) and family support (stress, fear). Most children with non-IgE FA do not need day care hospitalization, since they are not at risk of anaphylaxis. According to the Adverse Reactions to Food Committee of the American Academy of Allergy, Asthma & Immunology, "if a patient has a negative skin test, undetectable serum food specific IgE level, and no history of convincing symptoms of immediate FA (e.g., symptoms limited to behavioral changes or delayed/chronic gastrointestinal symptoms), gradual home introduction of the food in question may be attempted" [7]. Exceptions are made for patients with suspected food protein-induced enterocolitis syndrome (FPIES), who require hospitalized medical supervision for the OFC, as they are at risk of dehydration.

The diagnosis of some forms of non-IgE FA, and especially mixed FA, also requires an endoscopic examination to reveal any eosinophilic infiltration of gastrointestinal tissue.

Recent years have seen rising interest in non-IgE FA. A search for "non IgE mediated food allergy" on Pubmed revealed 9 articles published in 2000 and 11 in 2001, compared with 66 and 67 articles respectively for 2018 and 2019. This has resulted in a large increase in knowledge of many of its clinical and non-clinical aspects. The aim of this article is thus to provide an updated review on the different clinical pictures of non-IgE and mixed gastrointestinal FA in the first years of life. It also focuses on the role of both old and new diagnostic tools, including fecal biomarkers, atopy patch tests, endoscopy, Immunoglobulin G, Immunoglobulin G4 (IgG, IgG4), allergen specific lymphocyte stimulation test (ALST), clinical score, and other novel and future tests for the diagnosis of non-IgE and mixed FA. Celiac disease, although classified as a non-IgE-mediated food allergy, is not included in this review.

2. Materials and Methods

Search strategy. A comprehensive search was conducted in September 2020 using MEDLINE via PubMed (www.pubmed.gov) and Embase databases (www.embase.com). Searches were not restricted by language of publication, publication type or study design, but had to have been published in the last 10 years. However, we also considered earlier relevant studies and guidelines by looking through the references of the reviews and clinical studies published on this topic. This search found 2176 articles in PubMed and 2800 in Embase. Total non-overlapping record identified in PubMed and/or Embase for each category was 4145. Of these articles 189 were considered useful and are cited in the references. The search strategies and results in PubMed (MedLine) and EMBASE are detailed in Table S1.

The PICO (Patient, Intervention, Comparison/Intervention and Outcome) system was used to generate questions in regard to three of the topics: endoscopy and food challenge, atopy patch test and food challenge, and clinical score and food challenge. Further details on the methods and results are given in the respective paragraphs.

3. Clinical Features of Non-IgE Gastrointestinal FA

Non-IgE gastrointestinal FA present with specific, well-characterized clinical pictures, such as food protein-induced allergic proctocolitis (FPIAP), food protein-induced enterocolitis syndrome (FPIES) and food protein-induced enteropathy syndrome (FPE), eosinophilic gastrointestinal disorders (EGIDs) including eosinophilic esophagitis (EoE), allergic eosinophilic gastroenteritis (AEG) and eosinophilic colitis (EC) (Table 1), or with less specific clinical pictures. The latter come with nonspecific symptoms such as repeated regurgitation, vomiting, and watery or mucous hemorrhagic diarrhea, often in combination with other symptoms such as poor growth and crying crises (colic). These are not easily distinguishable from other childhood gastrointestinal diseases or functional disorders. Furthermore, FA itself can cause gastroesophageal reflux disease (GERD).

3.1. Food Protein-Induced Allergic Proctocolitis (FPIAP)

Notwithstanding a lack of prevalence studies, FPIAP is believed to be the most common non-IgE FA. It manifests in the first months of life with bloody stools in an otherwise seemingly well infant, in whom other causes of bleeding (constipation and/or anal fissures, infections, inflammatory intestinal diseases) have been excluded [8]. In a recent prospective population study in the USA, FPIAP was diagnosed from the presence of bloody stools or occult blood in 163 (18%) of 903 infants over a period of 3 years, and from the presence of occult blood alone in 63 (7%) [9]. The most frequent food trigger is cow's milk, followed by egg, soya and corn [10]. Diagnosis was most frequently made at the age of one month, and it seemed to be the most frequent cause of rectal bleeding in infants. FPIAP is believed to resolve rapidly: two studies reported that bloody stools or occult blood disappear in a few weeks, even without an elimination diet [11,12]. However, a meta-analysis by Lozinsky et al. found that an OFC was still positive in 34/47 patients (72.4%) after three months of an elimination diet and in 10/47 (21.2%) after 1 year [13].

A recent study of 257 infants with FPIAP showed even less optimistic data: only 60% of children developed tolerance in the first year of life, although 99% did so within 3 years [14].

SPT and sIgE tests for milk are usually negative. However, about 20% of children with FPIAP may show sensitization or develop IgE-mediated allergy to offending foods over time [15]. Endoscopy and rectal biopsy may prove inconclusive, with focal erythema ulceration, diffuse nodularity, or loss of vascular pattern, or they may be normal. For this reason, and because FPIAP is usually a patchy disease, multiple biopsies are necessary for diagnosis [16].

For these reasons, it has been suggested, for otherwise seemingly well infants with suspected FPIAP, to wait 2–4 weeks for spontaneous resolution without initiating an elimination diet [17,18]. If symptoms continue, an elimination diet is started; if the hematochezia stops, a specific IgE or SPT test for the suspected food may be useful. If these are negative, the food may be reintroduced at home, but if positive, an OFC is essential. If symptoms persist after elimination, other causes of rectal bleeding (fissures, infections, necrotizing enterocolitis, chronic inflammatory bowel diseases, coagulation defects, invagination, volvulus, Hirschsprung's disease) must be excluded to enable diagnosis [19].

An elimination diet (whether for the mother, as most cases arise during breastfeeding, or the infant) usually leads to regression of symptoms within 3–5 days, although some children may require a few weeks before any improvement is seen [8]. An OFC should be performed 2–4 weeks after the regression of symptoms. The first food excluded is usually cow's milk, or the food suspected by the mother. If this seems ineffective, other foods (egg, nuts, etc.) [14] should be excluded [20], following the same diagnostic procedure for each food. If the allergen is detected the elimination diet is usually continued until the age of 12 months. An OFC can then be proposed to assess the development of tolerance [2].

3.2. Food Protein-Induced Enterocolitis Syndrome (FPIES)

FPIES usually presents acutely, although a chronic form has also been described. Acute FPIES is characterized by bouts of vomiting 1–4 h after ingestion of the food responsible [21]. Repetitive emesis is associated with progressive lethargy, which may be associated with shock, dehydration and acidosis, hypotonia, and hypotension [6]. Episodes of diarrhea may occur, usually within 24 h. The severity of the symptoms often causes patients to seek emergency medical care [22]. SPT and sIgE tests for foods are usually negative. However, over 10% of patients have food-specific IgE (atypical FPIES) and associated IgE clinical features before or after the onset of FPIES [23,24].

Any food can induce FPIES, but the most common causes vary by age and location. Rice and oats have emerged as the most common triggers in the USA, followed by cow's milk, soya, egg, fish, fruits, and vegetables [22]. In Italy and Spain, fish is the most common solid food trigger [25,26]. According to a recent international consensus, the diagnosis of FPIES requires the major criterion and at least 3 minor criteria to be met [6] (Table 2). Infants presenting with a convincing history of FPIES likely do not require challenges to confirm their initial diagnosis. If only one episode has occurred, a diagnostic OFC is strongly recommended to confirm the diagnosis. Differential diagnosis includes sepsis, necrotizing enterocolitis, anaphylaxis, FPE, intussusception, pyloric stenosis, etc. [22].

A variety of protocols have been proposed in relation to OFCs. A recent International Consensus suggests administering a dose of 0.06 to 0.6 g (usually 0.3 g) of the food protein per kilogram of body weight, in three equal doses over 30 min [6]. Lower starting doses, longer observation periods between doses, or both should be considered in patients with a history of severe reactions [27]. It is generally recommended not to exceed a total dose of 3 g of protein or 10 g of total food (100 mL of liquid) for an initial feed and to observe the patient for 4 to 6 h [28].

The severity of chronic FPIES symptoms depends on the amount of food trigger continuously present in the diet. With low doses (e.g., solid foods or food allergens in breast milk), they manifest as intermittent vomiting and/or diarrhea and failure to thrive, without dehydration or metabolic acidosis. More regular intake (e.g., formula milk) is associated with intermittent but progressive vomiting and diarrhea (occasionally with blood), sometimes with dehydration and metabolic acidosis, and in about 50% of cases failure to thrive. Vomiting should regress within 3 days of excluding the responsible food, and its reintroduction may be followed by the sudden onset of acute clinical signs of FPIES. Without a confirmation challenge, the diagnosis of chronic FPIES remains presumptive [6].

3.3. Food Protein-Induced Enteropathy (FPE)

The incidence of FPE is unknown, although it seems to be less common than 20 years ago, and today is rather rare. It manifests with chronic diarrhea and consequent failure to thrive in the first 9 months of life [29]. It most frequently begins in the first two months of life, some weeks after the introduction of cow's milk to the diet. More than half of the affected infants also show vomiting, abdominal distention, poor growth, and lack of appetite [30]. In a minority of cases it leads to iron deficiency anemia, associated with the presence of occult blood in the stool. FPE causes a malabsorption similar to that of celiac disease, with which it has been found in association. It is usually caused by cow's milk, but may be due to soya or egg. Elimination of the responsible food leads to the regression of symptoms within 1–4 weeks, while the patchy villous atrophy it causes regresses several months after apparent clinical healing [29]. It is diagnosed by the reappearance of symptoms following the reintroduction of the food after 1–2 months [30].

3.4. Eosinophilic Gastrointestinal Disorders (EGIDs)

EGIDs are chronic diseases characterized by a range of gastrointestinal symptoms, eosinophilic infiltration of the gastrointestinal tract and, sometimes, peripheral eosinophilia. Diagnosis requires the exclusion of other causes of eosinophilic infiltration and the involvement of other organs.

3.4.1. Eosinophilic Esophagitis (EoE)

EoE may Start in the First Years of Life. In a multicentric study of 705 patients with EoE, about half were under 11 years of age. In this subgroup the median age of diagnosis was 3 years and median age of the onset of symptoms 1.1 years (interquartile range 0.4–3 years) [31]. In the early years of life EoE presents as GERD, and it is thought to be responsible for about 10% of cases of infants requiring treatment for GERD. The clinical picture includes regurgitation, vomiting, sometimes rumination, lack of appetite, burning, and pain, causing crying after feeding and sometimes immediately after starting to feed. This leads to refusal of food and sometimes abnormal posturing of the head and neck and severe arching of the spine, associated with melena and iron deficiency anemia (Sandifer syndrome) [32]. In these cases, failure to respond to proton pump inhibitor (PPI) should increase the suspicion of EoE.

The most common symptoms of EoE, such as dysphagia and food impaction, increase with age and are more common during adolescence. Concomitant atopic conditions should increase the suspicion of EoE [33]. It is diagnosed on the basis of symptoms of esophageal disfunction and >15 eosinophils per high power field (eos/hpf) on esophageal biopsy [34,35]. Other non-EoE disorders that cause or potentially contribute to esophageal eosinophilia should be excluded. An esophageal biopsy is necessary not only for diagnosis but also to monitor the results of treatment. The endoscopic signs of EoE include esophageal rings, longitudinal furrows, exudates, edema, strictures, or narrow caliber esophagus [2] Even in the absence of macroscopic lesions, multiple biopsies are needed for diagnosis: at least 4 biopsies (2 in the proximal and 2 in the distal esophagus) according to Dellon [36], and at least six biopsies from two different sites (typically the distal and proximal esophagus) according to Liacouras [37].

Food can be a trigger for children with EoE, especially in the early years of life. Studies showed that about 70% of children were allergic to one or more food, above all cow's milk, egg, wheat and soya. The younger the age, the more foods may be responsible [38,39].

3.4.2. Allergic Eosinophilic Gastroenteritis (AEG)

AEG is much less common than EoE. It affect both adults and children, and is rarely seen in the first year of life [40,41]. In young children it may cause abdominal pain, irritability, easy satiety, vomiting, diarrhea, weight loss, anemia and hypoalbuminemia, due to protein-losing enteropathy. However, symptoms are dependent not only on the patient's age but also on the organ affected, as well as the extent (invasion through bowel wall layers) [42]. Multiple food allergens are often implicated in this condition [42]. Peripheral eosinophilia is found in approximately 50% of patients with AEG. Serum tests for food-specific IgE antibodies or SPT reveal a food trigger in less than 50% of cases [43]. The diagnosis is made following endoscopic examination of the upper gastrointestinal tract, showing hyperemic edema and plaque in more than 50% of cases and, less frequently, erosion and ulceration [44]. The hallmark of AEG is marked eosinophilic infiltration of the gastric and/or duodenal mucosa, amounting to at least 30 eos/hpf [45]. Before initiating treatment of any AEG eosinophilic gastroenteritis, it is imperative to conduct a differential diagnosis to exclude other causes of hypereosinophilia with GI localization.

3.4.3. Eosinophilic Colitis (EC)

EC is the least common form of EGIDs [46], although like the other forms, its overall frequency seems to be increasing [43]. It is usually seen in adolescents in association with inflammatory bowel disease and/or celiac disease, and more rarely in infants associated with other atopic conditions and FA [47]. Its association with FA is unclear, but probably drops with increasing age. In a retrospective study of 69 children with colonic eosinophilia, Pensabene found that FA accounted for 10%, inflammatory bowel disease 32%, irritable bowel syndrome 33%, and other diagnoses 25% of cases of EC. In another retrospective study of 49 children aged over 3 years, Yang found sIgE for cow's milk and egg in 59.2% of cases. Elemental formula, simple elimination diet or combination therapy resulted in clinical improvement in 75%, 88.2% and 80% of patients, respectively [48].

Even if some studies found that more than half of cases of EC coexist with an allergy to cow's milk protein, soya, or peanuts, the elimination diet is not usually sufficient to treat it [47]. The IgE concentration associated with allergen stimulation does not reflect the tissue concentration at the location of the ongoing allergic inflammation. This suggests that most of the eosinophilic inflammation in the colon is associated with an IgE-independent mechanism [49]. The diagnosis is established by the presence of an increased eosinophilic infiltrate in the colon wall in symptomatic patients. However, this is problematic, as different studies have found different numbers of eos/hpf in healthy children, as well as a decrease in eos/hpf moving further down the colon [50]. Hurrel et al. suggested that more than 60 eos/10 hpfs in the lamina propria and eosinophilic infiltration in the epithelium or the muscularis mucosae are suggestive of eosinophilic proctocolitis [51]. However, there is no consensus on what comprises pathologic colonic eosinophilia versus normal variation in eosinophil levels [52].

3.5. Less Specific Clinical Features, Other Phenotypes and Associations

Non-IgE FA occur with less specific symptoms. These can include repeated regurgitation, vomiting, crises of crying, gas, poor growth, constipation or diarrhea, and it is not always possible to frame them in one of the clinical pictures listed above. There also seem to be some differences in the clinical features and laboratory findings in different ethnic groups and geographical regions [53,54]. In addition, many of these symptoms are also present in GERD, gastrointestinal functional disorders (which are much more common in the first year of life), and irritable bowel disease. Diagnosis is thus particularly difficult in these cases, not least because the different conditions can coexist in the same child. A relationship has been hypothesized between GERD and FA, in particular CMA [55,56]. According to Nielsen, 56% of children with severe GERD may also have CMA [57]. The same association (with percentages ranging from 16 to 55%) was also found in other studies [58,59]. In addition, response to diet may not help in the diagnosis of CMA, as extensive hydrolysis can also improve symptoms in functional disorders or GERD regardless of allergy: gastric emptying time is lower in children fed with extensive hydrolytes than in those fed with adapted milk [60].

Table 1. Clinical features of Non–IgE- or Mixed IgE/non–IgE-mediated Gastrointestinal Food Allergies.

	Food Protein Induced Allergic Proctocolitis (FPIPC)	Acute Food Protein Induced Enterocolitis (FPIES)	Chronic Food Protein Induced Enterocolitis (FPIES)	Food Protein Induced Enteropathy Syndrome (FPE)	Eosinophilic Esophagitis (EoE)	Allergic Eosinophilic Gastroenteritis (AEG)	Eosinophilic Colitis (EC)
Age	First months of life	First year, often after the first intake of allergenic food	Weeks or months after the first administrations of the responsible food	First year of life	About 10% of children with GERD who need medication	First years of life to adult	First years of life to adult
Food allergy	Cow's milk, egg, soya	Cow's milk, soya, grains, pulses, poultry, fish, variable in different countries	Cow's milk, soya, grains, pulses, poultry, fish, variable in different countries	Cow's milk, soya, egg, wheat	Cow's milk, soya, egg, wheat, peanut, walnut, fish	Cow's milk, egg, fish and seafood, soya, nuts, wheat	Cow's milk, egg
Food aversion	No	No	Sometimes	Sometimes	Yes	Sometimes	No
General condition	Good	Compromised	Compromised	Compromised	Good	Compromised	Compromised
Growth	Good at the beginning	Good at the beginning	Poor in 30% of cases	Poor	Sometimes poor	Poor	Sometimes poor
Vomiting	No	Immediate and repeated, 1-4 h after ingestion	Intermittent but progressive if the food is not withdrawn	More than half of cases	Yes	Yes	Sometimes
Regurgitation	No	No	No	No	Yes	Yes	No
Crying/colic/abdominal pain	No	No	No	No	Yes	Yes	Yes
Constipation	No	No	No	No	No	Sometimes	No
Watery diarrhea	No	in 20–50% after a few hours	Yes, chronic	Yes, chronic	Sometimes	Sometimes	Yes
Mucous diarrhea	Yes	No	No	No	No	Sometimes	Yes
Bloody diarrhea	Yes	No	Yes, in about 50% of cases	No	No	Sometimes	Yes
Abdominal distension	No	No	No	Yes	No	Yes	Yes
Acute symptoms	No	Yes	Only after the food is withdrawn	No	No	No	No
Fever	No	Sometimes	Only if acute onset after food withdrawal	No	No	No	No
Lethargy/Shock	No	Often	Only if acute onset after food withdrawal	No	No	No	No
Anemia	If not on a diet in severe forms	No	Sometimes	Yes	Sometimes	Sometimes	No
Hypoalbuminemia			Sometimes	Sometimes	Sometimes	Sometimes	No

No: almost never, Sometimes: less than 50%, Yes: more than 50%.

Table 2. Diagnostic tools for Non–IgE- or Mixed IgE/non–IgE-mediated Gastrointestinal Food Allergies.

	Food Protein Induced Allergic Proctocolitis (FPIPC)	**Acute Food Protein Induced Enterocolitis (FPIES)**	**Chronic Food Protein Induced Enterocolitis (FPIES)**	**Food Protein Induced Enteropathy Syndrome (FPE)**	**Eosinophilic Esophagitis (EoE)**	**Allergic Eosinophilic Gastroenteritis (AEG)**	**Eosinophilic Colitis (EC)**
Skin Prick Test/Specific IgEs	Usually negative	Positive only in 10–20% (atypical forms)	Positive only in 10–20% (atypical forms)	Usually negative	Positive for food in about 15–20% but not always related to the responsible food	Positive for food but not always related to the responsible food	Positive for food but not always related to the responsible food
Patch test	Usually negative	Positive in different percentages between studies (21–84%)	Positive in different percentages between studies (21–84%)	Not known	Positive for food in about 10% but not always related to the responsible food	Positive for food but not always related to the responsible food	Not known
	Shows lymphonodular hyperplasia or aphthous ulceration. Histologic examination shows focal aggregates of eosinophils in the large intestinal epithelium, lamina propria, crypt epithelium, and muscularis mucosa	Not indicated	Not indicated	If performed, it demonstrates damage to the intestinal mucosa with villi atrophy	Required for diagnosis. Shows eosinophilic infiltration (>15 per hpf)	Required for diagnosis. Shows eosinophilic infiltration (>30 per hpf)	Required for diagnosis. Shows eosinophilic infiltration (often >40 per hpf)
Response to diet	Within a few days (3–5, < 10)	Immediate	Within 72 h	Within 1–4 weeks	Clinical within weeks, histological within months	Clinical within weeks, histological within months	Clinical within weeks, histological within months
Diagnosis	Possible gradual home introduction after 1–2 months	If it does not meet the diagnostic criteria	OFC in absence of previous acute reaction	OFC or reintroduction after 1–2 months	Clinical and histological remission	Clinical and histological remission	Clinical and histological remission

4. Fecal Biomarkers

Non-IgE FA are characterized by intestinal inflammation and increased permeability, which leads to migration of granulocytes and eosinophils to the intestinal lumen. Due to the lack of reliable diagnostic tests, there is growing interest in finding fecal biomarkers. Several studies have investigated the use of various fecal biomarkers for diagnosis, such as fecal calprotectin (FC), α-1 antitrypsin (AT), β-defensin, tumor necrosis factor-α (TNF-α), fecal IgA, eosinophil-derived neurotoxin (EDN) and eosinophilic cationic protein (ECP).

4.1. FC in the Diagnosis of CMA

FC is an S-100 group cytosolic protein. This group comprises calcium- and zinc-binding proteins, thereby depriving microorganisms of these trace elements and inhibiting some zinc-dependent enzymes [61]. FC is immunomodulatory, antimicrobic, and antiproliferative and is present in the cytoplasm of neutrophils, in the membranes of macrophages, in activated monocytes and in mucosal epithelial cells [62]. It is a non-invasive marker of gastrointestinal inflammation, as its release into the intestine is correlated with the movement of neutrophils and mononuclear cells through the intestinal wall and their turnover and migration into the intestinal lumen [63,64]. Its concentration is correlated with the level of intestinal mucosal inflammation, as confirmed by endoscopic and histological examinations of intestinal inflammatory conditions [65–67]. It has been in use for several years in both the follow-up and remission monitoring of subjects with chronic intestinal conditions [68].

Other intestinal proteins (AT, β-defensin, TNF-α, fecal IgA, EDN and ECP), as well as FC, have been studied in non-IgE FA, offering surrogate markers of the cellular response.

To investigate the use of FC in FA (particularly CMA) in infants, we selected 6 studies conducted in children with non-IgE FA and 3 studies conducted in a population with both IgE-mediated and non-IgE-mediated forms (Tables 3 and 4). As was reported in a recent systematic review, some studies evaluated the use of FC as a biomarker for the diagnosis and monitoring of CMA, while several others investigated its use as a marker of intestinal response to OFC [69].

Baldassarre et al. reported significantly higher FC values in patients than controls, with values dropping by 50% after the elimination diet [70].

Beser et al. enrolled 32 infants under two years of age diagnosed with IgE and non-IgE mediated CMA by OFC. They found higher FC levels in the non-IgE mediated group, suggesting a possible use for this biomarker in the diagnosis and monitoring of non-IgE mediated gastrointestinal forms of CMA [71].

A prospective study conducted by Trillo Belizon et al. reported statistically higher FC values in infants aged 1 to 12 months diagnosed with non-IgE mediated CMA than in both infants for whom such a diagnosis had been excluded and healthy controls. Furthermore, there was a progressive decline in FC values after one to three months of a cow's milk elimination diet, with significant differences in patients with gastrointestinal symptoms such as diarrhea or rectal bleeding. The authors concluded that an FC value <138 μg/g permits the exclusion of a diagnosis of non-IgE mediated CMA, with a sensitivity of 95% and a specificity of 78.57% [72].

In a study of 46 children with allergic colitis suggestive of CMA, Lendvai-Emmert et al. found considerably lower FC values in children who had followed a strict cow's milk elimination diet for 3 months compared to their value at diagnosis, thereby indicating FC as a useful parameter for the diagnosis of CMA [73].

Prikhodchenko et al. monitored FC values in 18 children with FPE and 20 healthy age-matched children over the course of the disease. The mean FC concentration was higher in children with FPE than in the control group, but dropped significantly during the course of the disease. The authors concluded that FC shows promise for monitoring the course of FPE and evaluating treatment efficacy in children with FPE [74].

However, other studies reported results conflicting with those cited above. Some studies did not find any statistically significant difference between FC values on diagnosis of CMA and after a normal diet without cow's milk, or in comparison with healthy controls [75,76].

The use of FC as a biomarker of intestinal response to OFC has been investigated in very few studies. Berni-Canani et al. investigated the presence of subclinical intestinal inflammation in response to challenge testing of an amino acid-based formula under study in 60 infants aged $\leq$4 years with both IgE and non-IgE mediated CMA. FC and ECP were measured both before and 7 and 14 days after the challenge. Their values were unchanged in all patients, thereby demonstrating their optimal clinical tolerance of the formula [77].

Merras-Salmio et al. found higher FC values in patients following an elimination diet with a positive challenge than in patients with a negative challenge (39 children) towards cow's milk protein or in the controls (22 children), demonstrating the presence of mild inflammation of the intestinal mucosa during the challenge. The Mann-Whitney p values were significantly different between geometric means of FC values in non-IgE-mediated forms in comparison with IgE-mediated forms (18% versus 15%).

4.2. Other Fecal Biomarkers

Fecal biomarkers can indicate the degree of intestinal inflammation. Quantification of fecal eosinophils, above all EDN and ECP, reveals the extent of eosinophilic gastrointestinal inflammation, thus making them a non-invasive clinical biomarker. The feces of patients with FPIES, FPE and FPIAP show high levels of EDN, which remain stable at room temperature for at least 7 days, with matching histologic evidence of eosinophilic allergic colitis [78]. Kalach et al. determined various fecal markers (AT, TNF-α, β-defensin 2, secretory IgA, EDN and FC) and analyzed fecal microbiota and intestinal permeability in infants with digestive and non-digestive symptoms of CMA. A cow's milk challenge was performed in all children after an elimination diet, with a positive result in 11 patients. Eight patients presented non-IgE mediated CMA. The EDN cut-off level of 2818 ng/g gave a sensitivity of 55% and a specificity of 71% and the authors concluded that measurement of EDN in a single spot sample is promising in the diagnosis of non-IgE CMA [79].

In a child with FPIES, Wada et al. found an increase in TNF-α following sequential measurement prior to hydrolysate challenge and a paradoxical reduction during the challenge. After the challenge, it rose again for one month. Similar results were observed with fecal IgA, which dropped during the challenge, whilst fecal EDN rose during the challenge. The authors concluded that the sequential measurement of fecal TNF- α, together with other markers of intestinal inflammation, could offer a sensitive and non-invasive method to evaluate non-IgE mediated forms of CMA [80].

In a more recent study of eight patients with FPIES and 12 age-matched healthy infants, Wada et al. determined FC, EDN and fecal IgA levels before and after the OFC, finding a significant increase in all three fecal biomarkers in all patients after ingestion of the causative food. However, FC and fecal IgA levels were much lower than EDN, and the authors suggest that fecal EDN testing after ingestion of the causative food may serve as a useful diagnostic marker of FPIES [81].

Very recently, Rycyk et al. measured simultaneous FC, EDN and TNFα in 34 infants with gastrointestinal bleeding and 25 control group infants with functional gastrointestinal disorders. FPIAP was diagnosed by open OFC in 27 infants, and the offending food was identified as cow's milk in 23 and hen's eggs in 4 patients. Children with FA demonstrated significantly higher FC and EDN levels than the controls ($p < 0.05$). The authors found the best diagnostic performance in a combination of FC and EDN (88.9% and 84%) respectively and concluded that FC and EDN are reliable tools in differentiating between FPIAP and gastrointestinal functional disorders in infants [82].

Finally, a prospective case-control study carried out in Chile reported a sensitivity of 84%, a specificity of 66%, a positive predictive value (PPV) of 68% and a negative predictive value (NPV) of 83% for occult blood in the diagnosis of FPIAP [83].

In conclusion, the results from the available literature do not permit us to make any recommendations concerning the use of FC in the diagnosis of non-IgE FA. Further studies are necessary, involving an adequate number of participants with uniform characteristics such as age, nutrition, and duration of elimination diet, and, above all, the use of clearly defined reference values and FC cut-off times. The latter is a problem of great importance, given that whilst for adults and children over 4 years of age there is a well-defined cut-off value for FC (50 mg/kg), values in children under 4 years are considerably higher, and no cut-off values have been established for infants under one year of age [84–89]. EDN values too are higher in younger children, suggesting the activation and increased degranulation of intestinal eosinophils in this age group, given the immaturity of their epithelial barrier and reduced ability to regulate the intestinal microbiota.

Table 3. Fecal calprotectin in the diagnosis of CMA.

Author Year Ref	Study Design	Study Population and Sample Size	OFC	FC before Elimination Diet	FC after Elimination Diet	p	Comment
Baldassarre 2010 [70]	Prospective cohort study	30 (median age 8.57 months) with CMA 4 IgE mediated 26 non-IgE mediated vs. 32 (age-matched) healthy controls	No	325.89 ± 152.31 vs. 131.97 ± 37.98 $p < 0.001$	157.5 ± 149.13	$p < 0.001$	FC useful for diagnosis and monitoring of non-Ige mediated CMA
Beşer 2014 [71]	Prospective cohort study	32 (median age 12.5 ± 8.5 months) IgE mediated CMA 8 (median age 2.8 ± 1.7 months) non-IgE mediated CMA vs. 39 (median age 11.5 ± 7.6 months) healthy controls	Yes	392 ± 209 886 ± 278 vs. 296 ± 94 $p < 0.001$ $p = 0.142$	218 ± 90 359 ± 288	$p < 0.001$ $p = 0.025$	FC useful for diagnosis and monitoring of non-IgE mediated CMA
Trillo Belizon 2016 [72]	Prospective	40 (median age 3.68 months) with non-IgE mediated CMA vs. 12 (median age 3.25 months) without non-IgE mediated CMA vs. 30 (median age 3.8 months) healthy controls	Yes	442.65 vs. 268.58 vs. 100.30 $p < 0.0001$	228.51 ° 92.78 °°	$p < 0.001$	FC < 138 µg/g rules out non-IgE mediated CMA. FC > 138 µg/g offers sensitivity 95% specificity 78.57% PPV 80.9% NPV 94%
Ataee 2018 [75]	Prospective cohort study	29 (median age 117.2 days) with non-IgE mediated CMA	No	209.1 (SD 387.9)	189.9 § (SD 382.4) 125.2 §§ (SD 105.4)	$p = 0.741$ $p - 0.281$	FC not useful for diagnosis or follow-up of CMA
Lendvai/Emmert 2018 [73]	Prospective cohort study	46 (median age 7.28 years) with CMA of which 36 following a strict diet	No	61.17 (SD 63.72) 77	68.35 (SD 74.74) 41.69 (SD 34.68)	$p = 0.21$ $p < 0.001$	FC useful parameter in diagnosing CMA
Diaz 2018 [76]	Prospective cohort study	17 (13–23 months) with non-IgE mediated CMA vs. 10 (age-matched) healthy controls	Yes	47.25 (28.80–106.10) vs. 68.4 (30.38–76.73) $p = 1.0$			FC not useful
Prikhodchenko/Russia [74]	Prospective cohort study	18 (1-2 months) non IgE mediated vs. 20 (age matched) controls	No	384.41 ± 46.05 vs. 58.38 ± 8.05 $p < 0.001$	186.29 ± 14.16	$p < 0.001$	FC is the marker of intestinal inflammation in FPE and is useful for monitoring the disease course and evaluating the treatment

CMA = cow's milk allergy; OFC = oral food challenge; FC = fecal calprotectin; FPE = food protein enteropathy; IgE = immunoglobulin E; SD = standard deviation; PPV = positive predictive value; NPV = negative predictive value. ° 1 month after diet; °° 3 months after diet, § 2 months after diet; §§ 6 months after diet.

Table 4. Fecal calprotectin in OFC.

Author/Country/ Year/Ref	Study Design	Study Population and Sample Size	OFC	FC before Elimination Diet	FC after Elimination Diet		
BerniCanani/ Italy 2013 [77]	Prospective	60 (median age 37 months) with CMA 29 IgE-mediated 31 non-IgE mediated	Yes	36.3 ± 21.6	32.5 ± 23.8 * 33.5 ± 21.6 ^		FC useful for monitoring intestinal response to OFC in IgE and non-IgE mediated CMA
Merras-Salmio/ Finland 2014 [90]	Prospective cohort study	57 (median age 8.7 months) with non-IgE mediated CMA vs. 22 (13.2 months) healthy controls	Yes	18 OFC positive 52 (33–86) vs. 39 OFC negative 28 (24–44)	60(30–122) 33(24–44)	$p = 0.5$ $p = 0.4$	FC not useful for diagnosis in non-IgE mediated CMA

FC = fecal calprotectin; OFC = oral food challenge; CMA = cow's milk allergy; IgE = immunoglobulin E. * 7 days after OFC; ^ 14 days after OFC

5. IgG, IgG4, Allergen-Specific Lymphocyte Stimulation Test (ALST)

The measurement of food-specific IgG and IgG4 antibody levels is often proposed for the diagnosis of non-IgE mediated FA, but the results are currently still uncertain. We performed a search on PUBMED and Embase to establish the diagnostic usefulness of IgG and IgG4 testing in this type of allergy, particularly in pediatric age (see Table 1). None of the six articles selected confirmed the diagnostic usefulness of these tests. The lack of robust evidence leads to uncertainty over their use in childhood, [91,92] therefore this is not currently recommended [8,90].

Stapel et al. [93] pointed out that the presence of specific IgG and IgG4 antibodies against a given food is merely an indicator of the immune system's physiological response to repeated exposure to its components and a condition of immune tolerance, and it is logical to expect positive test results for specific IgG antibodies against food in healthy adults and children. Furthermore, the Canadian Society of Allergy and Clinical Immunology (CSACI) recently issued a position statement, in agreement with the American Academy of Allergy, Asthma and Immunology (AAAAI) and the European Academy of Allergy and Clinical Immunology (EAACI), declaring that there is no evidence of the usefulness of the IgG or IgG4 assay in identifying and/or predicting the presence of adverse reactions to foods [94].

The situation for EoE could be different. Recent data have highlighted the presence of high titers of specific food IgG4 antibodies in sera and esophageal tissue biopsy specimens from adults with EoE [95,96]. The clinical significance of these results is not yet clear, nor has the applicability of these findings to pediatric EoE, or their clinical functional significance in this population, been established. To further investigate this, we selected two studies.

Schuyler AJ et al. [97] demonstrated that high sIgG4 levels to cow's milk proteins are much more common in children with EoE than in the control group and sIgG4/sIgE ratios were often 10,000:1 or higher, with an OR > 20 to all 3 cow's milk proteins. Rosenberg CE et al. [98] reported that esophageal IgG subclasses were increased in pediatric subjects with EoE relative to controls; with IgG4 showing a 21-fold change, independently of age and duration of disease. Although more studies are needed, these data demonstrated that high specific sIgG4 or esophageal IgG4 levels could be useful biomarkers for the diagnosis or monitoring of EoE.

The allergen-specific lymphocyte stimulation test (ALST), also called lymphocyte proliferation or transformation test, has also recently been used to improve the diagnostic work up in non-IgE FA. We selected nine articles from the PubMed and Embase search (see Table S1).

The ALST analyzes lymphocyte proliferation and cytokine production in a culture of peripheral blood mononuclear cells after stimulation with food antigen for 3–5 days. Response is typically reported as the percentage of stimulated cells (stimulation index). Although this test has long been used in the diagnosis and research of disorders associated with immune diseases (immunodeficiencies, cancer, malnutrition, autoimmune diseases, etc.), its role in allergic diseases is still uncertain [99,100].

Two studies showed the diagnostic utility of ALST in neonates and infants with non-IgE GI symptoms after ingestion of cow's milk formula [101,102], but further evaluation of its sensitivity and specificity is needed in a larger population.

CMA can cause functional bowel disorders, which can create difficulty in managing pediatric surgical patients who also have CMA.

Ikeda K et al. [103] examined the effect of CMA on the management of 14 pediatric surgical patients in their institute, finding that a high LST index (normal range < 300%) was an important diagnostic tool for pediatric surgeons, who are in the front line for the treatment of neonates and infants with functional bowel symptoms.

Yagi H et al. [54] evaluated the relationship between the severity of non-IgE mediated gastrointestinal FA and both clinical and laboratory findings in neonates and infants, using a new symptom severity scale (grade 1–3). All patients tested positive to at least one milk component on ALST, with the most severely affected group (Grade 3) showing significantly higher positive levels than the other groups.

Kajita N et al. [104], in a recent case report of a 7-year-old Japanese girl with FPIES to quail egg, but not to chicken egg, reported that the ALST stimulation index (cut-off value > 180%) for quail egg yolk was higher than for other antigens, suggesting that the yolk might be a major allergen in quail-egg-induced FPIES.

Overall, all these studies showed that ALST could be a useful tool in the diagnosis of non-IgE mediated gastrointestinal FA, given that it can be performed regardless of the patient's clinical condition and hence enables early diagnosis. Nevertheless, there are a number of limitations to its use in children: the use of antigens that are not yet standardized, the significant amounts of peripheral blood necessary for the test and the relatively long culture times (5–7 days) [105].

To try to improve these limits, Yagi H et al. [106] evaluated a more rapid allergen-specific lymphocyte stimulation test (IPAST) that detects IL2 mRNA expression by quantitative reverse transcription polymerase chain reaction within 24 h, using only small amounts of blood. Peripheral blood mononuclear cells from 16 young children with non IgE-mediated gastrointestinal FA and 17 controls were incubated for 24 h with cow's milk proteins. All antigens, and especially α-casein, significantly increased IL2RA mRNA expression in patients with non-IgE-GI FA compared to the controls, with similar results to those obtained with conventional ALSTs. The authors concluded that IPAST may be a useful alternative to ALST in the diagnosis of non-IgE-GI FA, due to its high diagnostic value, small requirements for peripheral blood and rapid analysis.

In conclusion, the possibility of using specific biomarkers in the diagnosis of non-IgE mediated FA is still uncertain. While ALST and IPAST appear very promising in this regard, further studies are needed for both tests to improve standardization, to enable their use for as many antigens as possible and to better understand the mechanisms underlying the expression of cytokines and/or their receptors.

6. Accuracy of Atopy Patch Test Compared to OFC

The atopy patch test (APT) is an in vivo test that aims to reproduce the allergic reaction by application of the suspected allergen to the skin. It mimics the cell-mediated immune responses in which T cells play a prominent role, such as in non-IgE FA. APT has been included as a potential test to assess suspected FA in subjects with clinical signs of FPIES, FPIAP, FPE and EGID, as well as those with less specific symptoms [107].

APT is performed by applying the suspected food allergen to healthy untreated skin [108].The diagnostic accuracy of APT has been reported as higher with fresh foods than with freeze dried food extracts [109]. Any food can be assessed with patch testing, although cow's milk, hen's egg, wheat, and soya have been studied most extensively.

Reactions are traditionally classified as + in the presence of erythema, slight infiltration and, possibly, papules; ++ in the event of erythema, infiltration, vesicles and papules; and

+++ for intense erythema with infiltrate and coalescing vesicles. They are negative if the skin is unaffected, and doubtful in the event of faint erythema only [110].

APT is not recommended for the routine diagnosis of FA [111]. Certain factors need to be taken into consideration, such as the lack of standardized test substances and wide variability in the sensitivity and specificity of results in previous studies. Moreover, there is no consensus among experts regarding the appropriate reagents, methodology or interpretation of results. The recent EAACI FA and anaphylaxis guidelines [6] states that APT remains under study, and that to date its use has not been well established. In contrast, a systematic review for FA diagnosis published in 2014 by Sampson HA et al. [42] showed some evidence that APT may be valuable in assessing food triggers in pediatric EoE. [Strength of recommendation: Moderate; Evidence: C]

We evaluated the accuracy of APT compared to OFC using a PICO system. All eligible studies had to meet the inclusion criteria: pediatric patients, FA adequately confirmed by OFC, specific results in relation to the accuracy of APT. Data extraction was developed on the inclusion criteria, taking in consideration the best available evidence. Studies from which it was impossible to extract data on the specificity and sensitivity of APT were excluded. Two reviewers screened all abstracts and full-text articles independently. Any disagreement was resolved by a third party.

A total of 56 articles were identified by the literature search. Two of these were systematic reviews [112,113], and one a meta-analysis. As this kind of study is considered to be the highest quality evidence, we have provided an overview of the research published since then. The last comprehensive systematic review searches were conducted in September and November 2017; we continued the search up to October 2020.

Articles were screened, but no relevant studies were found in addition to those already included in the systematic reviews, and only two studies published after 2017 were eligible according to the inclusion criteria.

The methodological quality of one of the systematic reviews [113] is low. The authors include 37 studies without a specified reference list, the inclusion and exclusion criteria are not clearly indicated and they did not consider the quality of included studies. Given the lack of information on the included studies, we have not taken this review into account.

The second systematic review [112] showed a good quality assessment. The authors indicated the PICO research approach and criteria for selecting eligible studies, and they included estimates of likely bias to give quality weights. This review evaluated studies of the diagnostic value of APT compared to OFC in children with FA. A total of 41 studies were included and their quality was assessed by QUADAS-2.

Most of the included studies investigated both IgE-mediated and non-IgE FA. Subgroup analyses were conducted in relation to the patients' age and clinical signs. The gastrointestinal symptoms analyzed were: vomiting, regurgitation, diarrhea, abdominal pain, constipation, hematochezia and failure to thrive. In other cases, the enrolled patients had a specific diagnosis of enterocolitis, enteropathy, gastroesophageal reflux (GER) or FPIES. Subgroup analyses for gastrointestinal allergic symptoms indicated high specificity (91.5%) and low sensitivity (57.4%) for APT. The results showed that FA cannot be ruled out completely in the event of a negative APT, while its high specificity means that a positive APT indicates a high risk of FA.

It should perhaps be emphasized that four studies in the above systematic review recruited patients with a non-IgE mediated reaction, a negative SPT and negative specific IgE to the suspected foods. One of these was a retrospective study on FPIES, [114] while two were prospective studies conducted on FPIES [115] and on non-IgE-mediated CMA [116]. The prospective FPIES study demonstrated excellent sensitivity and negative predictive value (both 100%), while the study of children with non-IgE-mediated CMA confirmed that caution is needed before performing an OFC in children with a positive APT, given their good specificity and PPV (respectively 88.3% and 82.8%). The last of these four studies [117] was conducted on patients with non-IgE mediated rectal bleeding. Only 6 of the 31 subjects enrolled had confirmed food allergy, and none of them had a positive APT.

Our search identified two relevant studies published after the 2017 systematic review. In July 2017, Gonzaga TA et al. [118] evaluated the accuracy of APT in predicting the development of tolerance in non IgE-mediated CMA. The APTs were prepared with powdered skimmed cow's milk in isotonic saline solution or in petrolatum vehicle and with fresh cow's milk. With all preparation types, APT gave more false negatives than true positives. These data demonstrate the low sensitivity of APT and its low efficacy in predicting true negative patients and, hence, the development of tolerance, but also its good specificity in identifying subjects with a high risk of allergy. Cow's milk powder in isotonic saline solution was slightly superior to the other preparations, with 33.3% sensitivity and 96.1% specificity.

The second recent study, by Sirin Kose S et al., [119] aimed to determine the diagnostic efficacy of APT compared to OFC in 133 patients with gastrointestinal symptoms caused by cow's milk and hen's egg allergy. The authors retrospectively investigated APT reactions compared to OFC results. APT procedures were performed by applying fresh milk or egg white and yolk on the patient's back. The results demonstrated high specificity but low sensitivity. In patients with milk allergy APT had a specificity of 100%, sensitivity of 9.1%, PPV of 100% and NPV of 48.7%, and in patients with egg allergy APT had a specificity of 78.6%, sensitivity of 77.0%, PPV of 47.2% and NPV of 75.0%.

The discussed systematic review and our own search excluded studies without an appropriate diagnosis of FA, namely those that did not compare APT with OFC. For this reason, EoE was not included in the list of food allergies, because the diagnosis must be confirmed by the number of eosinophils in the esophageal biopsy specimen [36].

There is little literature evidence in relation to APT and EoE, with only two studies from the same group evaluated. The first, published in 2007, is a prospective study [120], while the second is a retrospective data collection [39]. These studies calculated PPV, NPV, specificity and sensitivity for different foods that caused increased eosinophils in biopsies. No other studies investigating the accuracy of APT reported specificity and sensitivity data. The prospective study [12] found PPVs ranging from 53.8% to 94.4%, depending on the food concerned, with NPVs ranging from 59% to 98.7%. It is important to point out that the last study [121] by the same authors, published in 2020 on the diagnosis and treatment of EoE, did not include the use of APT.

In conclusion, APT can be included in the diagnostic workup because it is a safe, specific diagnostic test that could point to a possible FA, especially in children with non IgE-mediated gastrointestinal symptoms (above all FPIES, FPIAP and FPE, and probably EGID too). The predictive capacity of APT can therefore be improved by combining it with negative sIgE or SPT measurement. However, several aspects require further investigation, especially to enable the better definition and standardization of the technique.

7. Accuracy of Endoscopy Compared to OFC

Diagnosis of non-IgE FA in clinical practice is challenging, due to the lack of pathognomonic non-invasive laboratory tests. Many non-IgE and mixed FA such as EoE and FPIAP have typical histological findings which confirm the diagnosis and point to the best treatment [122,123]. However, endoscopy with tissue sample collection can be difficult to perform, since it requires trained staff and resources. It can also be technically difficult, particularly in the first years of life, requiring general anesthesia.

Moreover, with the exception of EoE, these investigations supply data that are not easy to interpret, and hence do not change the patient's elimination diet, the timing of trigger food reintroduction or any strong suspicions of a different diagnosis, such as autoimmune enteropathy, tufting enteropathy, microvillus inclusion disease, or congenital disaccharide deficiencies, in the case of persistent symptoms.

We aimed to compare the diagnosis of non-IgE and mixed FA based on histology and elimination diet vs. OFC. Only a handful of studies satisfied the above-mentioned diagnostic work-up (Table S1). After excluding repeat results from the two databases or different searches, case reports, and literature reviews, 3 articles remained.

Rectal bleeding is common in not-sick newborns and infants [124]. A recent study by Jang et al. [125] aiming to clarify the etiology of small rectal bleeding in not-sick newborns demonstrated that FPIAP is a rare cause of small rectal bleeding, while idiopathic neonatal transient colitis (INTC) is far more prevalent. All 16 patients included in the study underwent endoscopy with biopsy. A food elimination test was performed in patients who did not improve spontaneously, and when rectal bleeding resolved an OFC was performed in order to confirm the diagnosis of FPIAP. Ten patients satisfied the histological criteria for FPIAP diagnosis but only two cases were confirmed as FPIAP by food elimination and OFC. One of these presented erosions on endoscopy and 141 eos/10 hpf within the lamina propria on histology, while the other had ulcers and 260 eos/hpf. Based on these results, the authors underlined that without OFC testing, INTC is often misdiagnosed as FPIAP. When the FPIAP diagnosis is based on clinical symptoms, the misdiagnosis rate is 88%, when based on clinical and pathological guidelines it is 80%, and when based on an elimination diet it is a little lower, at 67%. Most cases proved to be INTC, which has similar clinical symptoms and histopathological findings to FPIAP but resolves spontaneously without diet avoidance or medical treatment within the first week of life (average time 4 days). The authors thus suggest the usefulness of waiting for spontaneous remission of the hematochezia, in agreement with other authors, who suggest a "one month watch and wait (W&W) approach" [8,36,39,90–125].

With persistent bleeding, a diagnosis of FPIAIP should be confirmed by avoidance diet and oral food reintroduction at home, or OFC under supervision if SPT and sIgE tests for food are positive. However, in clinical practice if symptoms disappear and the infant is well, the confirmatory oral provocation test may be overcome in the first months of life. It should be periodically performed over the first year of life to test the acquisition of tolerance.

Given the age of presentation and the favorable course in the majority of cases, biopsies are generally not recommended, except in cases of unusual or abnormal symptoms such as constipation, diarrhea with mucus-streaked stools but without grossly visible bleeding, or severe rectal bleeding complicated with anemia despite a cow's milk elimination diet.

In EoE the problem is far more complex. EoE is a chronic esophageal inflammatory disease characterized clinically by symptoms of esophageal dysfunction and histologically by eosinophil-predominant inflammation [34]. When EoE is suspected, the first diagnostic test is upper gastrointestinal endoscopy. The role of allergy testing to identify triggering foods is limited in EoE, and such foods might only be identified by an elimination diet and reintroduction of single foods under biopsy control. Although no specific recommendations exist, it is reasonable to recommend sIgE testing prior to food reintroduction under biopsy control, due to the possible loss of tolerance during the avoidance diet [126].

The search strategy ("GERD and allergy and endoscopy and oral food challenge") identified only one study. Yukelsen et al [59]. investigated the relationship between refractory GERD (defined as the persistence of symptoms despite PPI treatment for at least 8 weeks) and allergy in 151 patients undergoing allergy testing and OFC. Of these, 28 had positive allergy tests to cow's milk protein and 7 to egg, and also reacted during cow's milk and egg OFC, respectively; 30 with negative allergy tests also reacted during OFC. All of them underwent endoscopy with sample collection: six patients in the first group and four in the second were diagnosed with EoE.

These results lead to various observations. First, this study showed the existence of a relationship between GERD and allergic disease. Second, it underlined that while OFC and allergy testing can identify many patients with allergic disease, endoscopy enables the diagnosis of EoE.

The detection of eosinophilic infiltration (>15/hpf) in at least one esophageal biopsy is the diagnostic hallmark of EoE (35). The recent Joint Recommendations of the North American Society for Pediatric Gastroenterology, Hepatology, and Nutrition and the European Society for Pediatric Gastroenterology, Hepatology, and Nutrition report a diagnostic algorithm addressing gastroesophageal reflux disease (GERD) management in clinical practice [127].

Upper GI endoscopy with biopsies should be performed in cases of persistent symptoms, such as crying, vomiting, anemia, feeding problems and/or failure to thrive, to identify and characterize esophagitis or enteropathy.

Poddar [128] et al. performed sigmoidoscopy and rectal biopsy in forty children presenting with a presumptive diagnosis of CMA based on clinical history of diarrhea, response to cow's milk withdrawal and exclusion of other disease. Aphthous ulcers were found on sigmoidoscopy, while rectal biopsy revealed eosinophilia without much change in the crypt architecture. There was a recurrence of histological lesions in all patients who underwent challenge after 6 months of exclusion diet, but only 42% were symptomatic. This study showed that the correlation between histology and clinical features can be slippery: while all the symptomatic patients had endoscopic/histologic alterations at baseline, after re-challenge there was a considerable difference between histological recurrence and clinical symptoms.

8. Accuracy of Clinical Score Compared to OFC

The diagnosis of CMA in the first year of life is often challenging because its presentation is non-specific, especially in non-IgE mediated and mixed forms. A clinical score, the CoMiSS (Cow's Milk related Symptom Score), has recently been proposed. According to the authors, the CoMiSS should be used as an awareness tool to help recognize the symptoms of CMA in infants and young children [129,130].

In order to review the diagnostic performance of CoMiSS and any other clinical scores, we carried out a PUBMED and Embase search using the terms listed in Table 1. We identified 363 and 328 articles in PUBMED and in Embase respectively, of which just 27 were eligible for our purposes. We found only one clinical score (CoMiSS) applicable to FA diagnosis in the first years of life. The CoMiSS is based on the presence and severity of five items investigating general clinical signs and dermatological, gastrointestinal and respiratory symptoms (Table 5).

In the first study, Vandenplas and colleagues evaluated the diagnostic performance of CoMiSS in relation to an open challenge performed after 1 month of elimination diet (extensive hydrolysate). A total of 116 infants with symptoms compatible with mild to moderate non-IgE CMA were included. The challenge was performed in 73% and was positive in 69% of the infants. The study showed a reduction in the CoMiSS during the elimination diet, and found that it was correlated to the challenge result. The average score reduction was 8.07 points; the challenge was positive in 80% of patients in whom the score was reduced to 6.0 points or less, but only in 48% if it remained ≥ 7 ($p = 0.001$). A more than 50% reduction in a baseline score ≥ 12.0 was the best predictor of a positive challenge [131].

In a later study, Vandenplas calculated the CoMiSS in 413 infants aged ≤ 6 months attending for vaccinations or growth examinations, in order to define normal values in apparently healthy infants and to establish the cut-off to identify those requiring further evaluation. The median and mean scores were 3.0 and 3.7 respectively, and the 95th percentile was 9.0, while only 1.5% had a score ≥ 12.0. Based on these results, a panel of allergologists, pediatric gastroenterologists and Belgian general pediatricians established a CoMiSS threshold of 12 or more to consider the diagnosis of CMPA likely [132].

The same authors published data on 333 healthy infants aged <6 months, documenting mean and median CoMiSS values of 2.77 and 2.83 respectively. These values rose to 3.88 and 4.00 when the analysis was extended to infants initially excluded due to incomplete data on gender and type of diet.

While no infant in the first sample had a score ≥ 12, 14 infants (1.9%) in this larger cohort did [132].

In another study of 226 healthy, mostly (exclusively or partially) breastfed Polish infants of the same age, the median and average scores were 4.0 and 4.7 respectively, while the 95th percentile was 11.0. Only 11 infants (4.9%) scored ≥ 12.0 [133].

Table 5. **CoMiSS®**: Cow's Milk-related Symptom Score.

Symptom	Score			
Crying (only considered if the child has been crying for 1 week or more, assessed by parents)	0		≤1 h/day	
	1		1 to 1.5 h/day	
	2		1.5 to 2 h/day	
	3		2 to 3 h/day	
	4		3 to 4 h/day	
	5		4 to 5 h/day	
	6		≥5 h/day	
Regurgitation	0		0 to 2 episodes/day	
	1		≥3 to ≤5 of small volumes	
	2		>5 episodes of >1 coffee spoon	
	3		>5 episodes of ± half of the feeds in half of the feeds	
	4		Continuous regurgitations of small volumes >30 min after each feed	
	5		Regurgitation of half to complete volume of a feed in at least half of the feeds	
	6		Regurgitation of the complete feed after each feeding	
Stools (Bristol scale)	4		Type 1 and 2 (hard stools)	
	0		Type 3 and 4 (normal stools)	
	2		Type 5 (soft stools)	
	4		Type 6 (liquid stools, if unrelated to infection)	
	6		Type 7 (watery stools)	
Skin	0 to 6		Atopic eczema head-neck-trunk arms-hands-legs-feet	
		Absent	0	0
		Mild	1	1
		Moderate	2	2
		Severe	3	3
Urticaria	0 or 6	YES	NO	
		6	0	
RespiratorySymptoms	0		No respiratory symptoms	
	1		Slight symptoms	
	2		Mild symptoms	
	3		Severe symptoms	

The most relevant papers dealing with CoMiSS are listed in Table 6. CoMISS: Coe's Milk Related Symptoms Score.

Table 6. Characteristics of studies dealing with CoMiSS.

Author (Year)	Study Design	Number (Age)	Cases with +ve IgE and/or SPT	CoMiSS vs. OFC	CoMiSS and Elimination Diet	Sensitivity/ Specificity PPV-NPV	Author's Conclusions
Vandenplas (2014) [131]	Cohort	116 (2 weeks–6 months)	sIgE>0.35 KU/L = 8% +ve SPT = 10%	OFC in 85/116 (73%) +ve in 59 (69%)	Basal score ≥12 If reduced to ≤6, 80% positivity of OFC.	ND	Score ≥12 useful for CMA diagnosis If reduction >50% with diet, high VPP for positive OFC
Chakrabarty (2017) [132]	Prospective	30 (24–136 days)	ND	OFC +ve in 8/10	Significant score reduction (from >12 to 6)	ND	Useful for early diagnosis and to monitor response to therapy
Rigley (2017) [133]	Prospective	58 (<1 year)	ND	OFC +ve in 2/2	Score reduction in all (from 16.5 to 3.4, average values)	ND	Useful for early diagnosis, may help reduce specialist consultations
Bajerova (2017) [134]	Cohort	121 (6 weeks–1 year)	ND	OFC +ve in 11/18	Performed in 21	ND	A cut-off of 8 reached much more frequently in allergic patients, but a lower threshold could increase sensitivity

Abbreviations: CoMISS: Coe's Milk Related Symptoms Score; ND = Not Done; +ve = positive; Sens = Sensitivity; Spec = Specificity; PPV = Positive Predictive Value; NPP = Negative Predictive Value; SPT = Skin Prick Test; OFC = Oral Food Challenge; CMA = Cow's Milk Allergy; FA = Food Allergy; GI = gastrointestinal.

Table 6. *Cont.*

Author (Year)	Study Design	Number (Age)	Cases with +ve IgE and/or SPT	CoMiSS vs. OFC	CoMiSS and Elimination Diet	Sensitivity/ Specificity PPV-NPV	Author's Conclusions
Prasad (2018) [135]	Observational Cross-sectional	83 (0–24 months)	+ve sIgE and/or SPT = 26/83 (31%)	Diagnosis confirmed in 70: by OFC in 56% of cases	ND	CoMiSS>12 Sens = 77% Spec = 66% PPV = 93% NPV = 33%	High PPV confirming the reliability of parameters included in CoMiSS
Armano (2017) [136]	Prospective	40 (3–41 months)	ND	OFC +ve in 40/40	38/40 score reduction >50%	Score ≥ 12 (in 17/40, 42.5%) predicted diet efficacy with 100% PPV and 9% NPV	Selection of candidate patients for diet
Salvatore (2019) [137]	Prospective	47 (1–12 months)	+ve SPT in 8/47 = 17%	OFC in 21/39 patients responsive to diet +ve in 6 (29%)	In 19/47 (40%) score reduction ≥50%	Best cut-off = 9 for response to diet: Sens = 84% Spec = 85% PPV = 80% NPV = 88%	To predict diet response in children with persistent GI symptoms.
Vandenplas (2013) [138]	Prospective /Multicentric	116 (80-64 days (median of two groups respectively)	sIgE>0.35 KU/L = 7.5% +ve SPT = 17% (rash); 10.5% (papule)	OFC in 85/116 (74%) +ve in 69%	Significant score reduction after 1 month diet	ND	CoMiSS useful for CMA diagnosis (OFC positive in 70% with score ≥12)
Vandenplas (2014) [139]	Prospective/ Multicentric	40 (3.4 months) (mean age)	SPT = 15/40 (37.5%) tested only 17 cases	OFC in 38/40 +ve in 38/40	Score significantly reduced after 1,3,6 months of diet	ND	ND
Vandenplas (2016) [140]	Prospective/ Multicentric	71 (6 months)	ND	OFC in 50/71 (70.4%) +ve in 34	After 1 month of diet, score significantly reduced in both confirmed and unconfirmed CMA (OFC not performed or negative)	ND	ND
Vandenplas (2017) [141]	Aggregate analysis of the previous 3 studies	See above	See above	See above	Both a score <5 (median) and a score reduction from 13 to 5 (median) after 1 month of diet increase likelihood of CMA (+ve OFC)	See above	See above
Kose (2018) [142]	Cohort	112 (5.6 months (mean)	sIgE and SPT +ve = 66/112 (59%).	OFC in 46/112 (41%)	Significant score reduction after 1 month of diet in infants allergic to milk, egg or both.	Score reduction after diet ≥50%: Sens = 83.7% 84.6%, 87.5% for milk, egg allergy or both respectively	Score reduction after diet ≥50% to be used for diagnosis of FA
Selbuz (2020) [143,144]	Prospective	168 (0–12 months)	+ve sIgE = 23/168 (13.8); +ve SPT = 20/168 (12%).	OFC in 154/168 (91.7%) +ve in 91/168 (54,2%)	After 4 weeks of diet, score reduced by ≥3 points in 154 (91.7%)	Cut-off 12.5: Sens = 64.8% Spec = 54.4%	Association of symptoms in CoMiSS helps in recognition of CM-allergic infants
Vandenplas (2020) [145,146]	Cohort	148 2.3 months (median) = Spanish cohort. 72 3 months (mean) = Belgian cohort.	ND	Spanish cohort: OFC in 13, score ≥10 +ve in 10/13 (76%), score>12 +ve in 7/8	ND	ND	ND
Kherkhheulidze (2017) [147]	Prospective	34/<1 year	ND	ND	Significant score reduction after 2 weeks of diet.	ND	ND

Abbreviations: CoMISS: Coe's Milk Related Symptoms Score; ND = Not Done; +ve = positive; Sens = Sensitivity; Spec = Specificity; PPV = Positive Predictive Value; NPP = Negative Predictive Value; SPT = Skin Prick Test; OFC = Oral Food Challenge; CMA = Cow's Milk Allergy; FA = Food Allergy; GI = gastrointestinal.

So, the next question was: what is the clinical usefulness of CoMiSS? In particular, what is its predictive value when compared to the OFC in non IgE mediated CMA diagnosis?

Chakrabarty and Rigley evaluated the diagnostic performance of CoMiSS in two small studies, comparing it with the outcome of the elimination diet (and, in some cases only, with OFC). Interestingly, significantly reduced values were found at the end of the observation period [134,135].

Bajerova et al. used CoMiSS to identify patients at risk of CMA. The authors suggested that lower cut-offs (threshold value = 8) would increase the sensitivity of the method in children with non-specific symptoms of milk protein allergy [136].

None of the above studies reported the number of cases with sIgE and/or SPT.

Prasad carried out a study on 83 patients aged between 0 and 24 months with symptoms suggestive of CMA. A score >12.0 was obtained in 60 patients (72%). CMA was confirmed in 70 patients by OFC (performed in only 56% of cases) or ImmunoCAP. In detail, in 78.6% of patients with CoMiSS >12.0 and in 15% of patients with a value $\leq$12.0, the CoMiSS showed a PPV of 93% and an NPV of 33%. According to the authors the low NPV was probably because many children were already on the elimination diet and this would have led to a reduction in the score [137].

In two Italian studies, CoMiSS was compared with response to the elimination diet [138]. In the first, the PPV and NPV for score $\geq$12 were 100% and 9% respectively. The second was a prospective open study that investigated 47 infants aged between 1 and 12 months (median 3 months) who were on a cow' milk protein-free diet due to the presence of persistent gastrointestinal symptoms. A significant response to the diet, defined as a $\geq$50% score reduction from the baseline value and below the median of the control population, was obtained in 40% of patients. The ROC curve identified a value of 9.0 as the best cut-off to predict diet response (sensitivity 84%, specificity 85% vs. 37% and 92% with a cut-off of 12; PPV 80%, NPV 88%) [139,140].

A meta-analysis of 3 studies to investigate the usefulness of CoMiSS as a predictor of CMA as confirmed by open challenge [141] found that a low score (median 5.0) after 1 month of elimination diet was associated with a higher risk of a positive challenge test (odds ratio = 0.83). Moreover, a median score reduction from 13.0 to 5.0 after a 1-month diet was predictive of the appearance of symptoms upon the introduction of cow's milk as confirmed by the result of the confirmatory OFC, which was positive in 69% of cases in the Nestlé Health Science study [142] and in 81% in the other two studies [143,144].

Kose and Seda evaluated the response to the elimination diet according to the CoMiSS score in 112 children diagnosed with CMA, egg allergy or both. OFC confirmed the diagnosis in 46 patients (41%), in whom the modification of the score during the 1-month elimination diet was assessed. A $\geq$50% reduction corresponded to a sensitivity of 83.7%, 84.6% and 87.5% for milk allergy, egg allergy and both, respectively. According to the authors, this value could be employed as a cut-off for the diagnosis of the corresponding allergies [145,146].

A very recent study evaluated 168 children with a baseline score $\geq$12 who were started on elimination diet for 4 weeks: children who responded to the diet also underwent the open challenge. This study has two important strengths: the large number of children enrolled and the diagnostic confirmation, in all the "responders" to the diet, by open challenge. The allergy was confirmed in 54.2% patients; the ROC curve showed that the best cut-off for CoMiSS was 12.5, which corresponded to a sensitivity of 64.8% and a specificity of 54.4%. The study also showed that some symptoms, such as skin involvement, were more frequently observed in children with confirmed allergy whose score was significantly higher. The authors therefore concluded that the systematic evaluation of symptoms associated with CoMiSS can aid the selection of infants who might benefit from an elimination diet [147,148].

A collaborative study by Belgian and Spanish authors on children aged <6 months assessed the variability of the score when calculated by a pediatrician and by parents, as well as day to day variability when evaluated by parents over 3 consecutive days. The data suggest that CoMiSS can be calculated by parents, before medical consultation, without the need for special training. In the Spanish arm of the study, the diagnostic performance of the score was also compared in relation to OFC: 10 out of 13 children (76%) with a score ≥10 and 7 out of 8 with a score >12 were diagnosed as allergic to CM by OFC [149,150].

Finally, CoMiSS was recently included in a computer-based algorithm in which a score ≥12 increases the likelihood of the diagnosis and supports a dietary prescription for babies, whether exclusively breast-fed or not [151]. The score was also employed to evaluate the effect of hydrolyzed formula therapy in a study conducted at the Central Hospital of Tbilisi (Georgia). After 2 weeks, there was already a significant score reduction in children fed with hydrolyzed formula, along with a significant decrease in crying and regurgitation scores and a significant rise in the percentage of children with normal stool consistency [152].

In conclusion, 14 studies have evaluated the diagnostic efficacy of the CoMiSS. Of these, 9 were prospective studies, and 5 enrolled less than 50 children. Only about half the studies reported the percentage of positive specific IgE and/or SPT in the study population; in these studies, the vast majority seemed to be non-IgE FA. The diagnostic efficacy of CoMiSS compared to OFC was evaluated in 13 studies. In the 3 studies (Prasad, Armano, Salvatore) which used a CoMiSS cut-off value of >12, the PPV was between 80 and 100%. The vast majority of studies found a reduction in CoMiSS after elimination diet and that a >50% reduction in CoMiSS was predictive of a subsequent positive OFC.

Although further studies are needed to validate CoMiSS in the diagnostic workup of CMA and, possibly, other types of FA, and to define the optimal cut-off values, it can already be considered a useful tool, especially for suspected non-IgE mediated FA. As also affirmed by other authors, it should also be used to monitor response to therapeutic interventions such as the elimination diet, but at present it is not sufficient in itself to diagnose FA and cannot replace the OFC [153].

9. Novel and Future Diagnostic Tests for Non IgE-Mediated Food Allergy

Recent years have seen great interest in the search for biomarkers that, supported by clinical evidence, could facilitate the diagnostic path for non IgE-mediated and mixed FA. However, results have been poor [154]. Laboratory tests such as blood count, C-reactive protein (CRP), serum electrolytes and protein profile offer little help with the differential diagnosis in the presence of diseases with symptoms similar to non IgE-mediated FA (e.g., sepsis, gastroenteritis). However, testing for specific cytokines produced by cells involved in the immune response may be useful. A recent study in patients with FPIES identified TARC (thymus and activation-regulated chemokine) as a potential biomarker. TARC is produced by eosinophils when stimulated by TNFα and IL4, and it promotes the expression by Th2 cells of cytokine receptor type 4, which is involved in cell migration to the inflammation site [155]. TARC was initially proposed as a marker of severity and for treatment monitoring in atopic dermatitis [156]. It was recently reported that some patients with FPIES showed an increase in TARC about 24 h after being exposed to the trigger food, whether accidentally or during OFC. This increase only appears alongside gastrointestinal symptoms, suggesting that changes in serum TARC levels are likely linked to allergy reactions in intestinal epithelium cells [157,158]. This study is an example of how the measurement of cytokines and changes in their levels following OFC may help in the diagnosis of non IgE-mediated FA.

There is growing evidence that the microbiome contributes to the development and presentation of allergic diseases. It seems that gut dysbiosis likely precedes the development of food allergy, and the timing of dysbiosis appears to be critical [159]. Specific microbiome signatures have been observed in non-IgE food allergies, such as eosinophilic esophagitis and FPIAIP and FPIES [160]. This suggests that the microbiome may offer a simple and non-invasive diagnostic marker for these disorders [159].

Other studies have shown that activation of the innate immune response underlies the pathogenetic signs of these diseases. Mehr et al. used RNA sequencing and bioinformatic approaches to analyze whole blood from children with FPIES before OFC and during any acute reactions [161]. Patients reacting to the OFC showed an increased expression of the genes that activate monocytes, neutrophiles and their receptors, which are responsible for the observed reactions. In contrast, this was not observed in patients showing no reaction to OFC. In this case too, a better knowledge of the basic pathogenic mechanisms of delayed FA may contribute to the development of future new diagnostic techniques.

Similarly, Schouten et al. [162] found a high concentration of immunoglobulin free light chains (Ig-fLC) in patients with non IgE-mediated CMA who, in any case, showed a type I immediate clinical response. Increased Ig-fLC levels are normally found in chronic inflammatory diseases such as intestinal diseases, rheumatoid arthritis, Sjogren's syndrome, systemic erythematous lupus and multiple sclerosis. Besides confirming that a chronic inflammatory state underlies allergy, this result may suggest the use of this immunoglobulin subpopulation for the diagnosis of non IgE-mediated CMA.

A recently proposed ALST measures interleukin 2 α-receptor mRNA expression within 24 h, using a small amount of peripheral blood. However, tests like these need further study to adapt their use to as many allergens as possible and to better understand the mechanisms underlying the expression of both cytokines and their receptors [106].

The basophil activation test (BAT) is a laboratory test for the in vitro simulation of an in vivo allergenic challenge, using cytofluorometric evaluation of basophil activation markers (CD63 and CD203c). The BAT is mainly used for IgE-mediated FA, but can also be used for non IgE-mediated allergies, albeit with a lower diagnostic accuracy; it is in fact one of the few in vitro techniques currently available for this kind of allergic reaction [163]. Indeed, basophil activation occurs not only through IgE signal transduction, but also as a consequence of non IgE-mediated reactions [164]. The main advantages of the BAT are its reliability, the small amount of peripheral blood required (1 mL), and its high specificity and sensitivity, enabling it to replace OFC for some patients (especially in the case of tests requiring high allergen dosages). Its limitations are related to possible basophil anergy, which is responsible for a lack of response in about 10% of cases; in addition, it requires specialized training and is still not commercially available. Furthermore, it must be carried out within 24 h (ideally within 4 h) of sampling and, last but not least, large scale validation is needed [165].

Some promising results are also arriving from the instrumental diagnostics field. To support the diagnosis of non IgE-FA in symptomatic individuals, a recent study proposed the use of abdominal ultrasound and Doppler imaging to evaluate intestinal vessel density (VD) [166]. The authors evaluated the VD of patients with a history of delayed food allergy and compared it with the VD observed in patients with gastroenteritis and in case controls. All patients with non IgE-FA showed thickening of the small intestinal wall and reduced peristalsis, and most also showed thickening of the mesentery and gastric wall. These findings suggest that non IgE-FA is characterized by a relatively severe involvement of the gastroenteric segment, as in the case of acute abdomen, gastroenteric perforation and Crohn's disease.

In contrast, infectious diseases do not produce the same ultrasound evidence. Moreover, patients with delayed allergy showed a larger VD in the ileum and jejunum than did the other two groups. These parameters could therefore be used to distinguish a non IgE-mediated FA from a severe infection. This non-invasive examination is suitable for use in children, but as with all ultrasound procedures, it is operator-dependent. In any case, given the low number of cases included in the clinical study and the variation in the participants' ages, further investigation is needed [166].

Finally, genetics could be used to identify individuals affected by or at risk of numerous disorders, including allergic diseases. A large number of genes have been identified by genome-wide association studies for food allergy [167]. Allergic diseases are the result of a complex interplay between genetic and environmental factors [168]. Epigenetic mechanisms may explain how the environment influences gene expression, modulating immune responses throughout life, especially early life [169,170].

Classical epigenetic mechanisms, including DNA methylation and histone modifications, have been shown to be involved in the development of IgE-mediated food allergies such as CMA [171–173]. Differences in DNA methylation in different gene pathways have been observed in children who subsequently developed an IgE-mediated food allergy [174], suggesting their possible role as potential biomarkers. Although epigenetic mechanisms have mostly been investigated in IgE-mediated food allergies, they are also likely to play a role in non-IgE mediated food allergy. The symptoms of non-IgE-mediated allergy to food proteins are mostly gastrointestinal and the pathogenetic mechanisms are probably cell-mediated [175].

In recent years, a common inflammatory pathway has been hypothesized for allergic diseases, characterized by a "type-2 inflammation" involving different cells besides the classic Th2 cells. These cells, from both the innate and adaptive systems, produce a unique Th set of so-called 'type-2′ cytokines, the effectors of the allergic response [172–176]. This inflammatory pathway has been implicated in a wide range of allergic diseases, including atopic dermatitis, asthma and eosinophilic esophagitis [176].

Interestingly, a different DNA methylation profile of Th1 and Th2 cytokine genes and achievement of tolerance has been demonstrated in children with IgE mediated food allergy [177]. This suggests that epigenetic modifications may be potential biomarkers for predicting tolerance. The role of epigenetics in this field has been specifically demonstrated for IgE food allergies, but a similar effect might also be hypothesized for non-IgE food allergies. Different DNA methylation profiles were in fact recently demonstrated between patients with EoE who responded to treatment in comparison with non-responders [178], suggesting that epigenetic modifications may also be biomarkers of treatment response in some non-IgE mediated food allergies.

In addition to the classic mechanisms discussed above, other epigenetic mechanisms have also been proposed in non-IgE-mediated food allergy. It has been reported that post-transcriptional control elements such as miRNAs may be involved in the pathogenesis of non-IgE delayed cow's milk hypersensitivity [179], suggesting the possibility that this reaction could be downregulated.

Although epigenetic research is still in its infancy, especially in the field of non-IgE mediated food allergies, it may have several promising clinical applications, ranging from prevention to early prediction of the success of a given therapeutic strategy. Finally, ongoing progress in molecular biology and omics sciences (e.g., genomics, proteomics, epigenomics, metabolomics, and metagenomics) may offer new insights into non-IgE food allergies [180].

10. Conclusions

Non-IgE mediated and mixed FA constitute a heterogeneous group of diseases arising through immunological mechanisms that are not yet well understood. In clinical practice, diagnosis generally relies on a compatible clinical history and the resolution of symptoms upon the elimination of the presumed triggering antigens. Diagnostic confirmation, however, requires a different approach in the different clinical pictures. An OFC or home reintroduction of food may be attempted in many cases, while some cases, endoscopy and biopsy of the affected intestinal tract is also essential for diagnosis. Promising new diagnostic tools to facilitate diagnosis are being studied, with encouraging results in some cases, such as CoMiSS, LSTs and IPAST. Further studies are still necessary to fully understand the physiopathology of these diseases and, consequently, improve their diagnosis and prognosis.

Supplementary Materials: The following are available online at https://www.mdpi.com/2072-6643/13/1/226/s1, Table S1: title. Search strategies and results in PubMed (MedLine) and EMBASE.

Author Contributions: Conceptualization, M.C.; writing—original draft preparation, M.C., C.A., B.C., E.D., F.D., G.C.I., V.M., M.U.A.S., A.S.; E.V.; writing—review and editing, M.C., C.A., B.C., E.D., F.D., G.C.I., V.M., M.U.A.S., A.S.; E.V., G.M. All authors have read and agreed to the published version of the manuscript.

Funding: This research received no external funding.

Institutional Review Board Statement: Not applicable.

Informed Consent Statement: Not applicable.

Data Availability Statement: Data sharing not applicable.

Acknowledgments: We wish to thank Marie-Hélène Hayles MITI for her editing assistance.

Conflicts of Interest: The authors declare no conflict of interest.

References

1. Boyce, J.; Assa'ad, A.; Burks, A.W.; Jones, S.M.; Sampson, H.A.; Wood, R.A.; Plaut, M.; Cooper, S.F.; Fento, M.J.; Arshad, S.H.; et al. Guidelines for the diagnosis and manage-ment of food allergy in the United States: Report of the NIAID-sponsored expert panel. *J. Allergy Clin. Immunol.* **2010**, *126*, 1–58. [CrossRef] [PubMed]
2. Cianferoni, A. Non-IgE Mediated Food Allergy. *Curr. Pediatr. Rev.* **2020**, *16*, 95–105. [CrossRef] [PubMed]
3. Saarinen, K.M.; Savilahti, E. Infant feeding patterns affect the subsequent immunological features in cow's milk allergy. *Clin. Exp. Allergy* **2000**, *30*, 400–406. [CrossRef] [PubMed]
4. Burks, W.; Tang, M.; Sicherer, S.; Muraro, A.; Eigenmann, P.A.; Ebisawa, M.; Fiocchi, A.; Chiang, W.; Beyer, K.; Wood, R.; et al. ICON: Food allergy. *J. Allergy Clin. Immunol.* **2012**, *129*, 906–920. [CrossRef] [PubMed]
5. Labrosse, R.; Graham, F.; Caubewt, J.C. Non-IgE-Mediated Gastrointestinal Food Allergies in Children: An Update. *Nutrients* **2020**, *12*, 2086. [CrossRef]
6. Nowak–Wegrzyn, A.; Chehade, M.; Groetch, M.; Spergel, J.M.; Wood, R.A.; Allen, K.; Atkins, D.; Bahna, S.; Barad, A.V.; Berin, C.; et al. International consensus guidelines for the diagnosis and management of food protein–induced enterocolitis syndrome: Executive summary-Workgroup Report of the Adverse Reactions to Foods Committee, American Academy of Al-lergy, Asthma & Immunology. *J. Allergy Clin. Immunol.* **2017**, *139*, 1111–1126.
7. Nowak-Wegrzyn, A.; Assa'ad, A.H.; Bahna, S.L.; Bock, S.A.; Sicherer, S.H.; Teuber, S.S. Adverse Reactions to Food Commit-tee of American Academy of Allergy, Asthma & Immunology. Work Group report: Oral food challenge testing. *J. Allergy Clin. Immunol.* **2009**, *123*, S365–S383.
8. Nowak-Wegrzyn, A.; Katz, Y.; Mehr, S.S.; Koletzko, S. Non-IgE-mediated gastrointestinal food allergy. *J. Allergy Clin. Immunol* **2015**, *135*, 1114–1124. [CrossRef]
9. Martin, V.M.; Virkud, Y.V.; Seay, H.; Hickey, A.; Ndahayo, R.; Rosow, R.; Southwick, C.; Michael Elkort, M.; Gupta, B.; Kramer, E.; et al. Prospective Assessment of Pediatrician-Diagnosed Food Pro-tein-Induced Allergic Proctocolitis by Gross or Occult Blood. *J. Allergy Clin. Immunol. Pract.* **2020**, *8*, 1692–1699. [CrossRef]
10. Nowak-Węgrzyn, A. Food protein-induced enterocolitis syndrome and allergic proctocolitis. *Allergy Asthma Proc.* **2015**, *36*, 172–184. [CrossRef]
11. Arvola, T.; Ruuska, T.; Keränen, J.; Hyöty, H.; Salminen, S.; Isolauri, E. Rectal bleeding in infancy: Clinical, allergological, and mi-crobiological examination. *Pediatrics* **2006**, *117*, 760–768. [CrossRef]
12. Elizur, A.; Cohen, M.; Goldberg, M.R.; Rajuan, N.; Cohen, A.; Leshno, M.; Katz, Y. Cow's milk associated rectal bleeding: A popula-tion based prospective study. *Pediatr. Allergy Immunol.* **2012**, *23*, 766–770. [CrossRef] [PubMed]
13. Lozinsky, A.C.; Morais, M. Eosinophilic colitis in infants. *Jornal Pediatr.* **2014**, *90*, 16–21. [CrossRef] [PubMed]
14. Buyuktiryaki, B.; KulhasCelik, I.; Erdem, S.B.; Capanoglu, M.; Civelek, E.; Guc, B.U.; Guvenir, H.; Cakir, M.; Misirlioglu, E.; Akcal, O. Risk Factors Influencing Tolerance and Clinical Features of Food Protein-induced Allergic Proctocolitis. *J. Pediatr. Gastroenterol. Nutr.* **2020**, *70*, 574–579. [CrossRef] [PubMed]
15. Cetinkaya, P.G.; Kahveci, M.; Sahiner, U.M.; Sekerel, B.E.; Soyer, O. Food protein-induced allergic proctocolitis may have distinct phenotypes. *Ann. Allergy Asthma Immunol.* **2021**, *126*, 75–82. [CrossRef] [PubMed]
16. Diaz Del Arco, C.; Taxonera, C.; Olivares, D.; Fernandez Acenero, M.J. Eosinophilic colitis: Case series and literature review. *Pathol Res. Pract.* **2018**, *214*, 100–104. [CrossRef]
17. Miceli Sopo, S.; Monaco, S.; Bersani, G.; Romano, A.; Fantacci, C. Proposal for management of the infant with suspected food pro-tein-induced allergic proctocolitis. *Pediatr. Allergy Immunol.* **2018**, *29*, 215–218. [CrossRef]
18. Mennini, A.; Fiocchi, A.G.; Cafarotti, A.; Montesano, M.; Mauro, A.; Villa, M.P.; Di Nardo, G. Food protein-induced allergic proctocol-itis in infants: Literature review and proposal of a management protocol. *World Allergy Organ. J.* **2020**, *13*, 100–471. [CrossRef]
19. Fox, V.L. Gastrointestinal bleeding in infancy and childhood. *Gastroenterol. Clin. N. Am.* **2000**, *29*, 37–66. [CrossRef]
20. Kaya, A.; Toyran, M.; Civelek, E.; Misirlioglu, E.; Kirsaclioglu, C.; Kocabas, C. Characteristics and prognosis of allergic proctocolitis in infants. *J. Pediatr Gastroenterol. Nutr.* **2015**, *61*, 69–73. [CrossRef]
21. Leonard, S.A.; Nowak-Wegryn, A. Food Protein–Induced Enterocolitis Syndrome. *Pediatr. Clin. N. Am.* **2015**, *62*, 1463–1477. [CrossRef] [PubMed]
22. Nowak-Wegrzyn, A.; Berin, M.C.; Mehr, S. Food Protein-Induced Enterocolitis Syndrome. *J. Allergy Clin. Immunol. Pr.* **2020**, *8*, 24–35. [CrossRef] [PubMed]

23. Caubet, J.C.; Ford, L.S.; Sickles, L.; Jarvinen, K.M.; Sicherer, S.H.; Sampson, H.A.; Nowak-Węgrzyn, A. Clinical features and resolution of food protein-induced enterocolitis syndrome: 10-year experience. *J. Allergy Clin. Immunol.* **2014**, *134*, 382–389. [CrossRef] [PubMed]
24. Sicherer, S.H.; Eigenmann, P.A.; Sampson, H.A. Clinical features of food protein–induced enterocolitis syndrome. *J. Pediatr.* **1998**, *133*, 214–219. [CrossRef]
25. Diaz, J.J.; Espin, B.; Segarra, O.; Dominguez-Ortega, G.; Blasco-Alonso, J.; Cano, B.; Rayo, A.; Moreno, A. Food protein-induced enter-ocolitis syndrome: Data from a multicenter retrospective study in Spain. *J. Pediatr. Gastroenterol. Nutr.* **2019**, *68*, 232–236. [CrossRef]
26. Miceli Sopo, S.; Monaco, S.; Badina, L.; Barni, S.; Longo, G.; Novembre, E.; Viola, S.; Monti, G. Food protein-induced enterocolitis syn-drome caused by fish and/or shellfish in Italy. *Pediatr. Allergy Immunol.* **2015**, *26*, 731–736. [CrossRef]
27. Sicherer, S.H. Food protein-induced enterocolitis syndrome: Case presentations and management lessons. *J. Allergy Clin. Immunol.* **2005**, *115*, 149–156. [CrossRef]
28. American College of Allergy, Asthma, and Immunology. Food allergy: A practice parameter. *Ann. Allergy Asthma Immunol.* **2006**, *96* (Suppl. 2), S1–S68. [CrossRef]
29. Savilahti, E. Food-Induced Malabsorption Syndromes. *J. Pediatr. Gastroenterol. Nutr.* **2000**, *30*, S61–S66. [CrossRef]
30. Sampson, H.A.; Aceves, S.; Bock, S.A.; James, J.; Jones, S.; Lang, D.; Nadeau, K.; Nowak-Wegrzyn, A.; Oppenheimer, J.; Perry, T.T.; et al. Food allergy: A practice parameter update 2014. J. *Allergy Clin. Immunol.* **2014**, *134*, 1016–1025. [CrossRef]
31. Chehade, M.; Jones, S.M.; Pesek, R.D.; Burks, A.W.; Vickery, B.P.; Wood, R.A.; Leung, D.; Furuta, G.T.; Fleischer, D.M.; Henning, A.; et al. Phenotypic Characterization of Eosinophilic Esophagitis in a Large Multicenter Pa-tient Population from the Consortium for Food Allergy Research. *J. Allergy Clin. Immunol. Pract.* **2018**, *6*, 1534–1544. [CrossRef] [PubMed]
32. Cianferoni, A.; Spergel, J. Eosinophilic Esophagitis: A Comprehensive Review. *Clin. Rev. Allergy Immunol.* **2016**, *50*, 159–174. [CrossRef] [PubMed]
33. Muir, A.B.; Brown-Whitehorn, T.; Godwin, B.; Cianferoni, A. Eosinophilic esophagitis: Early diagnosis is the key. *Clin. Exp. Gastroenterol.* **2019**, *12*, 391–399. [CrossRef] [PubMed]
34. Lucendo, A.J.; Molina-Infante, J.; Arias, Á.; von Arnim, U.; Bredenoord, A.J.; Bussmann, C.; Dias, J.A.; Bove, M.; González-Cervera, J.; Larsson, H.; et al. Guidelines on eosinophilic esophagitis: Evidence based statements and recommendations for diagnosis and management in children and adults. *United Eur. Gastroenterol. J.* **2017**, *5*, 335–358. [CrossRef] [PubMed]
35. Spergel, J.M.; Andrews, T.; Brown-Whitehorn, T.F.; Beausoleil, J.L.; Liacouras, C.A. Treatment of eosinophilic esophagitis with spe-cific food elimination diet directed by a combination of skin prick and patch tests. *Ann. Allergy Asthma Immunol.* **2005**, *95*, 336–343. [CrossRef]
36. Dellon, E.S.; Liacouras, C.A.; Molina Infante, J.; Furuta, G.T.; Spergel, J.M.; Zevit, N.; Spechler, S.J.; Attwood, S.E.; Straumann, A.; Aceves, S.; et al. Updated International Consensus Diagnostic Criteria for Eosinophilic Esophagitis: Proceed-ings of the AGREE Conference. *Gastroenterology* **2018**, *155*, 1022–1033. [CrossRef] [PubMed]
37. Liacouras, C.A.; Furuta, G.T.; Hirano, I.; Atkins, D.; Attwood, S.E.; Bonis, P.A.; Burks, A.W.; Chehade, M.; Collins, M.H.; Dellon, E.S.; et al. Eosinophilic esophagitis: Updated consensus recommendations for children and adults. *J. Allergy Clin. Immunol.* **2011**, *128*, 3–20. [CrossRef]
38. Kagalwalla, A.F.; Shah, A.; Li, B.U.; Sentongo, T.A.; Ritz, S.; Manuel-Rubio, M.; Jacques, K.; Wang, D.; Melin-Aldana, H.; Nelson, S.P. Identification of Specific Foods Responsible for Inflammation in Children With Eosinophilic Esophagitis Successfully Treated With Empiric Elimination Diet. *J. Pediatr. Gastroenterol. Nutr.* **2011**, *53*, 145–149. [CrossRef]
39. Spergel, J.M.; Brown Witehorn, T.F.; Cianferoni, A.; Shuker, M.; Wang, M.L.; Verma, R.; Liacouras, C.A. Identification of causative foods in children with eosinophilic esophagitis treated with an elimination diet. *J. Allergy Clin. Immunol.* **2012**, *130*, 461–467. [CrossRef]
40. Shetty, V.; Daniel, K.E.; Kesavan, A. Hematemesis as Initial Presentation in a 10-Week-Old Infant with Eosinophilic Gastroenter-itis. *Case Rep. Pediatr.* **2017**, *2017*, 2391417.
41. Costa, C.; Pinto Pais, I.; Rios, E.; Costa, C. Milk-sensitive eosinophilic gastroenteritis in a 2-month-old boy. *BMJ Case Rep.* **2015**, *13*, 2015210157. [CrossRef] [PubMed]
42. Gonsalves, N.P. Eosinophilic Gastrointestinal Disorders. *Clin. Rev. Allergy Immunol.* **2019**, *57*, 272–285. [CrossRef] [PubMed]
43. Pesek, R.D.; Reed, C.C.; Muir, A.B.; Fulkerson, P.C.; Menard-Katcher, C.; Falk, G.W.; Kuhl, J.; Martin, E.K.; Magier, A.Z.; Ahmed, F.; et al. Increasing Rates of Diagnosis, Substantial Co-Occurrence, and Variable Treatment Patterns of Eosinophilic Gastritis, Gastroenteritis, and Colitis Based on 10-Year Data Across a Multicenter Consortium. *Am. J. Gastroenterol.* **2019**, *114*, 984–994. [CrossRef] [PubMed]
44. Han, X.X.; Guan, D.X.; Zhou, J.; Yu, F.H.; Wang, G.L.; Mei, T.L.; Guo, S.; Fu, L.B.; Zhang, J.; Shen, H.Q.; et al. Clinical analysis of eosinophilic gastroenteritis in 71 children. *Zhonghua Er Ke Za Zhi* **2018**, *56*, 500–504.
45. Sampson, H.A.; Anderson, J.A. Summary and recommendations: Classification of gastrointestinal manifestations due to immu-nologic reactions to foods in infants and young children. *J. Pediatr. Gastroenterol. Nutr.* **2000**, *30*, S87Y94. [CrossRef]
46. Alfadda, A.A.; Storr, M.A.; Shaffer, E. Eosinophilic colitis: An update on pathophysiology and treatment. *Br. Med. Bull.* **2011**, *100*, 59–72. [CrossRef]
47. Grzybowska-Chlebowczyk, U.; Horowska-Ziaja, S.; Kajor, M.; Więcek, S.; Chlebowczyk, W.; Woś, H. Eosinophilic colitis in chil-dren. *Adv. Dermatol. Allergol.* **2017**, *34*, 52–59. [CrossRef]

48. Pensabene, L.; Brundler, M.-A.; Bank, J.M.; Di Lorenzo, C. Evaluation of mucosal eosinophils in the pediatric colon. *Dig. Dis. Sci.* **2005**, *50*, 221–229. [CrossRef]
49. Yang, M.; Geng, L.; Chen, P.; Wang, F.; Xu, Z.; Liang, C.; Li, H.; Fang, T.; Friesen, C.; Gong, S.; et al. Effectiveness of Dietary Allergen Exclusion Therapy on Eosinophilic Colitis in Chinese Infants and Young Children ≤ 3 Years of Age. *Nutrients* **2015**, *7*, 1817–1827. [CrossRef]
50. Yan, B.M.; Shaffer, E.A. Primary eosinophilic disorders of the gastrointestinal tract. *Gut* **2008**, *58*, 721–732. [CrossRef]
51. Saad, A.G. Normal Quantity and Distribution of Mast Cells and Eosinophils in the Pediatric Colon. *Pediatr. Dev. Pathol.* **2011**, *14*, 294–300. [CrossRef] [PubMed]
52. Hurrell, J.M.; Genta, R.M.; Melton, S.D. Histopathologic Diagnosis of Eosinophilic Conditions in the Gastrointestinal Tract. *Adv. Anat. Pathol.* **2011**, *18*, 335–348. [CrossRef] [PubMed]
53. Mark, J.; Fernando, S.D.; Masterson, J.C.; Pan, Z.; Capocelli, K.E.; Furuta, G.T.; Furuta, G.T.; de Zoeten, E.F. Clinical implications of pedi-atric colonic eosinophilia. *J. Pediatr. Gastroenterol. Nutr.* **2018**, *66*, 760–766. [CrossRef] [PubMed]
54. Yagi, H.; Takizawa, T.; Sato, K.; Inoue, T.; Nishida, Y.; Ishige, T.; Tatsuki, M.; Hatori, R.; Kobayashi, Y.; Yamada, Y.; et al. Sever-ity scales of non-IgE-mediated gastrointestinal food allergies in neonates and infants. *Allergol. Int.* **2019**, *68*, 178–184. [CrossRef]
55. Nomura, I.; Morita, H.; Ohya, Y.; Saito, H.; Matsumoto, K. Non–IgE-Mediated Gastrointestinal Food Allergies: Distinct Differences in Clinical Phenotype BetweenWestern Countries and Japan. *Curr. Allergy Asthma Rep.* **2012**, *12*, 297–303. [CrossRef]
56. Salvatore, S.; Vandenplas, Y. Gastroesophageal reflux and cow milk allergy: Is there a link? *Pediatrics* **2002**, *110*, 972–984. [CrossRef]
57. Nielsen, R.G.; Bindslev-Jensen, C.; Kruse-Andersen, S.; Husby, S. Severe gastroesophageal reflux disease and cow milk hy-persensitivity in infants and children: Disease association and evaluation of a new challenge procedure. *J. Pediatr. Gastroenterol. Nutr.* **2004**, *39*, 383–391. [CrossRef]
58. Rybak, A.; Pesce, M.; Thapar, N.; Borrelli, O. Gastro-Esophageal Reflux in Children. *Int. J. Mol. Sci.* **2017**, *18*, 1671. [CrossRef]
59. Yukselen, A.; Celtik, C. Food allergy in children with refractory gastroesophageal reflux disease. *Pediatr. Int.* **2015**, *58*, 254–258. [CrossRef]
60. Staelens, S.; Driessche, M.V.D.; Barclay, D.; Carrié-Faessler, A.-L.; Haschke, F.; Verbeke, K.; Vandebroek, H.; Allegaert, K.; Van Overmeire, B.; Van Damme, M.; et al. Gastric emptying in healthy newborns fed an intact protein formula, a partially and an extensively hydrolysed formula. *Clin. Nutr.* **2008**, *27*, 264–268. [CrossRef]
61. Johne, B.; Fagerhol, M.K.; Lyberg, T.; Prydz, H.; Brandtzaeg, P.; Naess-Andresen, C.F.; Dale, I. Functional and clinical aspects of the myelomonocyte protein calprotectin. *Mol. Pathol.* **1997**, *50*, 113–123. [CrossRef] [PubMed]
62. Striz, I.; Trebichavský, I. Calprotectin-a pleiotropic molecule in acute and chronic inflammation. *Physiol. Res.* **2004**, *53*, 245–253. [PubMed]
63. Berni Canani, R.; De Horatio, L.T.; Terrin, G.; Romano, M.T.; Miele, E.; Staiano, A.; Rapacciuolo, L.; Polito, G.; Bisesti, V.; Manguso, F.; et al. Combined Use of Noninvasive Tests is Useful in the Initial Diagnostic Approach to a Child with Suspected Inflammatory Bowel Disease. *J. Pediatr. Gastroenterol. Nutr.* **2006**, *42*, 9–15. [CrossRef] [PubMed]
64. von Roon, A.C.; Karamountzos, L.; Purkayastha, S.; Reese, G.E.; Darzi, A.W.; Teare, J.P.; Paraskeva, P.; Tekkis, P.P. Diagnostic precision of fecal calprotectin for inflammatory bowel disease and colorectal malignancy. *Am. J. Gastroenterol.* **2007**, *102*, 803–813. [CrossRef] [PubMed]
65. Chang, M.H.; Chou, J.W.; Chen, S.M.; Tsai, M.C.; Sun, Y.S.; Lin, C.C.; Lin, C.P. Faecal calprotectin as a novel biomarker for differentiat-ing between inflammatory bowel disease and irritable bowel syndrome. *Mol. Med. Rep.* **2014**, *10*, 522–526. [CrossRef]
66. Degraeuwe, P.L.J.; Beld, M.P.A.; Ashorn, M.; Berni Canani, R.; Day, A.S.; Diamanti, A.; Fagerberg, U.L.; Henderson, P.; Kolho, K.-L.; Van De Vijver, E.; et al. Faecal Calprotectin in Suspected Paediatric Inflammatory Bowel Disease. *J. Pediatr. Gastroenterol. Nutr.* **2015**, *60*, 339–346. [CrossRef]
67. Berni Canani, R.; Rapacciuolo, R.M.T.; Tanturri de Horatio, L.; Terrin, G.; Manguso, F.; Cirillo, P.; Paparo, F.; Troncone, R. Di-agnostic value of faecal calprotectin in pediatric gastroenterology clinical practice. *Dig. Liver Dis.* **2004**, *36*, 467–470. [CrossRef]
68. Bunn, S.K.; Bisset, W.M.; Main, M.J.; Golden, B.E. Fecal Calprotectin as a Measure of Disease Activity in Childhood Inflammatory Bowel Disease. *J. Pediatr. Gastroenterol. Nutr.* **2001**, *32*, 171–177. [CrossRef]
69. Xiong, L.-J.; Xie, X.-L.; Li, Y.; Deng, X.-Z. Current status of fecal calprotectin as a diagnostic or monitoring biomarker for cow's milk protein allergy in children: A scoping review. *World J. Pediatr.* **2020**. [CrossRef]
70. Baldassarre, M.E.; Laforgia, N.; Fanelli, M.; Laneve, A.; Grosso, R.; Lifschitz, C. Lactobacillus GG improves recovery in in-fants with blood in the stool and presumptive allergic colitis compared with extensively hydrolyzed formula alone. *J. Pediatr.* **2010**, *156*, 397–401. [CrossRef]
71. Beşçr, O.F.; Sancak, S.; Erkan, T.; Kutlu, T.; Cokugraş, H.; Cokugraş, F.C. Can fecal calprotectin level be used as a markers of inflammation in the diagnosis and follow-up of cow's milk protein allergy? *Allergy Asthma Immunol. Res.* **2014**, *6*, 33–38. [CrossRef] [PubMed]
72. Trillo Belizón, C.; Ortega Paez, E.; Medina Claros, A.F.; Rodriguez Sanchez, I.; Reina Gonzalez, A.; Vera Medialdea, R.; Ra-mon Salguero, J.M. Fecal calprotectin as an aid to the diagnosis of non-IgE mediated cow's milk protein allergy. *An. Pediatr.* **2016**, *84*, 318–323. [CrossRef] [PubMed]
73. Lendvai-Emmert, D.; Emmert, V.; Fusz, K.; Premusz, V.; Boncz, I.; Toth, G.P. Quantitative assay of fecal calprotectin in chil-dren with cow's milk protein allergy. *Value Health* **2018**, *21*, s128. [CrossRef]

74. Prikhodchenko, N.G.; Shumakova, T.A.; Nee, A.N.; Katenkova, E.U.; Zernova, E.S.; Grigoryan, L.A. Noninvasive marker of the intestinal epithelial barrier state in infants with Food Protein-Induced Enteropathy. *Res. J. Pharm. Biol. Chem. Sci.* **2018**, *9*, 625–628.
75. Ataee, P.; Zoghali, M.; Nikkhoo, B.; Ghaderi, E.; Mansouri, M.; Nasiri, R.; Eftekhari, K. Diagnostic Value of Fecal Calprotectin in Response to Mother's Diet in Breast-Fed Infants with Cow's Milk Allergy Colitis. *Iran. J. Pediatr.* **2018**, *28*, 66172. [CrossRef]
76. Diaz, M.; Guadamuro, L.; Espinosa-Martos, I.; Mancabelli, L.; Jiménez, S.; Molinos-Norniella, C.; Pérez Solis, D.; Milani, C.; Rodriguez, J.M.; Ventura, M.; et al. Microbiota and derived parameters in fecal samples of infants with non-IgE cow's milk protein allergy under a restricted diet. *Nutrients* **2018**, *10*, 1481. [CrossRef]
77. Berni Canani, R.; Nocerino, R.; Leone, L.; Di Costanzo, M.; Terrin, G.; Passariello, A.; Cosenza, L.; Troncone, R. Tolerance to a new free amino acid-based formula in children with IgE or non-IgE-mediated cow's milk allergy: A randomized controlled clinical trial. *BMC Pediatr.* **2013**, *13*, 24. [CrossRef]
78. Yamada, Y. Unique features of non-IgE-mediated gastrointestinal food allergy during infancy in Japan. *Curr. Opin. Allergy Clin. Immunol.* **2020**, *20*, 299–304. [CrossRef]
79. Kalach, N.; Kapel, N.; Waligora-Dupriet, A.; Castelain, M.C.; Cousin, M.O.; Sauvage, C.; Ba, F.; Nicolis, I.; Campeotto, B.; Butel, M.J.; et al. Intestinal permeability and fecal eosinophilic-derived neurotoxin are the best diagnosis tools for digestive non-IgE-mediated cow's milk allergy in toddlers. *Clin. Chem. Lab. Med.* **2013**, *51*, 351–361. [CrossRef]
80. Wada, H.; Horisawa, T.; Inoue, M.; Yoshida, T.; Toma, T.; Yachie, A. Sequential measurement of fecal parameters in a case of non-immunoglobulin E-mediated milk allergy. *Pediatr. Int.* **2007**, *49*, 109–111. [CrossRef]
81. Wada, T.; Toma, T.; Muraoka, M.; Matsuda, Y.; Yachie, A. Elevation of fecal eosinophil-derived neurotoxin in infants with food protein-induced enterocolitis syndrome. *Pediatr. Allergy Immunol.* **2014**, *25*, 617–619. [CrossRef] [PubMed]
82. Rycyk, A.; Cudowska, B.; Lebensztejn, D.M. Eosinophil-Derived Neurotoxin, Tumor Necrosis Factor Alpha, and Calprotectin as Non-Invasive Biomarkers of Food Protein-Induced Allergic Proctocolitis in Infants. *J. Clin. Med.* **2020**, *9*, 3147. [CrossRef] [PubMed]
83. Concha, S.; Cabalin, C.; Iturriaga, C.; Pérez-Mateluna, G.; Gomez, C.; Cifuentes, L.; Harris, P.R.; Gana, J.C.; Borzutzky, A. Diagnostic validity of fecal occult blood test in infants with food protein-induced allergic proctocolitis. *Rev. Chil. Pediatr.* **2018**, *89*, 630–637. [PubMed]
84. Tøn, H.; Brandsnes, Ø.; Dale, S.; Holtlund, J.; Skuibina, E.; Schjønsby, H.; Johne, B. Improved assay for fecal calprotectin. *Clin. Chim. Acta* **2000**, *292*, 41–54. [CrossRef]
85. Fagerberg, U.L.; Loof, L.; Merzoug, R.D.; Hansson, L.O.; Finkel, Y. Fecal calprotectin levels in healthy children studied with an im-proved assay. *J. Pediatr. Gastroenterol. Nutr.* **2003**, *37*, 468–472. [CrossRef]
86. Li, F.; Ma, J.; Geng, S.; Wang, J.; Liu, J.; Zhang, J.; Sheng, X. Fecal Calprotectin Concentrations in Healthy Children Aged 1–18 Months. *PLoS ONE* **2015**, *10*, e0119574. [CrossRef]
87. Zhu, Q.; Li, F.; Wang, J.; Shen, L.; Sheng, X. Fecal Calprotectin in Healthy Children Aged 1-4 Years. *PLoS ONE* **2016**, *11*, e0150725. [CrossRef]
88. Oord, T.; Hornung, N. Fecal calprotectin in healthy children. *Scand. J. Clin. Lab. Investig.* **2014**, *74*, 254–258. [CrossRef]
89. Şahin, B.S.G.; Keskindemirci, G.; Özden, T.A.; Durmaz, Ö.; Gökçay, G. Faecal calprotectin levels during the first year of life in healthy children. *J. Paediatr. Child. Health* **2020**, *56*, 1806–1811. [CrossRef]
90. Merras-Salmio, L.; Kolho, K.L.; Pelkonen, A.S.; Kuitunen, M.; Mäkelä, M.J.; Savilahti, E. Markers of gut mucosal inflammation and cow's milk specific immunoglobulins in non-IgE cow's milk allergy. *Clin. Transl. Allergy* **2014**, *4*, 8. [CrossRef]
91. Meyer, R.; ChebarLozinsky, A.; Fleischer, D.M.; Vieira, M.C.; Du Toit, G.; Vandenplas, Y.; Dupont, C.; Knibb, R.; Uysal, P.; Cavkaytar, O.; et al. Diagnosis and management of Non-IgE gastrointestinal allergies in breastfed infants-An EAACI Position Paper. *Allergy* **2020**, *75*, 14–32. [CrossRef] [PubMed]
92. Hochwallner, H.; Schulmeister, U.; Swoboda, I.; Twaroch, T.E.; Vogelsang, H.; Kazemi-Shirazi, L.; Kundi, M.; Balic, N.; Quirce, S.; Rumpold, H.; et al. Patients suffering from non-IgE-mediated cow's milk protein intolerance cannot be diagnosed based on IgG subclass or IgA responses to milk allergens. *Allergy* **2011**, *66*, 1201–1207. [CrossRef] [PubMed]
93. Stapel, S.O.; Asero, R.; Ballmer-Weber, B.K.; Knol, E.F.; Strobel, S.; Vieths, S.; Kleine-Tebbe, J.; EAACI Task Force. Testing for IgG4 against foods is not recommended as a diagnostic tool: EAACI Task Force Report. *Allergy* **2008**, *63*, 793–796. [CrossRef]
94. Carr, S.; Chan, E.S.; LaVine, E.; Moote, D.W. CSACI Position statement on the testing of food-specific IgG. *Allergy, Asthma Clin. Immunol.* **2012**, *8*, 12. [CrossRef] [PubMed]
95. Clayton, F.; Fang, J.C.; Gleich, G.J.; Lucendo, A.J.; Olalla, J.M.; Vinson, L.A.; Lowichik, A.; Chen, X.; Emerson, L.; Cox, K.; et al. Eosinophilic esophagitis in adults is associated with IgG4 and not mediated by IgE. *Gastroenterology* **2014**, *147*, 602–609. [CrossRef]
96. Wright, B.L.; Kulis, M.; Guo, R.; Orgel, K.A.; Wolf, W.A.; Burks, A.W.; Vickery, B.P.; Dellon, E.S. Food-specific IgG4 is associated with eosinophilic esophagitis. *J. Allergy Clin. Immunol.* **2016**, *138*, 1190–1192. [CrossRef]
97. Schuyler, A.J.; Wilson, J.M.; Tripathi, A.; Commins, S.P.; Ogbogu, P.U.; Kruzsewski, P.G.; Barnes, B.H.; McGowan, E.C.; Workman, L.J.; Lidholm, J.; et al. Specific IgG4 antibodies to cow's milk proteins in pediatric patients with eosinophilic esophagitis. *J. Allergy Clin. Immunol.* **2018**, *142*, 139–148. [CrossRef]
98. Rosenberg, C.E.; Mingler, M.K.; Caldwell, J.M.; Collins, M.H.; Fulkerson, P.C.; Morris, D.W.; Mukkada, V.A.; Putnam, P.E.; Shoda, T.; Wen, T.; et al. Esophageal IgG4 levels correlate with histopathologic and transcriptomic features in eosinophilic esopha-gitis. *Allergy* **2018**, *73*, 1892–1901. [CrossRef]

99. Giavi, S.; Megremis, S.; Papadopoulos, N.G. Lymphocyte stimulation test for the diagnosis of non-IgE-mediated cow's milk allergy: A step closer to a noninvasive diagnostic tool? *Allergy Immunol.* **2012**, *157*, 1–2. [CrossRef]
100. Morita, H.; Nomura, I.; Matsuda, A.; Saito, H.; Matsumoto, K. Gastrointestinal Food Allergy in Infants. *Allergol. Int.* **2013**, *62*, 297–307. [CrossRef]
101. Kimura, M.; Oh, S.; Narabayashi, S.; Taguchi, T. Usefulness of Lymphocyte Stimulation Test for the Diagnosis of Intestinal Cow's Milk Allergy in Infants. *Int. Arch. Allergy Immunol.* **2012**, *157*, 58–64. [CrossRef] [PubMed]
102. Miyazawa, T.; Itabashi, K.; Imai, T. Retrospective Multicenter Survey on Food-Related Symptoms Suggestive of Cow's Milk Allergy in NICU Neonates. *Allergol. Int.* **2013**, *62*, 85–90. [CrossRef] [PubMed]
103. Ikeda, K.; Ida, S.; Kawahara, H.; Kawamoto, K.; Etani, Y.; Kubota, A. Importance of evaluating for cow's milk allergy in pediatric surgical patients with functional bowel symptoms. *J. Pediatric. Surgery* **2011**, *46*, 2332–2335. [CrossRef]
104. Kajita, N.; Yoshida, K.; Furukawa, M.; Akasawa, A. High Stimulation Index for Quail Egg Yolk After an Allergen-Specific Lym-phocyte Stimulation Test in a Child With Quail Egg Induced Food Protein Induced Enterocolitis Syndrome. *J. Investig. Allergol. Clin. Immunol.* **2019**, *29*, 66–68. [CrossRef] [PubMed]
105. Matsumoto, K. Non-IgE-Related Diagnostic Methods (LST, Patch Test). *Chem. Immunol. Allergy* **2015**, *101*, 79–86.
106. Yagi, H.; Takizawa, T.; Sato, K.; Inoue, T.; Nishida, Y.; Yamada, S.; Ishige, T.; Hatori, R.; Inoue, T.; Yamada, Y.; et al. Interleukin 2 recep-tor-alpha expression after lymphocyte stimulation for non-IgE-mediated gastrointestinal food allergies. *Allergol. Int.* **2020**, *69*, 287–289. [CrossRef]
107. Turjanmaa, K.; Darsow, U.; Niggemann, B.; Rancé, F.; Vanto, T.; Werfel, T. EAACI/GA2LEN Position paper: Present status of the atopy patch test. *Allergy* **2006**, *61*, 1377–1384. [CrossRef]
108. Niggemann, B.; Ziegert, M.; Reibel, S. Importance of chamber size for the outcome of atopy patch testing in children with atopic dermatitis and food allergy. *J. Allergy Clin. Immunol* **2002**, *110*, 515–516. [CrossRef]
109. Berni Canani, R.; Ruotolo, S.; Auricchio, L.; Caldore, M.; Porcaro, F.; Manguso, F.; Terrin, G.; Troncone, R. Diagnostic accuracy of the atopy patch test in children with food allergy-related gastrointestinal symptoms. *Allergy* **2007**, *62*, 738–743. [CrossRef]
110. Johansen, J.D.; Aalto-Korte, K.; Agner, T.; Andersen, K.E.; Bircher, A.J.; Bruze, M.; Cannavó, A.; Giménez-Arnau, A.; Gonçalo, M.; Goossens, A.; et al. European Society of Contact Dermatitis guideline for diagnostic patch testing-recommendations on best practice. *Contact Dermat.* **2015**, *73*, 195–221. [CrossRef]
111. Muraro, A.; Werfel, T.; Hoffmann-Sommergruber, K.; Roberts, G.; Beyer, K.; Bindslev-Jensen, C.; Cardona, V.; Dubois, A.; duToit, G.; Eigenmann, P.; et al. EAACI food allergy and ana-phylaxis guidelines: Diagnosis and management of food allergy. *Allergy* **2014**, *69*, 1008–1025. [CrossRef] [PubMed]
112. Luo, Y.; Zhang, G.-Q.; Li, Z.-Y. The diagnostic value of APT for food allergy in children: A systematic review and meta-analysis. *Pediatr. Allergy Immunol.* **2019**, *30*, 451–461. [CrossRef] [PubMed]
113. Gayam, S.; Zinn, Z.; Chelliah, M.; Teng, J. Patch testing in gastrointestinal diseases-a systematic review of the patch test and atopy patch test. *J. Eur. Acad. Dermatol. Venereol.* **2018**, *32*, e349–e351. [CrossRef] [PubMed]
114. Dávila, G.; Zapatero, L.; Alonso, E.; Rojas, P.; Martin, E.; Martínez-Molero, I. Food Protein-Induced Enterocolitis Syndrome Caused by Fish. *J. Allergy Clin. Immunol.* **2006**, *117*, S304. [CrossRef]
115. Fogg, M.I.; Brown-Whitehorn, T.A.; Pawlowski, N.A.; Spergel, J.M. Atopy patch test for the diagnosis of food protein-induced en-terocolitis syndrome. *Pediatr. Allergy Immunol.* **2006**, *17*, 351–355. [CrossRef]
116. Nocerino, R.; Granata, V.; Di Costanzo, M.; Pezzella, V.; Leone, L.; Passariello, A.; Terrin, G.; Troncone, R.; BerniCanani, R. Atopy patch tests are useful to predict oral tolerance in children with gastrointestinal symptoms related to non-IgE-mediated cow's milk allergy. *Allergy* **2013**, *68*, 246–248. [CrossRef]
117. Alves, F.A.; Cheik, M.F.A.; De Nápolis, A.C.R.; Rezende, É.R.M.D.A.; Barros, C.P.; Segundo, G.R.S. Poor utility of the atopy patch test in infants with fresh rectal bleeding. *Ann. Allergy Asthma Immunol.* **2015**, *115*, 161–162. [CrossRef]
118. Gonzaga, T.A.; Alves, F.A.; Cheik, M.F.A.; de Barros, C.P.; Rezende, E.R.M.A.; Segundo, G.R.S. Low efficacy of atopy patch test in predicting tolerance development in non-IgE-mediated cow's milk allergy. *Allergol. Immunopathol. (Madr.)* **2018**, *46*, 241–246. [CrossRef]
119. Sirin Kose, S.; Asilsoy, S.; Tezcan, D.; Atakul, G.; Al, S.; Atay, O.; KangalliBoyacioglu, O.; Uzuner, N.; Anal, O.; Karaman, O. Atopy patch test in children with cow's milk and hen's egg allergy: Do clinical symptoms matter? *Allergol. Immunopathol. (Madr.)* **2020**, *48*, 323–331. [CrossRef]
120. Spergel, J.M.; Brown-Whitehorn, T.; Beausoleil, J.L.; Shuker, M.; Liacouras, C.A. Predictive values for skin prick test and atopy patch test for eosinophilic esophagitis. *J. Allergy Clin. Immunol.* **2007**, *119*, 509–511. [CrossRef]
121. Spergel, J.M.; Brown-Whitehorn, T.; Muir, A.B.; Liacouras, C.A. Medical algorithm: Diagnosis and treatment of eosinophilic esophagitis in children. *Allergy* **2020**, *75*, 1522–1524. [CrossRef] [PubMed]
122. Iyngkaran, N.; Robinson, M.J.; Prathap, K.; Sumithran, E.; Yadav, M. Cows' milk protein-sensitive enteropathy. Combined clinical and histological criteria for diagnosis. *Arch. Dis. Child.* **1978**, *53*, 20–26. [CrossRef]
123. Katzka, D.A. Eosinophilic Esophagitis. *Ann. Intern. Med.* **2020**, *172*, Itc65–Itc80. [CrossRef]
124. Lawrence, W.W.; Wright, J.L. Causes of rectal bleeding in children. *Pediatr. Rev.* **2001**, *22*, 394–395. [PubMed]
125. Jang, H.-J.; Kim, A.S.; Hwang, J.-B. The etiology of small and fresh rectal bleeding in not-sick neonates: Should we initially suspect food protein-induced proctocolitis? *Eur. J. Nucl. Med. Mol. Imaging* **2012**, *171*, 1845–1849. [CrossRef] [PubMed]

126. Fleischer, D.M.; Conover-Walker, M.K.; Christie, L.; Burks, A.W.; Wood, R.A. Peanut allergy: Recurrence and its management. *J. Allergy Clin. Immunol.* **2004**, *114*, 1195–1201. [CrossRef] [PubMed]
127. Rosen, R.; Vandenplas, Y.; Singendonk, M.; Cabana, M.; DiLorenzo, C.; Gottrand, F.; Gupta, S.; Langendam, M.; Staiano, A.; Thapar, N.; et al. Pediatric Gastroesophageal Reflux Clinical Practice Guidelines: Joint Recommendations of the North American Society for Pediatric Gastroenterology, Hepatology, and Nutrition and the European Society for Pediatric Gastro-enterology, Hepatology, and Nutrition. *J. Pediatr. Gastroenterol. Nutr.* **2018**, *66*, 516–554. [PubMed]
128. Poddar, U.; Yachha, S.K.; Krishnani, N.; Srivastava, A. Cow's milk protein allergy: An entity for recognition in developing countries. *J. Gastroenterol. Hepatol.* **2010**, *25*, 178–182. [CrossRef]
129. Vandenplas, Y.; Dupont, C.; Eigenmann, P.; Host, A.; Kuitunen, M.; Ribes-Koninckx, C.; Shah, N.; Shamir, R.; Staiano, A.; Szajewska, H.; et al. A workshop report on the development of the Cow's Milk-related Symptom Score awareness tool for young children. *Acta Paediatr.* **2015**, *104*, 334–339. [CrossRef]
130. Vandenplas, Y.; Mukherjee, R.; Dupont, C.; Eigenmann, P.; Høst, A.; Mikael Kuitunen, M.; Ribes-Koninkx, C.; Shah, N.; Szajewska, H.; von Berg, A.; et al. Protocol for the validation of sensitivity and specificity of the Cow's Milk-related Symptom Score (CoMiSS) against open food challenge in a single-blind, prospective, multicentre trial in infants. *BMJ Open* **2018**, *8*, 1–9. [CrossRef]
131. Vandenplas, Y.; Steenhout, P.; Grathwohl, D.; Althera Study Group. A pilot study on the application of a symptom-based score for the diagnosis of cow's milk protein allergy. *SAGE Open Med.* **2014**, *2*, 1–5. [CrossRef] [PubMed]
132. Vandenplas, Y.; Salvatore, S.; Ribes-Koninckx, C.; Carvajal, E.; Szajewska, H.; Huysentruyt, K. The Cow Milk Symptom Score (CoMiSSTM) in presumed healthy infants. *PLoS ONE* **2018**, *13*, e0200603. [CrossRef] [PubMed]
133. Bigorajska, K.; Filipiak, Z.; Winiarska, P.; Adamiec, A.; Trent, B.; Vandenplas, Y.; Ruszczyński, M.; Szajewska, H. Cow's Milk-Related Symptom Score in Presumed Healthy Polish Infants Aged 0–6 Months. *Pediatr. Gastroenterol. Hepatol. Nutr.* **2020**, *23*, 154–162. [CrossRef] [PubMed]
134. Chakrabarty, A.; Sahay, K. The use of cow's milk related Symptom Score (CoMiSS) as a predictor of cow's milk protein allergy. *Clin. Exp. Allergy* **2017**, *47*, 12.
135. Rigley, L.; Cribb, E. Using a symptom scoring tool aid diagnosis and management of cow's milk protein allergy. *J. Pediatr. Gastroenterol. Nutr.* **2017**, *64* (Suppl. 1), 958–959.
136. Bajerova, K.; Bajer, M. CoMiSS usage in practice-our first experience. *J. Pediatr. Gastroenterol. Nutr.* **2017**, *64* (Suppl. 1), 919.
137. Prasad, R.; Venkata, R.S.A.; Ghokale, P.; Chakravarty, P.; Anwar, F. Cow's Milk-related Symptom Score as a predictive tool for cow's milk allergy in Indian children aged 0–24 months. *Asia Pac. Allergy* **2018**, *8*, e36. [CrossRef]
138. Armano, C.; Ottaviano, G.; Fumagalli, L.; Luini, C.; Salvatore, S. CoMiSS score and the response to cow's milk free diet in infants and children. *Dig. Liver Dis.* **2017**, *49*, e269. [CrossRef]
139. Salvatore, S.; Bertoni, E.; Bogni, F.; Bonaita, V.; Armano, C.; Moretti, A.; Baù, M.; Luini, C.; D'Auria, E.; Marinoni, M.; et al. Testing the Cow's Milk-Related Symptom Score (CoMiSSTM) for the Response to a Cow's Milk-Free Diet in Infants: A Pro-spective Study. *Nutrients* **2019**, *11*, 2402. [CrossRef]
140. Moretti, A.; Bertoni, E.; Armano, C.; Marinoni, M.; Bogni, F.; Agosti, M.; Salvatore, S. Test and validation of the Cow's Milk-related Symptom Score (CoMiSS®) in a population of symptomatic and healthy infants. *J. Pediatr. Gastroenterol. Nutr.* **2019**, *68* (Suppl. 1), 1103.
141. Vandenplas, Y.; Steenhout, P.; Järvi, A.; Garreau, A.S.; Mukherjee, R. Pooled Analysis of the Cow's Milk-related-Symptom-Score (CoMiSSTM) as a Predictor for Cow's Milk Related Symptoms. *Pediatr. Gastroenterol. Hepatol. Nutr.* **2017**, *20*, 22–26. [CrossRef] [PubMed]
142. Vandenplas, Y.; Steenhout, P.; Planoudis, Y.; Grathwohl, D.; Althera Study Group. Treating cow's milk protein allergy: A dou-ble-blind randomized trial comparing two extensively hydrolysed formulas with probiotics. *Acta Paediatr.* **2013**, *102*, 990–998. [CrossRef] [PubMed]
143. Vandenplas, Y.; De Greefe, E.; Hauser, B.; Paradice Study Group. Safety and tolerance of a new extensively hydrolyzed rice pro-tein-based formula in the management of infants with cow's milk protein allergy. *Eur. J. Pediatr.* **2014**, *17*, 1209–1216. [CrossRef] [PubMed]
144. Vandenplas, Y.; De Greef, E.; Xinias, I.; Vrani, O.; Mavroudi, A.; Hammoud, M.; Al Refai, F.; Khalife, M.C.; Sayad, A.; Noun, P.; et al. Safety of a thickened extensively casein hydrolysate formula. *Nutrition* **2016**, *32*, 206–212. [CrossRef]
145. Kose, S.S.; Atakul, G.; Asilsoy, S.; Uzuner, N.; Anal, O.; Karaman, O. The efficiency of the symptom-based score in infants diagnosed with cow's milk protein and hen's egg allergy. *Allergol. Immunopathol.* **2019**, *47*, 265–271. [CrossRef]
146. Seda, S.K.; Gizem, A.; Suna, A.; Nevin, U.; Ozkan, K.; Ozden, A. The effectiveness of symptom-based score on the diagnosis and follow-up of food allergy. *Clin. Transl. Allergy* **2018**, *8* (Suppl. 2), P61.
147. Selbuz, S.K.; Astuntaş, C.; Kansu, A.; Kırsaçlıoğlu, C.T.; Kuloğlu, Z.; İlarslan, N.; Doğulu, N.; Günay, F.; Topçu, S.; Ulukol, B. Assessment of cow's milk-related symptom scoring awareness tool in young Turkish children. *J. Paediatr. Child. Health* **2020**, *56*, 1799–1805. [CrossRef]
148. Kaymak, S.; Kirsaçlioğlu, C.T.; Kansu, A.; Kuloğlu, Z.; Altuntaş, C.; Doğulu, N.; Ilarslan, N.E.C.; Günai, F.; Topçu, S.; Ulukol, B. Assessment of Cow's milk-related symptom scoring awareness tool in Turkish infants. *J. Pediatr. Gastroenterol. Nutr.* **2018**, *66* (Suppl. 2), 1030.
149. Vandenplas, Y.; Carvajal, E.; Peeters, S.; Balduck, N.; Jaddioui, Y.; Ribes-Koninckx, C.; Huysentruyt, K. The Cow's Milk-Related Symptom Score (CoMiSSTM): Health Care Professional and Parent and Day-to-Day Variability. *Nutrients* **2020**, *12*, 438. [CrossRef]

150. Huysentruyt, K.; Peeters, S.; Balduck, N.; Vandenplas, Y. Day to Day and inter-rater variability of the Cow's Milk-related Symptom Score (CoMiSSTM). *J. Ped. Gastr. Nutr.* **2019**, *68* (Suppl. 1), 1089.
151. Mukherjee, S.; Harding, J.; Daniel, E.; Natarajan, S.; Farmer, H.; Allan-Smith, D.; Holland, S. Computer based diagnosis and man-agement algorithm for cow's milk protein allergy (CMPA) in primary care in United Kingdom-a proof of concept study. *Allergy Eur. J. Allergy Clin. Immunol.* **2019**, *74*, 357.
152. Kherkheulidze, M.; Kavlashvili, N.; Kandelaki, E. Clinical symptoms and development of infants with cow's milk allergy fed with hydrolyzed formula. *World Allergy Organ. J.* **2017**, *10* (Suppl. 1), 25.
153. Zeng, Y.; Zhang, J.; Dong, G.; Liu, P.; Xiao, F.; Li, W.; Wang, L.; Wu, Q. Assessment of Cow's milk-related symptom scores in early identification of cow's milk protein allergy in Chinese infants. *BMC Pediatr.* **2019**, *19*, 191. [CrossRef] [PubMed]
154. Devonshire, A.L.; Durrani, S.; Assa'ad, A. Non-IgE-mediated food allergy during infancy. *Curr. Opin. Allergy Clin. Immunol.* **2020**, *20*, 292–298. [CrossRef]
155. Sugaya, M. Chemokines and Skin Diseases. *Arch. Immunol. Ther. Exp.* **2014**, *63*, 109–115. [CrossRef]
156. Fujisawa, T.; Nagao, M.; Hiraguchi, Y.; Katsumata, H.; Nishimori, H.; Iguchi, K.; Kato, Y.; Higashiura, M.; Ogawauchi, I.; Tamaki, K. Serum measurement of thymus and activation-regulated chemokine/CCL17 in children with atopic dermatitis: Elevated normal levels in infancy and age-specific analysis in atopic dermatitis. *Pediatr. Allergy Immunol.* **2009**, *20*, 633–641. [CrossRef]
157. Hamano, S.; Yamamoto, A.; Fukuhara, D.; Yan, K. Serum thymus and activation regulated chemokine level as a potential bi-omarker for food protein induced enterocolitis syndrome. *Pediatr. Allergy Immunol.* **2019**, *30*, 387–389. [CrossRef]
158. Aparicio, M.; Alba, C.; Proctocolitis Study Group of CAM Public Health Area 6 Proctocolitis Study Group of CAM Public Health Area 6; Rodríguez, J.M.; Fernández, L. Microbiological and Immunological Markers in Milk and Infant Feces for Common Gastrointestinal Disorders: A Pilot Study. *Nutrients* **2020**, *12*, 634. [CrossRef]
159. Dong, P.; Feng, J.-J.; Yan, D.-Y.; Lyu, Y.-J.; Xu, X. Early-life gut microbiome and cow's milk allergy—A prospective case-control 6-month follow-up study. *Saudi J. Biol. Sci.* **2018**, *25*, 875–880. [CrossRef]
160. Zhao, W.; Ho, H.E.; Bunyavanich, S. Faculty Opinions recommendation of The gut microbiome in food allergy. *Fac. Opin. Post-Publication Peer Rev. Biomed. Lit.* **2020**, *122*, 276–282. [CrossRef]
161. Mehr, S.; Lee, E.; Hsu, P.; Anderson, D.; de Jong, E.; Bosco, A.; Campbell, D.E. Innate immune activation occurs in acute food pro-tein-induced enterocolitis syndrome reactions. *J. Allergy Clin. Immunol.* **2019**, *144*, 600–602. [CrossRef] [PubMed]
162. Schouten, B.; Van Esch, B.C.; Van Thuijl, A.O.; Blokhuis, B.R.; Kormelink, T.G.; Hofman, G.A.; Moro, G.E.; Boehm, G.; Arslanoglu, S.; Sprikkelman, A.B.; et al. Contribution of IgE and immunoglobulin free light chain in the allergic reaction to cow's milk proteins. *J. Allergy Clin. Immunol.* **2010**, *125*, 1308–1314. [CrossRef] [PubMed]
163. Brusca, I.; Bizzaro, N.; Pesce, G.; Caponi, L.; Caruso, B.; Villalta, D. Recommendations for the use of the basophil activation test in the diagnosis of allergic diseases. La Rivista Italiana della Medicina di Laboratorio-Italian. *J. Lab. Med.* **2015**, *11*, 225–231.
164. Ansotegui, I.J.; Melioli, G.; Canonica, G.W.; Caraballo, L.; Villa, E.; Ebisawa, M.; Passalacqua, G.; Savi, E.; Ebo, D.; Gómez, R.M.; et al. IgE allergy diagnostics and other relevant tests in allergy, a World Allergy Organization position paper. *World Allergy Organ. J.* **2020**, *13*, 100080. [CrossRef] [PubMed]
165. Lopes, J.P.; Sicherer, S.H. Food allergy: Epidemiology, pathogenesis, diagnosis, prevention, and treatment. *Curr. Opin. Immunol.* **2020**, *66*, 57–64. [CrossRef]
166. Jimbo, K.; Ohtsuka, Y.; Kono, T.; Arai, N.; Kyoudo, R.; Hosoi, K.; Aoyagi, Y.; Kudo, T.; Arai, N.; Shimizu, T. Ultrasonographic study of intestinal Doppler blood flow in infantile non-IgE-mediated gastrointestinal food allergy. *Allergol. Int.* **2019**, *68*, 199–206. [CrossRef]
167. Lack, G. Update on risk factors for food allergy. *J. Allergy Clin. Immunol.* **2012**, *129*, 1187–1197. [CrossRef]
168. Renz, H.; Autenrieth, I.B.; Brandtzaeg, P. Gene environment interaction in chronic disease: A European Science Foundation Forward Look. *J. Allergy Clin. Immunol.* **2011**, *128* (Suppl. 6), S27–S49. [CrossRef]
169. Simmons, R.A. Epigenetics and maternal nutrition: Nature v. nurture. *Proc. Nutr. Soc.* **2010**, *70*, 73–81. [CrossRef]
170. Feil, R.; Fraga, M.F. Epigenetics and the environment: Emerging patterns and implications. *Nat. Rev. Genet.* **2012**, *13*, 97–109. [CrossRef]
171. Alhamwe, B.A.; Meulenbroek, L.A.P.M.; Veening-Griffioen, D.H.; Wehkamp, T.M.D.; Alhamdan, F.; Miethe, S.; Harb, H.; Hogenkamp, A.; Knippels, L.M.J.; Von Strandmann, E.P.; et al. Decreased Histone Acetylation Levels at Th1 and Regulatory Loci after Induction of Food Allergy. *Nutrients* **2020**, *12*, 3193. [CrossRef] [PubMed]
172. Potaczek, D.P.; Harb, H.; Michel, S.; Alhamwe, B.A.; Renz, H.; Tost, J. Epigenetics and allergy: From basic mechanisms to clinical applications. *Epigenomics* **2017**, *9*, 539–571. [CrossRef] [PubMed]
173. van Esch, B.C.A.M.; Porbahaie, M.; Abbring, S.; Garssen, J.; Potaczek, D.P.; Savelkoul, H.F.J.; van Neerven, R.J.J. The Impact of Milk and Its Components on Epigenetic Programming of Immune Function in Early Life and Beyond: Implications for Allergy and Asthma. *Front Immunol.* **2020**, *11*, 2141. [CrossRef] [PubMed]
174. Martino, D.; Joo, J.E.; Sexton-Oates, A. Epigenome wide association study reveals longitudinally stable DNA methylation differences in CD4+ T cells from children with IgE-mediated food allergy. *Epigenetics* **2014**, *9*, 998–1006. [CrossRef]
175. Hill, D.J.; Ball, G.; Hosking, C.S. Clinical manifestations of cows' milk allergy in childhood. I. Associations with in-vitro cellular immune responses. *Clin. Exp. Allergy* **1988**, *18*, 469–479. [CrossRef] [PubMed]
176. Gandhi, N.A.; Bennett, B.L.; Graham, N.M.; Pirozzi, G.; Stahl, N.; Yancopoulos, G.D. Targeting key proximal drivers of type 2 in-flammation in disease. *Nat. Rev. Drug Discov.* **2016**, *15*, 35–50. [CrossRef]

177. Berni Canani, R.; Paparo, L.; Nocerino, R. Differences in DNA methylation profile of Th1 and Th2 cytokine genes are asso-ciated with tolerance acquisition in children with IgE mediated cow's milk allergy. *Clin. Epigenet.* **2015**, *7*, 38. [CrossRef]
178. Jensen, E.T.; Langefeld, C.D.; Zimmerman, K.D.; Howard, T.D.; Dellon, E.S. Epigenetic methylation in Eosinophilic Esophagitis: Mo-lecular ageing and novel biomarkers for treatment response. *Clin. Exp. Allergy* **2020**, *500*, 1372–1380. [CrossRef]
179. Wąsik, M.; Nazimek, K.; Nowak, B.; Askenase, P.W.; Bryniarski, K. Delayed-Type Hypersensitivity Underlying Casein Allergy Is Suppressed by Extracellular Vesicles Carrying miRNA-150. *Nutrients* **2019**, *11*, 907. [CrossRef]
180. Volpicella, M.; Leoni, C.; Dileo, M.C.G.; Ceci, L.R. Progress in the Analysis of Food Allergens through Molecular Biology Approaches. *Cells* **2019**, *8*, 1073. [CrossRef]

MDPI

Review

Cow's Milk Allergy or Gastroesophageal Reflux Disease—Can We Solve the Dilemma in Infants?

Silvia Salvatore [1,*], Massimo Agosti [1], Maria Elisabetta Baldassarre [2], Enza D'Auria [3], Licia Pensabene [4], Luana Nosetti [1] and Yvan Vandenplas [5]

1 Department of Medicine and Surgery, Pediatric Unit, "F. Del Ponte" Hospital, University of Insubria, 21100 Varese, Italy; massimo.agosti@uninsubria.it (M.A.); luana.nosetti@uninsubria.it (L.N.)
2 Department of Biomedical Sciences and Human Oncology-Neonatology and NICU Section, "Aldo Moro" University of Bari, 70124 Bari, Italy; mariaelisabetta.baldassarre@uniba.it
3 Department of Pediatrics, Vittore Buzzi Children's Hospital, University of Milan, 20154 Milan, Italy; enza.dauria@unimi.it
4 Department of Medical and Surgical Sciences, Pediatric Unit, University "Magna Graecia" of Catanzaro, 88100 Catanzaro, Italy; licia.pensabene@gmail.com
5 Kidz Health Castle, Universitair Ziekenhuis Brussel, Vrije Universiteit Brussel, 1090 Brussels, Belgium; yvan.vandenplas@uzbrussel.be
* Correspondence: silvia.salvatore@uninsubria.it; Tel.: +39-0332-299-247

Citation: Salvatore, S.; Agosti, M.; Baldassarre, M.E.; D'Auria, E.; Pensabene, L.; Nosetti, L.; Vandenplas, Y. Cow's Milk allergy or Gastroesophageal Reflux Disease—Can We Solve the Dilemma in Infants?. *Nutrients* **2021**, *13*, 297. https://doi.org/10.3390/nu13020297

Academic Editor: Sara Manti
Received: 24 December 2020
Accepted: 18 January 2021
Published: 21 January 2021

Abstract: Cow's milk allergy (CMA) and gastro-esophageal reflux disease (GERD) may manifest with similar symptoms in infants making the diagnosis challenging. While immediate reaction to cow's milk protein indicate CMA, regurgitation, vomiting, crying, fussiness, poor appetite, sleep disturbances have been reported in both CMA and GERD and in other conditions such as functional gastrointestinal disorders, eosinophilic esophagitis, anatomic abnormalities, metabolic and neurological diseases. Gastrointestinal manifestations of CMA are often non-IgE mediated and clinical response to cow's milk free diet is not a proof of immune system involvement. Neither for non-IgE CMA nor for GERD there is a specific symptom or diagnostic test. Oral food challenge, esophageal pH impedance and endoscopy are recommended investigations for a correct clinical classification but they are not always feasible in all infants. As a consequence of the diagnostic difficulty, both over- and under- diagnosis of CMA or GERD may occur. Quite frequently acid inhibitors are empirically started. The aim of this review is to critically update the current knowledge of both conditions during infancy. A practical stepwise approach is proposed to help health care providers to manage infants presenting with persistent regurgitation, vomiting, crying or distress and to solve the clinical dilemma between GERD or CMA.

Keywords: reflux; GER; GERD; cow's milk allergy; CMA; eosinophilic esophagitis; infants; hydrolyzed formula; alginate; thickened formula

1. Introduction

Gastroesophageal reflux (GER) and cow milk allergy (CMA) occur frequently in the first year of life [1–4]. The pathogenesis of these two conditions is complex and involves multiple mechanisms of nutrition, motility, immunology and hypersensitivity. A number of papers discussed the overlapping symptoms or simultaneous occurrence of CMA and GERD [1,4–29] Nonetheless, discrimination between both disorders is still challenging due to the similarity of the symptoms and the lack of accurate and handy diagnostic tests [1,27]). Although the response to a CM elimination diet and oral challenge are essential to confirm the diagnosis of CMA [30–33], a positive challenge test does not proof the involvement of the immune system. Moreover, delayed reactions as occurring in non-IgE mediated allergy, may be insufficiently recognized with an oral challenge test. Upper endoscopy and biopsies and esophageal pH-impedance are the recommended diagnostic investigations for GERD [34]. However, a normal endoscopy and histology does not rule out GERD, as is the case in non-erosive GERD. Normal ranges for pH-impedance are missing and parameters such as symptom association probability have not been validated in children. Performance of pH-impedance is also hampered by cost and investment of time [34,35]. As a consequence, under- or over-diagnosis of CMA and GERD are likely to occur. CM protein elimination diet and treatment with acid inhibitors are often empirically initiated and are, sometimes, excessively protracted.

The aim of this review is to critically update the current knowledge of both conditions during infancy and to provide clinicians a practical stepwise diagnostic and therapeutic approach for infants presenting with persistent regurgitation, vomiting, crying or distress.

2. CMA and GERD: A Pathogenic Twist

GER and other persistent gastrointestinal symptoms in allergic patients are predominantly associated with cellular immune mechanisms and delayed reactions. In non-IgE mediated CMA, activated mast-cells, eosinophils and Th2 lymphocytes, release histamine, tryptase, IL-4, IL-5, IL-13, eotaxin and other chemokines that lead to increased permeability, epithelial dysfunction, inflammatory infiltration in the mucosal, submucosal and, in some cases, muscle layers and nociception [25,27,28,36].

A migration of activated mast cells in proximity of enteric nervous system has been demonstrated in allergic children exposed to CM proteins and may determine gastrointestinal dysmotility and related symptoms [37].

GER and regurgitation are commonly related to overfeeding, short length of the (intra-abdominal) esophagus, obtuse His angle, horizontal position of the infant. Inappropriate relaxations of the lower esophageal sphincter (LES), ineffective clearance and the impaired resistance of the esophageal mucosa contribute to GERD [34].

Crying and pain in infants and children are determined by interplaying factors such as esophageal and gastrointestinal distension, dysmotility, visceral hyperalgesia, genetics, early life events, inflammatory and microbiota components, increased permeability, stress, parental and individual coping and perception [4,38,39].

GER and CMA can coexist in the same patient and it has been reported that CMA can induce GER and also be a predisposing factor for gastrointestinal functional disorders [22,27]. Conversely, treatment with acid inhibitors for GERD increase the risk of allergy later in life [40,41].

3. Functional Disorder, CMA or GERD: The Clinical Enigma

3.1. Definition and Epidemiological Data of Infant Regurgitation and Colic

Infant regurgitation and colic are defined by the Rome IV criteria as functional gastrointestinal disorders (FGIDs) of infancy [42]. Diagnostic criteria for infant regurgitation must include at least due episodes of regurgitation per day for at least three weeks in an otherwise healthy infant 3 weeks to 12 months of age without retching, hematemesis, aspiration, apnea, failure to thrive, feeding or swallowing difficulties or abnormal posturing [42]. Infant colic is defined by recurrent or prolonged periods of crying, fussing or irritability that occur without an obvious cause, that cannot be prevented or resolved by caregivers in an infant younger than 5 months with no failure to thrive, fever or illness [42]. For clinical research purposes, to fulfill the definition of colic these episodes of crying or fussiness should last at least 3 h per days, for a minimum of one day when measured by a prospectively kept 24 h behavior diary or 3 days per week according to a caregiver's interview [42]. They affect, alone or in combination and depending on selection and inclusion criteria around 20 to 25% of infants all over the world [4,39,43,44]. Neonates born preterm, small for gestational age or exposed to early life antibiotics have been recently reported to be at increased risk of infantile regurgitation and colic [45,46]. One fifth to one third of parents are concerned about their infant's health condition and consult health care providers because of regurgitation, fussiness and crying [3,4,39,41,47]. Regurgitation and infantile colic occur mostly during the first three to four months of life, with a natural resolution in the vast majority of cases around 4 to 5 months for colic and from 6 months onwards for regurgitation [3,42,48–50]. When the onset of regurgitation is in the first two weeks of life or when projectile vomiting is the predominant symptom, secondary GER related to anatomic malformations or conditions such as CMA are more likely [42].

3.2. Symptoms and Prevalence of GERD in Infants

When GER is associated with troublesome, persistent severe symptoms or complications (e.g., respiratory problems or esophagitis) it is referred to as GERD [34]. As the definition of troublesome is subjective, the distinction between GER and GERD is challenging in infants and the two terms are often misused interchangeably [34].

The most frequently reported symptom of GER in infants is regurgitation but the latter is neither sensitive nor specific to diagnose GERD, neither if associated with crying or fussiness [14,15,34,38,47,51–53]. Thus, acid inhibitors should not be started in these infants unless an investigation-based diagnosis of GERD is established [34]. The exact prevalence of GERD in infants is difficult to define because symptoms are not specific, empirical treatment is often started, many infants are not submitted to pH-impedance and/or endoscopy and prospective data are limited. The only report in which healthy infants (N = 509), screened for risk of sudden infant death syndrome, underwent pH-monitoring dates from 1991 [54]. Using a glass microelectrode to detect acid pH, the 95th percentile of esophageal acid exposure rate, during the first 12 months of life, was about 10% [54]. Hence, 5% of healthy infants, would present a pathological oesophageal acid exposure when the threshold is fixed to 10%. In the last 30 years, for ethical reasons, only symptomatic infants suspected to have GERD were investigated. When 151 infants with persistent crying underwent pH-monitoring, 17.9% infants had pathological acid exposure time (>10%) and no association with total crying duration was noted [15]. Regurgitation occurring more than 5 times daily was the most specific GERD symptom (specificity 70.9%) but had a poor positive predictive value (22%). In the absence of frequent regurgitation or feeding difficulties, pathological GERD according to pH monitoring results was unlikely (negative predictive value 87–90%) [15]. In another study evaluating 100 infants, suspected of having GERD, a pathological pH tracing was found in 21% of cases and esophagitis was identified in 17 out of 44 infants (39%) underwent endoscopy, with poor correlation between clinical symptoms, histology and pH results [51]. In a multicenter retrospective cross-sectional study in the United States using an Endoscopy Database System, emerged that 5.5% of children aged 0 to 1 year had erosive esophagitis [55]. In another cohort of 245 infants with symptoms of reflux submitted to endoscopy and esophageal biopsy, 62 cases (25%) had histological esophagitis [56]. In 8 out of 40 infants (20%) referred for persisting symptoms attributed to GERD (regurgitation and/or vomiting and inconsolable crying, fussiness, irritability, sleeping difficulties or respiratory problems for at least 2 weeks, in the absence of any other identifiable cause) a pathological acid exposure (defined as $\geq$7%, as measured by an antimony electrode) was found by pH-impedance [57]. More recently, our group analyzed impedance-pH tracings of 62 children (ages 15 days to 23 months, median age 3.5 months) with persistent unexplained fussiness or distress and 19% showed an acid reflux exposure time >7% [58].

3.3. Symptoms and Prevalence of CMA in Infants

The prevalence of hospital based diagnosed CMA in the first year of life ranges from 0.5% to 3% of infants, with the lowest rate when breast feeding and food challenge are considered [25,28,36,59]. Nonetheless, in a Finnish study, of the 824 exclusively breast-fed infants, 2.1% had CMA, verified by a CM elimination-challenge test [60].

In the EuroPrevall birth cohort study, 12,049 children with symptoms possibly related to CMA were enrolled and 77.5% were followed up to 2 years of age. Clinical evaluation included CM-specific IgE antibodies (IgE), skin prick test and double-blind, placebo-controlled food challenge. CMA was suspected in 358 (3%) children and confirmed by the food challenge in 55 cases (0.54%, 95% CI 0.41–0.70). Of all children with CMA, 23.6%

had negative specific serum IgE and all of them tolerated CM one year after diagnosis compared to 57% of those children with IgE-associated CMA [59].

According to these epidemiological data, the expected casual coexistence of CMA and GERD would occur, by far, in less than 1% of the breastfed or formula fed infants. In breastfed infants, reflux and infantile colic as single manifestations are only seldom caused by CMA [61].

GERD may be the cause of regurgitation, vomiting, feeding disorders, day and night crying [34]. Similar symptoms may also be present in CMA and make it difficult to understand which condition is responsible for the clinical picture, especially in the absence of other signs of allergy, such as atopic dermatitis or otherwise unexplained rectal bleeding in the first months of life [1,4,30,31,61,62].

Prolonged crying during or after a meal or in the evening and night are often erroneously attributed to both CMA and GERD which seem to be responsible for only 5–10% of cases of infantile colic [25,27,38].

Repeated episodes of incoercible vomiting, with possible severe dehydration, lethargy and diarrhea occurring within a few hours from CM intake, can be classified as food protein induced enterocolitis syndrome (FPIES) [63,64]. Diarrhea, poor feeding, vomiting, failure to thrive and malabsorption are reported in food protein enteropathy. Food protein induced allergic proctocolitis typically shows he presence of blood and mucous in the stools and mild diarrhea in otherwise well-appearing, often breastfed infants [28,31–33,64].

3.4. Literature Data on the Association of CMA and GERD

A number of studies examined the presence of CMA in infants with symptoms attributed to GERD (Table 1).

Table 1. Summary of the studies evaluating the association of cow's milk allergy (CMA) and gastro-esophageal reflux disease (GERD) (modified from Ferreira 2014 [23]).

Author, Year	Population	Investigation	Main Results
Forget, 1985 [5]	15 children with recurrent vomiting	Contrast X-ray, small bowel biopsy	All children showed GER on X-ray. 3/15 (20%) had enteropathy with IgE plasmatocytes, reported no improvement with GER treatment but disappearance on symptoms on CM free diet
McLain, 1994 [6]	10 infants with GERD who failed to respond to reflux treatment	pH-monitoring	Symptoms improved in 2/10 (20%) infants on CM free diet. No infant showed significant improvement in pH monitoring indices
Staiano, 1995 [11]	25 infants with recurrent vomiting	Endoscopy and small bowel biopsies, permeability test	Primary GERD in 16/25 (64%), GERD + CMA in 4/25 (16%), CMA alone in 4/25 (16%). Enteropathy in 19% GERD, 67% CMA. Abnormal permeability test in 6% GERD, 100% CMA
Iacono, 1996 [9]	204 infants (median age, 6.3 months) with GERD	pH-monitoring, upper endoscopy, allergy tests, CM challenge	93 (45%) had positive allergy tests, 85 (42%) improved with hydrolyzed formula and reappeared on challenge. GER + CMA significantly associated with the presence of diarrhea or atopic dermatitis

Table 1. *Cont.*

Author, Year	Population	Investigation	Main Results
Cavataio, 1996 [8]	96 infants with suspected GERD, CMA and controls	Serum specific IgE and IgG, blood eosinophils, pH-monitoring, endoscopy, CM challenge	14 out of 47 (30%) infants with GERD had CMA These infants had similar symptoms to those with primary GERD but significantly higher concentrations of total IgE, circulating eosinophils and IgG anti-beta lactoglobulin. A specific phasic pH pattern, with progressive decrease in pH tracing, occurred in 24/25 infants with CMA, 12/14 GERD + CMA and 0 controls. CM free diet improved only in the ones with CMA
Milocco, 1997 [10]	112 infants with GERD	pH-monitoring, CM challenge	18 infants (16%) had CMA, 10/18 had failure to thrive. A phasic pH-pattern was present in 1/18 with CMA and in 3 with only GERD
Hill, 2000 [14]	19 infants with persistent distress and GER symptoms with no response to eHF and GERD treatment	Endoscopy, pH-monitoring, CM challenge	Nine infants had histologic evidence of esophagitis and 9 had inflammatory changes in the stomach and/or duodenum. Symptoms remitted in all infants within 2 weeks of starting AAF. On double blind challenge, after a median period of 3 months of AAF, 12 infants were still intolerant to CM
Ravelli, 2001 [21]	26 vomiting infants (7 CMA, 9, GER, 10 controls)	Electrogastrography electrical impedance tomography, CM challenge	Children with CMA showed more gastric dysrythmia (67% vs. 29.4% GER and 30.4% controls) and delayed gastric emptying (89 ± 26 min) compared to infants with GERD (54 ± 13 min) and controls (62 ± 13 min). 7/7 CMA patients had regurgitation and/or vomiting, colic and positive family history of allergy
Garzi, 2002 [12]	10 infants with GER symptoms, 10 controls	Ultrasonography to measure gastric emptying time-with CM formula and protein hydrolysate	All infants with a clinical diagnosis for GER showed delayed gastric emptying vs. normal subjects (205 vs. 124 min, $p = 0.000$). With eHF there was a significant improvement in gastric emptying time and symptoms especially in infants with positive skin-test and RAST
Nielsen, 2004 [17]	18 infants and children (median age 8.7 years; range 2 months to 14.8 years) with GERD	Endoscopy, 48-h pH-metry (Day 1-elimination diet, Day 2-challenge test), 2nd CM challenge	10 (56%) infants had CMA + GERD (higher acid exposure time vs. primary GERD), responded to CM free diet and had a positive challenge which was not associated with a significant increase in the esophageal acid exposure in the simultaneous pH monitoring
Nielsen, 2006 [18]	17 infants and children (aged 2–178 months) (mean age of 7.8 years) with GERD	Endoscopy and biopsies, pH-monitoring, allergy tests, CM challenge	10/17 (59%) were classified as CMA-GERD. Two patients showed >15 eosinophils at biopsies (=EoE) No differences in the number of eosinophils, mast cells or T cells were found between children with CMA and those with primary GERD

Table 1. *Cont.*

Author, Year	Population	Investigation	Main Results
Semeniuk, 2007 [19] and 2008 [20]	264 children with suspected GERD (mean age 21 ± 17 months) or CMA	Esophageal manometry, pH-monitoring, allergy tests and CM challenge	138 children with GERD: 76 only GERD, 62 (23.5%) GER + CMA/FA, 32 only CMA/FA. No differences between primary GERD and GERD+ CMA in reflux parameters, in the mean values of resting LES pressure and LES length at baseline and during 2 years of follow-up
Farahmand, 2011 [13]	81 children (aged 1mo-2 yrs, median 12.5 mo) with supsected GERD.	Clinical study	54 (66%) responded to PPI, 27 (33%) to CM elimination diet
Borrelli, 2012 [22]	17 children (median age: 14 months) with proven f CMA and suspected GERD	48-h pH-impedance. Day 1-amino acid formula Day 2-challenge with cow's milk	The total reflux episodes and the number of weakly acidic episodes were higher during CM challenge compared with the amino acid-based formula period. No differences were found for either acid or weakly alkaline reflux
Vandenplas, 2014 [24]	72 Infants with suspected CMA	Clinical study comparing a thickened and non-thickened eHF casein formula: results after one month.	Regurgitation was reduced in all infants (from 6.4 ± 3.2 to 2.8 ± 2.9, $p < 0.001$) but fell more with the thickened hydrolyzed formula (−4.2 ± 3.2 regurgitations/day) vs. non thickened formula, especially in infants with a negative challenge (−3.9 ± 4.0 vs. −1.9 ± 3.4, ns). In the group with positive challenge the two formulas showed a similar decrease (−4.4 ± 2.6 vs. 4.7 ± 5.6). The global reduction of a symptom-based score was −7.4 points and the non-thickened hydrolysate was more effective in the group with a positive challenge (−9.2 vs. −5.7 points)
Yukselen, 2016 [26]	151 children (aged 3–60 mo) with GERD resistant to 8 wks PPI treatment	skin prick test, specific serum IgE, eosinophil count, atopy patch test and CM challenge	58 children (38.4%) had positive CM challenge and 28 (48%) of them had positive skin prck tests or IgE, 16 (28%) had positive patch tests. Bloody stools, atopic dermatitis and recurrent wheezing episodes were significantly more common in these children Vomiting and diarrhea were more common in non-IgE children. Ten children who had positive challenge were finally diagnosed as EoE
Omari, 2020 [29]	50 infants with persistent crying, vomiting and/or food refusal (suspected to be GERD and or CMA related)	48 h cry-fuss chart, I-GERQ-R, allergy tests, blinded milk elimination-challenge sequence, pH-impedance before and after CM elimination, ^{13}C-octanoate breath test for gastric emptying, dual-sugar intestinal permeability, fecal calprotectin	14 (28%) were diagnosed as non-IgE-mediated CMA, 17 (34%) had negative challenge, 19 were excluded for equivocal findings or incomplete data. No baseline differences in any of the tests or GERD parameters between infants with and without CMA. In the CMA group, CM elimination significantly reduced reflux symptoms, esophageal acid exposure, acid clearance time and increased impedance baseline

The association of CMA-GERD was reported in 16–56% of cases with persistent gastrointestinal symptoms and suspicion of GERD, irrespective of breast or formula feeding [1,17,23,27–29,45,62]. The percentage of infants with persistent GER symptoms with clinical improvement on diet and worsening on challenge is extremely variable depending on the population recruited, design of the study and follow up data [27]. In one study, out of 19 infants with persistent distress and GER symptoms with no response to eHF and acid suppressive agents, 9 infants had esophagitis, 9 had inflammatory changes in the stomach and/or duodenum and all 19 improved on amino acid-based formula [14].

4. The Stepwise Approach to Infants with Regurgitation, Vomiting and Crying

In each infant, alarm signals indicative of other conditions such as infectious, neurological, anatomic, surgical, genetic or metabolic pathologies should be excluded throughout an accurate medical history and full physical examination (Figure 1).

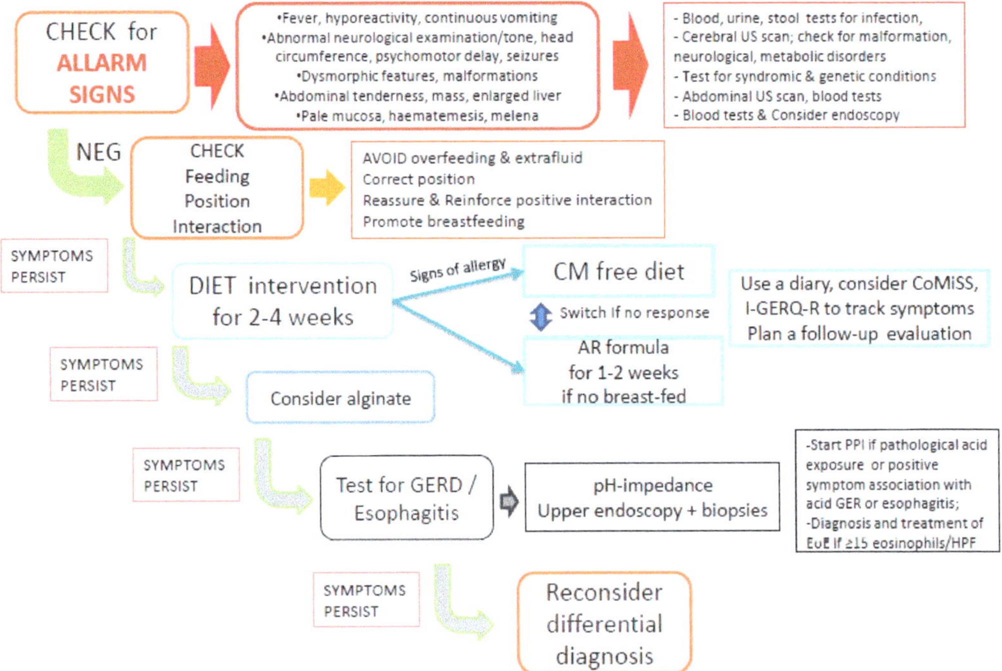

Figure 1. Simplified stepwise approach and ACTION PLAN for infants with persistent (≥1 week) regurgitation, vomiting and crying. See text for complete explanation and further details. Legend: US = ultrasound; CM = cow's milk protein; AR = thickened; PPI = proton pump inhibitors; GER = gastroesophageal reflux; EoE = eosinophilic esophagitis; HPF = high power field.

Onset of symptoms in the first week or beyond six months of life is not typical of GER. The presence of seizure, psychomotor delay, lethargy or hyporeactivity, abnormal head circumference, abnormal posturing, prolonged inconsolable crying/irritability, muscle hypo/hypertonia or impaired reflexes should alert for neurological or neuromotor or metabolic diseases. Fever, recurrent infections, prolonged apneas, recurrent brief resolved unexplained events (BRUE) or apparent life-threatening events (ALTE), jaundice, pallor, dehydration, bulging or depressed fontanelle, cyanosis, gastrointestinal bleeding, bilious vomiting, abdominal mass or tenderness, hepato/splenomegaly, multiple bruising or hematomas, weight loss or severe failure to thrive should be promptly investigated [34,65].

Abnormal growth, malformations and dysmorphic features should be considered for syndromes and genetic disorders.

Differential diagnosis and specific investigations for these different diseases will not be discussed in this review.

4.1. Management of CMA and GER in Infants

In the absence of warning signs, the first step in the management of infants presenting with infantile colic and regurgitation fulfilling the Rome IV criteria is to avoid overfeeding by checking infant's growth and feeding modalities regarding frequency and duration of feeding and preparation and volume in formula-fed infants. Parental education and information on their infant's symptoms mechanisms and evolution are of outmost importance [4,34,39]. Reassurance and positive interaction between parents and babies need empathy and patience and should be reinforced [39,65–67].

4.2. Nutrition, Dietary Modification and Diagnosis of CMA in Infants

Breastfeeding should always be promoted and continued in all infants, even in CMA, functional gastrointestinal disorders and GERD, as human milk represents the best nutritional option. In formula-fed infants, feeding volume and frequency should be progressively adapted according to age and weight and formula changing should be considered in cases with persistent (distressing) symptoms and/or poor weight gain. Commercial thickened formulas provide controlled concentration of various (locust bean gum/carob flour, tapioca, potato, rice, corn starch) thickening agents and nutritional requirements and is likely to decrease the daily episodes of regurgitation by half [66] within the first week.

The National Institute for Health and Care Excellence (NICE) GER- guidelines suggest a greater likelihood of CMA in the presence of regurgitation associated with chronic diarrhea or blood in the stool, other atopic manifestations (eczema) or a positive family history of allergy. In the ESPGHAN guidelines [30,34] the involvement of symptoms in different organ systems in association with the regurgitation increases the likelihood of CMA. Both regurgitation and atopic dermatitis are common disorders in the first months of life and their relation (overlapping age, coincidence or comorbidity) still needs to be further clarified, especially in infants with severe eczema.

Nonetheless, skin prick tests and specific IgE dosage are positive in only a minority of patients with gastrointestinal symptoms [1,28]. Atopy patch tests and the dosage of specific IgG antibodies are not well standardized and thus not recommended for diagnosing CMA [27,30,31,33]. As a consequence, elimination of CM proteins during 2 to 4 weeks is the recommended approach [30,34].

In breastfed infants, maternal CM free diet can be considered if symptoms are severe enough. In non-breastfed infants with CMA, formulas with CM based extensively hydrolyzed proteins is indicated as first choice, rice hydrolysates are second options and amino acid based formulas (AAF) should be reserved for more severe clinical reactions [28,30–33,68]. Soy infant formula could be considered in some cases, particularly in infants older than six months and in the absence of severe IgE mediated reactions (e.g., anaphylaxis) and gastrointestinal symptoms [31]. Other milk substitutes (from other mammalian species or plant-based beverages) are not recommended because of possible cross-reactivity, limited studies and scarce evidence of efficacy and nutritional adequacy [69]. Noteworthy, hydrolyzed formulas may vary considerable in terms of source of proteins, method and degree of hydrolysis, macro and micronutrients, additional components (i.e., pre- and probiotics) and proof of clinical benefit [70]. Thus, the results of one particular formula cannot be transferred to a "new" or "similar" one.

In one study, the effect of a thickened and non-thickened casein extensive hydrolyzed formula was analyzed in 72 formula-fed infants (younger than 6 months) with suspected CMA (including persistent unexplained distress or colic, respiratory and/or dermatological symptoms, diarrhea or constipation or blood in the stools and troublesome regurgitation/vomiting of more than five episodes a day) with no previous anaphylactic reactions [24]. The challenge was performed in 52/72 (72%) of the enrolled population and was positive in 65.4%. All cases tolerated both study-formulas and regurgitation was reduced in all infants (6.4 ± 3.2–2.8 ± 2.9, $p < 0.001$). The thickened hydrolysate showed a higher reduction of episodes of regurgitation (−4.2 ± 3.2 regurgitations/day) in infants with both a positive and a negative (−3.9 ± 4.0 regurgitations/day) CM challenge after one month of treatment compared to a minimal effect of the non-thickened hydrolysate (−1.9 ± 3.4 episode of regurgitation) in the group with a negative challenge [24]. The global reduction of a symptom-based score (assessing crying time, number and volume of episodes of regurgitation, consistency of stools, presence and severity of respiratory and dermatological symptoms unrelated to infections), was −7.4 points, with the highest efficacy for the non-thickened hydrolysate in the group with a positive challenge compared to the negative challenge (−9.2 vs. −5.7 points) and versus the thickened formulas between the two groups (−8.1 and −7.1 points) [24]. To better target and assess the effect of a CM free diet, based also on the previous study, a Cow's Milk Related symptom score (CoMiSS) has been proposed as an "awareness tool" for CMA [71]. This is based on scoring daily duration of crying, number and volume of regurgitation episodes, stool pattern, presence and severity of cutaneous and respiratory manifestations, unrelated to infections. The score ranges from 0 to 33 points [71]. A pooled analysis showed that infants with a CoMiSS > 12 had a 75 % chance to have a positive challenge test [72] and a 89% probability to respond to CM free diet according to another report [73]. In a presumed healthy population of infants, the P95 of the CoMiSS was >9 [73,74]. Despite CoMiSS is an easy tool to help identifying infants who can benefit from CM free diet, it does not replace the need for a diagnostic challenge and still requires further validation studies.

The importance of a clinical re-evaluation after a 2–4 weeks is emphasized both to evaluate the clinical benefit and programming the oral challenge in infants who improved or consider other diagnostic steps for the non-responders (Figure 1). The oral challenge test is required for diagnostic confirmation of CMA, proving a reaction to CM proteins after a clinical response to the exclusion diet [31–33,36]. Given the common acquisition of tolerance in the first year of life, particularly in infants with non-IgE allergy [59], diet re-evaluation and reintroduction of CM proteins should be considered and scheduled in order not to prolong unnecessary dietary restrictions. Supervised CM protein challenges are required; hospital setting and time frame, (after 2, 6 or 12 months of diet) should depend on the clinical scenario [30], including symptoms at onset and results of allergic tests [28,31–33,36].

The role of food allergy and the benefit of CM free diet in persistent unexplained crying classified as infantile colic are still controversial [3,25,75–77]. In an early small trial enrolling 21 colicky infants, CM free diet was superior to parental education and counseling [78]. In another study, enrolling 267 colicky babies, a partially hydrolyzed whey-based formula, containing fructo- and galacto-oligosaccharides and reduced lactose, showed a significant decrease in crying episodes compared to a standard formula after two weeks [79]. In 2010 a systematic review did not report evidence of diet efficacy in colicky infants and highlighted that in most studies data on the reintroduction of normal protein were lacking [75]. However, in 2012, another systematic review analyzed the eleven randomized controlled trials considered to be of good quality and concluded that both breast-fed and formula-fed colicky infants benefited from CM elimination diet [76]. According to the 2018 Cochrane review on dietary modification for infantile colic, including 15 randomized controlled trials and 1121 infants (aged 2 to 16 weeks), a greater reduction in crying time in the intervention group compared to normal CM protein intake was noted in 25% of infants with moderate or severe symptoms in many but not all studies [77]. However, the available studies had small sample sizes and most had a significant risk of bias [77].

Furthermore, symptoms such as vomiting, regurgitation and crying can decrease and disappear because of the natural evolution or a placebo effect. Nonetheless, symptoms may reappear when a formula with whole proteins (and normal lactose content) is reintroduced for mechanisms other than the immunological ones of allergy, such as the facilitating effect of gastric emptying of the (partial and extensively) hydrolyzed proteins or less fermentation in the case of a formula with reduced lactose [1,4,27]. Bradigastria and tachigastria have been more frequently detected in patients with CMA than in patients with GER or healthy children [21]. In allergic patients, dysrhythmia, mainly determined by an interaction between eosinophils, mast cells and nerve fibers [37], can impair gastric emptying causing vomiting, increasing reflux and possible pain.

4.3. Diagnosis and Treatment of GER and GERD

In both breast-fed and formula-fed infants with persistent regurgitation and distress, aluminum free alginate-based formulations have been reported to significantly reduce the number of episodes of GER and regurgitation and associated symptoms [57,67], with no adverse effects reported in short term trials.

No symptom or cluster of symptoms or questionnaire showed a high sensitivity and specificity for GERD in infants and young children [34,51]. The revised infant GER questionnaire (I-GERQ-R) has a controversial diagnostic value for GERD [29,51,80] but it provides a validated tool to monitor the evolution of symptoms during an intervention trial [80].

The infants who continue to present inconsolable crying and distress, with insufficient improvement after parental reassurance, behavioral and dietetic approaches should be submitted to investigations to identify GERD [34,65].

In some children with GER due to CMA, a particular pH-metric esophageal pattern with a gradual drop in pH after the meal was noted [8]. However, this finding is not present in all infants who respond to the diet and has not been confirmed by other authors [10]. The pH-impedance analysis showed that patients with CMA have predominantly a non-acid GER component [22] that can be even more painful than acid GER [58] but do not benefit from therapy with acid inhibitors.

Several clinical trials, two systematic reviews [47,81], one meta-analysis [82] and pediatric guidelines on GERD [34,83] have shown that treatment with acid inhibitors is not significantly effective in infants with regurgitation or vomiting and/or protracted crying without instrumental evidence of GERD. However, proton pump inhibitors are often empirically prescribed [84] while should be reserved to infants with pathological acid exposure time or significant temporal association between symptoms and acid GER during pH-impedance [34,35] or with evidence of esophagitis [34].

Upper endoscopy is indicated for cases with persistent crying, vomiting, anemia, feeding problems and failure to thrive to identify and characterize esophagitis or enteropathy. Quantification of eosinophils in esophageal biopsies help to differentiate GERD from eosinophilic esophagitis. The presence of villous atrophy and inflammatory infiltrate in the lamina propria on duodenal biopsies is characteristic of patients with CMA [11]. Intestinal permeability tests are also abnormal in these patients [11] but they are not performed in many hospitals, are non-specific and are of limited sensitivity for cases without enteropathy. Contrast X-ray is useful to detect anatomical abnormalities but has no role in diagnosis of GERD [34]. Video fluoroscopy and laryngeal examination by ENT pediatric specialist may identify abnormal swallowing and malformation determining respiratory manifestations. Nevertheless, the presence of laryngeal edema and hyperemia has a limited correlation with pH-impedance results in infants and children [85].

In a recent study 50 infants with persisting crying, vomiting and/or food refusal attributed to CMA and/or GERD were extensively investigated including atopy patch test for CM, milk specific serum IgE antibodies, 48 h cry-fuss diary, I-GERQ-R questionnaire, blinded milk elimination-challenge sequence, 24h pH-impedance monitoring before and after CM elimination, ^{13}C-octanoate breath testing for gastric emptying, dual-sugar intestinal permeability, fecal calprotectin and serum vitamin D level, Fourteen infants (28%) were finally diagnosed as CMA. No test or parameter at baseline differentiate infants with and without CMA. Only one infant had positive atopy patch test, none had positive serum IgE and, surprisingly, permeability test was higher in non-CMA infants. In the group with CMA, elimination diet significantly improved GERD symptoms, esophageal clearance and baseline, indirect parameters of esophageal function and mucosal integrity [29].

To quantify the evolution of symptoms and the benefit to the individual patient of any diet or therapeutic intervention, a follow-up visit after 2 weeks should be planned and the evaluation of a daily diary reporting pattern of stools, duration of inconsolable crying, episodes of regurgitation, feeding and sleeping disturbs, CoMiSS and I-GERQ-R scores would be useful to track symptoms.

A simplified stepwise approach and action plan for infants with persistent regurgitation, vomiting and crying is shown in Figure 1.

A correct diagnostic classification is essential to avoid the possible mislabel of "disease" in a "functional" condition or the use of protracted or unnecessary diets [31] or drugs with possible adverse effects [84].

5. The Third Wheel: Eosinophilic Esophagitis

The first report of eosinophilic esophagitis (EoE) dates back 1995 [86]. Ten children (median age 5 years, range 8 months–12.5 years), with intractable symptoms attributed to GERD but not responsive to reflux treatment (included Nissen fundoplication in 6 of them), showed improvement (in two patients) or complete resolution (in 8 children) of clinical picture when fed with an amino acid based formula (for at least 6 weeks) and relapsed on challenge. The striking feature was the detection of a high eosinophilic infiltrate (median, 41; range, 15–100) in the esophagus in all cases, with mucosal healing on elemental diet (median, 0.5; range, 0–22) [86]. Since then, EoE has been increasingly recognized at all ages throughout the world. While in children and adolescents dysphagia, bolus impaction, vomiting, epigastric pain and selective feeding can be indicators of EoE, in infants symptoms include regurgitation, vomiting, feeding difficulties, crying, fussiness and poor growth [86,87].

The overlap with CMA not only results from the clinical picture but also from the presence of positive family history of allergy, atopic manifestations and positive allergy tests in about 50% of EoE cases, with a response to a CM and/or other food elimination diet in 70–90% of patients [88].

The similarity with GERD is mainly based on the possible reduction of symptoms, acid exposure and esophageal inflammation with PPI [82] (Figure 2). Furthermore, patients with EoE may present a pathological pH-impedance, esophageal dysfunction and stricture [82].

Moreover, CMA, GERD and EoE can all occur with acute, chronic and relapsing manifestations which are difficult to differentiate between the three conditions, particularly in infants and young children [18,26,87,88].

The endoscopic finding of EoE is very variable and can range from normal appearance (particularly in infants) [86] to one or more of the following suggestive but not specific features: food bolus impaction, edema, linear furrows, friability, erosions, ulcerations, concentric rings (up to appearance of trachealization of the lumen), whitish exudates and stricture. The detection of a marked eosinophilic infiltration (>15 by high magnification field, HPF) in at least one esophageal biopsy is the diagnostic hallmark of EoE [87].

The exact prevalence of EoE in both breast- and formula-fed infants [61] is difficult to determine because few infants have endoscopy and esophageal biopsies before been attempted CM free diet or PPI treatment. Moreover, pediatric EoE case series did not provide a subgroup analysis of infants [89] and one large report on infant esophagitis did not detail eosinophilic infiltration [56]. Noteworthy, several early-life factors, including maternal fever, preterm labor, cesarean delivery, esophageal atresia, antibiotic or acid suppressant use in the first months of life, dysbiosis, other atopic conditions and celiac disease have been associated with risk of pediatric EoE [90,91].

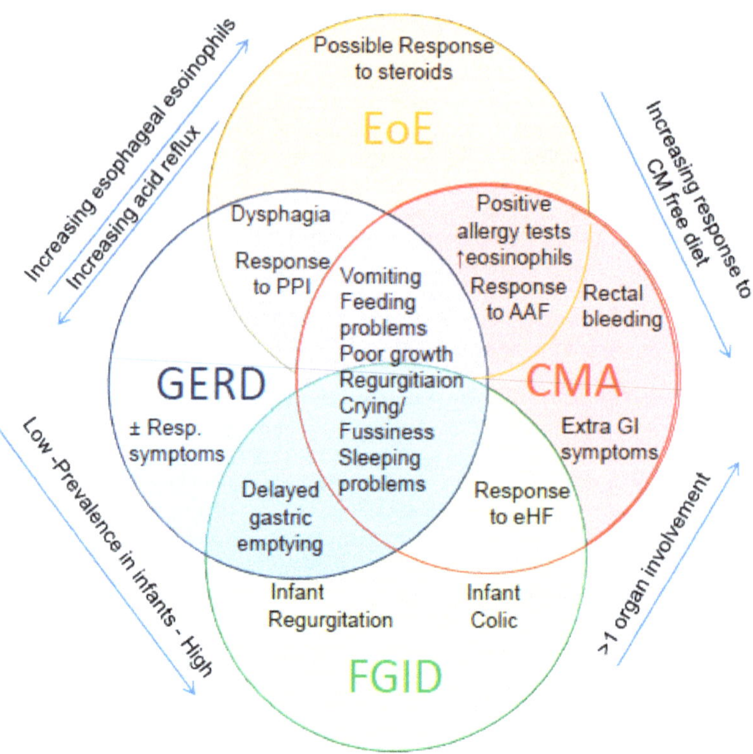

Figure 2. The challenging clinical overlap among Functional Gastrointestinal Disorder (FGID), GERD, CMA and eosinophilic esophagitis (EoE) in infants (modified from Nielsen 2006 [18].

The natural history and disease progression of EoE, as well as of CMA and GERD, are not yet well defined because pathogenesis is complex and not fully understood [92]. Therapeutic options for EoE include, as single or sequential intervention: proton pump inhibitors, elimination or elementary diet to avoid allergenic exposure and related inflammation [88]; topical steroids for anti-inflammatory effects; endoscopic dilations for severe stenosis. Immunosuppressive agents and immunomodulators have also been proposed, especially in non-responders adolescents and adults but need further validation [92]. To date, there is no specific, universally accepted and effective treatment for EoE in all patients; consequently, in clinical practice, the therapeutic approach is often individually adapted, especially as regards the choice between dietary or steroid treatment [82,87,92]. On the contrary, in infants the first and, in almost all cases, the only treatment needed for EoE is CM free diet with recommended amino acid formula [31,86,87]. In a recent review of ten studies, enrolling 462 EoE patients (mean age 6.7 years, range, 4 months–20 years), elemental diet resulted in clinical and histological remission (defined as $\leq$10 eosinophils/hpf) in 75–100% of children [89]. Despite diagnostic difficulties an early recognition of EoE is important to resolve or reduce clinical manifestations and possible long term esophageal complications.

6. Conclusions

Persistent regurgitation, vomiting, distress and crying are common symptoms in the first year of life, often coexist in the same patient and can be related to CMA, functional gastrointestinal disorders, GERD, eosinophilic esophagitis and also other different diseases. The real prevalence and the mechanisms underlying the association between CMA and

GERD are not yet fully clarified. The lack of an accurate test for non-IgE mediated CMA and for GERD determines the difficulty of a correct diagnostic classification and carries the risk of both delayed recognition and overtreatment. After exclusion of alarm signs for other organic pathologies, a stepwise approach, starting from behavioral and nutritional intervention moving to selected investigations in infants with persistent symptoms could better select infants to start diet and drugs. Because the response to elimination diet, alginate or acid inhibitors may be due to the natural evolution of underlying condition or other than immune or reflux-related mechanisms, periodic reassessment of the patient is essential to avoid misdiagnosis and excessive use of the proposed intervention.

Author Contributions: S.S., M.A. and Y.V. contributed to conception and design of the review, interpretation of data, drafting the article and final approval of the version to be published; M.E.B., E.D., L.N., L.P. contributed to interpretation of data, drafting the article and final approval of the version to be published. All authors have read and agreed to the published version of the manuscript.

Funding: This review received no external funding.

Institutional Review Board Statement: Not applicable.

Informed Consent Statement: Not applicable.

Conflicts of Interest: S.S. has participated as advisory board member, and/or consultant, and/or speaker for Bioproject, Danone, D.M.G., Mellin, Nestlé, Novalac; M.E.B. has participated as consultant for Aurora Biofarma, Y.V. has participated as a clinical investigator, and/or advisory board member, and/or consultant, and/or speaker for Abbott Nutrition, Biocodex, Biogaia, CHR Hansen, Danone, Nestle Health Science, Nestle Nutrition Institute, Nutricia, Mead Johnson Nutrition, United Pharmaceuticals, Wyeth. All the above manufacturers and companies have had no input or involvement in any aspect of this study. The other authors have no conflict of interest to disclose.

Abbreviations

AAF	Amino acid-based formula
CM	cow's milk
CMA	cow's milk protein allergy
CoMiSS	cow's milk related symptom score
EoE	eosinophilic esophagitis
FGIDs	functional gastrointestinal disorders
GER	gastroesophageal reflux
GERD	gastroesophageal reflux disease
I-GERQ-R	revised infant GER questionnaire
PPI	proton pump inhibitors

References

1. Salvatore, S.; Vandenplas, Y. Gastroesophageal reflux and cow milk allergy: Is there a link? *Pediatrics* **2002**, *110*, 972–984. [CrossRef] [PubMed]
2. Nwaru, B.I.; Hickstein, L.; Panesar, S.S.; Roberts, G.; Muraro, A.; Sheikh, A. EAACI Food Allergy and Anaphylaxis Guidelines Group Prevalence of common food allergies in Europe: A systematic review and meta-analysis. *Allergy* **2014**, *69*, 992–1007. [CrossRef] [PubMed]
3. Vandenplas, Y.; Abkari, A.; Bellaiche, M.; Benninga, M.A.; Chouraqui, J.P.; Çokuðraþ, F.; Harb, T.; Hegar, B.; Lifschitz, C.H.; Ludwig, T.; et al. Prevalence and Health Outcomes of Functional Gastrointestinal Symptoms in Infants From Birth to 12 Months of Age. *J. Pediatr. Gastroenterol. Nutr.* **2015**, *61*, 531–537. [CrossRef] [PubMed]
4. Vandenplas, Y.; Benninga, M.; Broekaert, I.; Falconer, J.; Gottrand, F.; Guarino, A.; Lifschitz, C.; Lionetti, P.; Orel, R.; Papadopoulou, A.; et al. Functional gastro-intestinal disorder algorithms focus on early recognition, parental reassurance and nutritional strategies. *Acta Paediatr.* **2015**, *105*, 244–252. [CrossRef]
5. Forget, P.; Arends, J.W. Cow's milk protein allergy and gastro-oesophageal reflux. *Eur. J. Pediatr.* **1985**, *144*, 298–300. [CrossRef]
6. McLain, B.I.; Cameron, D.J.S.; Barnes, G.L. Is cow's milk protein intolerance a cause of gastro-oesophageal reflux in infancy? *J. Paediatr. Child. Health* **1994**, *30*, 316–318. [CrossRef]
7. Cavataio, F.; Iacono, G.; Montalto, G.; Soresi, M.; Tumminello, M.; Campagna, P.; Notarbartolo, A.; Carroccio, A. Gastroesophageal reflux associated with cow's milk allergy in infants, which diagnostic examinations are useful? *Am. J. Gastroenterol.* **1996**, *91*, 1215–1220.

8. Cavataio, F.; Iacono, G.; Montalto, G.; Soresi, M.; Tumminello, M.; Carroccio, A. Clinical and pH-metric characteristics of gastro-oesophageal reflux secondary to cows' milk protein allergy. *Arch. Dis. Child.* **1996**, *75*, 51–56. [CrossRef]
9. Iacono, G.; Carroccio, A.; Cavataio, F.; Montalto, G.; Kazmierska, I.; Lorello, D.; Soresi, M.; Notarbartolo, A. Gastroesophageal reflux and cow's milk allergy in infants, a prospective study. *J. Allergy Clin. Immunol.* **1996**, *97*, 822–827.
10. Milocco, C.; Torre, G.; Ventura, A. Gastro-oesophageal reflux and cows' milk protein allergy. *Arch. Dis. Child.* **1997**, *77*, 183. [CrossRef]
11. Staiano, A.; Troncone, R.; Simeone, D.; Mayer, M.; Finelli, E.; Cella, A.; Auricchio, S. Differentiation of cows' milk intolerance and gastro-oesophageal reflux. *Arch. Dis. Child.* **1995**, *73*, 439–442. [CrossRef] [PubMed]
12. Garzi, A.; Messina, M.; Frati, F.; Carfagna, L.; Zagordo, L.; Belcastro, M.; Parmiani, S.; Sensi, L.; Marcucci, F. An extensively hydrolysed cow's milk formula im-proves clinical symptoms of gastroesophageal reflux and reduces the gastric emptying time in infants. *Allergol. Immunopathol.* **2002**, *30*, 36–41.
13. Farahmand, F.; Najafi, M.; Ataee, P.; Modarresi, V.; Shahraki, T.; Rezaei, N. Cow's milk allergy among children with gastroesophageal reflux disease. *Gut Liver.* **2011**, *5*, 298–301. [CrossRef] [PubMed]
14. Hill, D.J.; Heine, R.G.; Cameron, D.J.; Cairo-Smith, A.G.; Chow, C.W.; Francis, D.E.; Hosking, C.S. Role of food protein intolerance in infants with persistent distress attributed to reflux esophagitis. *J. Pediatr.* **2000**, *136*, 641–647. [CrossRef] [PubMed]
15. Heine, R.G. Gastroesophageal reflux disease, colic and constipation in infants with food allergy. *Curr. Opin. Allergy Clin. Immunol.* **2006**, *6*, 220–225. [CrossRef] [PubMed]
16. Heine, R.G.; Jordan, B.; Lubitz, L.; Meehan, M.; Catto-Smith, A.G. Clinical predictors of pathological gastro-oesophageal reflux in infants with persistent distress. *J. Paediatr. Child. Health* **2006**, *42*, 134–139. [CrossRef] [PubMed]
17. Nielsen, R.G.; Bindslev-Jensen, C.; Kruse-Andersen, S.; Husby, S. Severe Gastroesophageal Reflux Disease and Cow Milk Hypersensitivity in Infants and Children: Disease Association and Evaluation of a New Challenge Procedure. *J. Pediatr. Gastroenterol. Nutr.* **2004**, *39*, 383–391. [CrossRef]
18. Nielsen, R.G.; Fenger, C.; Bindslev-Jensen, C.; Husby, S. Eosinophilia in the upper gastrointestinal tract is not a characteristic fea-ture in cow's milk sensitive gastro-oesophageal reflux disease Measurement by two methodologies. *J. Clin. Pathol.* **2006**, *59*, 89–94. [CrossRef]
19. Semeniuk, J.; Kaczmarski, M. 24-hour esophageal pH monitoring in children with pathological acid gastroesophageal reflux: Primary and secondary to food allergy. Part, I. Intraesophageal pH values in distal channel; preliminary study and control studies–after 1, 2, 4 and 9 years of clinical observation as well as dietary and pharmacological treatment. *Adv. Med. Sci.* **2007**, *52*, 199–205.
20. Semeniuk, J.; Kaczmarski, M.; Uścinowicz, M. Manometric study of lower esophageal sphincter in children with primary acid gastroesophageal reflux and acid gastroesophageal reflux secondary to food allergy. *Adv. Med. Sci.* **2008**, *53*. [CrossRef]
21. Ravelli, A.M.; Tobanelli, P.; Volpi, S.; Ugazio, A.G. Vomiting and gastric motility in infants with cow's milk allergy. *J. Pediatr. Gastroenterol. Nutr.* **2001**, *32*, 59–64. [CrossRef] [PubMed]
22. Borrelli, O.; Mancini, V.; Thapar, N.; Giorgio, V.; Elawad, M.; Hill, S.; Shah, N.; Lindley, K.J. Cow's Milk Challenge Increases Weakly Acidic Reflux in Children with Cow's Milk Allergy and Gastroesophageal Reflux Disease. *J. Pediatr.* **2012**, *161*, 476–481.e1. [CrossRef] [PubMed]
23. Ferreira, C.T.; de Carvalho, E.; Sdepanian, V.L.; Morais, M.B.; Vieira, M.C.; Silva, L.R. Gastroesophageal reflux disease, exaggerations, evidence and clinical practice. *J. Pediatr.* **2014**, *90*, 105–118. [CrossRef] [PubMed]
24. Vandenplas, Y.; De Greef, E.; ALLAR Study Group. Extensive protein hydrolysate formula effectively reduces regurgitation in infants with positive and negative challenge tests for cow's milk allergy. *Acta Paediatr.* **2014**, *103*, e243–e250. [CrossRef]
25. Nocerino, R.; Pezzella, V.; Cosenza, L.; Amoroso, A.; Di Scala, C.; Amato, F.; Iacono, G.; Canani, R.B. The Controversial Role of Food Allergy in Infantile Colic: Evidence and Clinical Management. *Nutrients* **2015**, *7*, 2015–2025. [CrossRef]
26. Yukselen, A.; Celtik, C. Food allergy in children with refractory gastroesophageal reflux disease. *Pediatr. Int.* **2015**, *58*, 254–258. [CrossRef]
27. Pensabene, L.; Salvatore, S.; D'Auria, E.; Parisi, F.; Concolino, D.; Borrelli, O.; Thapar, N.; Staiano, A.; Vandenplas, Y.; Saps, M. Cow's Milk Protein Allergy in Infancy: A Risk Factor for Functional Gastrointestinal Disorders in Children? *Nutrients* **2018**, *10*, 1716. [CrossRef]
28. D'Auria, E.; Salvatore, S.; Pozzi, E.; Mantegazza, C.; Sartorio, M.U.A.; Pensabene, L.; Baldassarre, M.E.; Agosti, M.; Vandenplas, Y.; Zuccotti, G.V. Cow's Milk Allergy: Immunomodulation by Dietary Intervention. *Nutrients* **2019**, *11*, 1399. [CrossRef]
29. Omari, T.; Tobin, J.M.; McCall, L. Characterization of Upper Gastrointestinal Motility in Infants with Persistent Distress and Non-IgE-mediated Cow's Milk Protein Allergy. *J. Pediatr. Gastroenterol. Nutr.* **2020**, *70*, 489–496. [CrossRef]
30. Koletzko, S.; Niggemann, B.; Arato, A.; Dias, J.A.; Heuschkel, R.; Husby, S.; Mearin, M.L.; Papadopoulou, A.; Ruemmele, F.M.; Staiano, A.; et al. Diagnostic approach and management of cow's-milk protein allergy in infants and children, ESPGHAN gastrointestinal committee practical guidelines. *J. Pediatr. Gastroenterol. Nutr.* **2012**, *55*, 221–229. [CrossRef]
31. Muraro, A.; Werfel, T.; Hoffmann-Sommergruber, K.A.; Roberts, G.; Beyer, K.; Bindslev-Jensen, C.; Cardona, V.; Dubois, A.; Dutoit, G.; Eigenmann, P.; et al. EAACI Food Allergy and Anaphylaxis Guidelines: Diagnosis and management of food allergy. *Allergy* **2014**, *69*, 1008–1025. [CrossRef] [PubMed]
32. Luyt, D.; Ball, H.; Makwana, N.; Green, M.R.; Bravin, K.; Nasser, S.M.; Clark, A. BSACI guideline for the diagnosis and management of cow's milk allergy. *Clin. Exp. Allergy* **2014**, *44*, 642–672. [CrossRef] [PubMed]

33. Fiocchi, A.G.; Dahda, L.; Dupont, C.; Campoy, C.; Fierro, V.; Nieto, A. Cow's milk allergy: Towards an update of DRACMA guidelines. *World Allergy Organ. J.* **2016**, *9*, 35. [CrossRef] [PubMed]
34. Rosen, R.; Vandenplas, Y.; Singendonk, M.; Cabana, M.; Di Lorenzo, C.; Gottrand, F.; Gupta, S.; Langendam, M.; Staiano, A.; Thapar, N.; et al. Pediatric Gastroesophageal Reflux Clinical Practice Guidelines, Joint Recommendations of the North American Society for Pediatric Gastroenterology, Hepatology, and Nutrition (NASPGHAN) and the European Society for Pediatric Gastroenterology, Hepatology, and Nutrition (ESPGHAN). *J. Pediatr. Gastroenterol. Nutr.* **2018**, *66*, 516–554. [PubMed]
35. Quitadamo, P.; Tambucci, R.; Mancini, V.; Cristofori, F.; Baldassarre, M.; Pensabene, L.; Francavilla, R.; Di Nardo, G.; Caldaro, T.; Rossi, P.; et al. Esophageal pH-impedance monitoring in children: Position paper on indications, methodology and interpretation by the SIGENP working group. *Dig. Liver Dis.* **2019**, *51*, 1522–1536. [CrossRef] [PubMed]
36. Sicherer, S.H.; Sampson, H.A. Food allergy, A review and update on epidemiology, pathogenesis, diagnosis, prevention, and management. *J. Allergy Clin. Immunol.* **2018**, *141*, 41–58. [CrossRef]
37. Schäppi, M.G.; Borrelli, O.; Knafelz, D.; Williams, S.; Smith, V.V.; Milla, P.J.; Lindley, K.J. Mast Cell–Nerve Interactions in Children with Functional Dyspepsia. *J. Pediatr. Gastroenterol. Nutr.* **2008**, *47*, 472–480. [CrossRef]
38. Shamir, R.; St James-Roberts, I.; Di Lorenzo, C.; Burns, A.J.; Thapar, N.; Indrio, F.; Riezzo, G.; Raimondi, F.; Di Mauro, A.; Francavilla, R.; et al. Infant crying, colic, and gastrointestinal discomfort in early childhood, a review of the evidence and most plausible mechanisms. *J. Pediatr. Gastroenterol. Nutr.* **2013**, *57*, S1.
39. Salvatore, S.; Abkari, A.; Cai, W.; Catto-Smith, A.; Cruchet, S.; Gottrand, F.; Hegar, B.; Lifschitz, C.; Ludwig, T.; Shah, N.; et al. Review shows that parental reassurance and nutritional advice help to optimise the management of functional gastrointestinal disorders in infants. *Acta Paediatr.* **2018**, *107*, 1512–1520. [CrossRef]
40. Yadlapati, R.; Kahrilas, P.J. The "dangers" of chronic proton pump inhibitor use. *J. Allergy Clin. Immunol.* **2018**, *141*, 79–81. [CrossRef]
41. Levy, E.J.; Vandenplas, Y. Proton pump inhibitors, microbiota and micronutrients. *Acta Paediatr.* **2020**, *109*, 1531–1538. [CrossRef] [PubMed]
42. Benninga, M.A.; Nurko, S.; Faure, C.; Hyman, P.E.; Roberts, I.S.J.; Schechter, N.L. Childhood Functional Gastrointestinal Disorders: Neonate/Toddler. *Gastroenterology* **2016**, *150*, 1443–1455.e2. [CrossRef]
43. Wolke, D.; Bilgin, A.; Samara, M. Systematic review and meta-analysis, fussing and crying durations and prevalence of colic in infants. *J. Pediatr.* **2017**, *185*, 55–61. [CrossRef] [PubMed]
44. Bellaiche, M.; Oozeer, R.; Gerardi-Temporel, G.; Faure, C.; Vandenplas, Y. Multiple functional gastrointestinal disorders are frequent in formula-fed infants and decrease their quality of life. *Acta Paediatr.* **2018**, *107*, 1276–1282. [CrossRef]
45. Salvatore, S.; Baldassarre, M.E.; Di Mauro, A.; Laforgia, N.; Tafuri, S.; Bianchi, F.P.; Dattoli, E.; Morando, L.; Pensabene, L.; Meneghin, F.; et al. Neonatal Antibiotics and Prematurity Are Associated with an Increased Risk of Functional Gastrointestinal Disorders in the First Year of Life. *J. Pediatr.* **2019**, *212*, 44–51. [CrossRef] [PubMed]
46. Baldassarre, M.E.; Di Mauro, A.; Salvatore, S.; Tafuri, S.; Bianchi, F.P.; Dattoli, E.; Morando, L.; Pensabene, L.; Meneghin, F.; DiLillo, D.; et al. Birth Weight and the Development of Functional Gastrointestinal Disorders in Infants. *Pediatr. Gastroenterol. Hepatol. Nutr.* **2020**, *23*, 366–376. [CrossRef]
47. Salvatore, S.; Barberi, S.; Borrelli, O.; Castellazzi, A.; Di Mauro, D.; Di Mauro, G.; Doria, M.; Francavilla, R.; Landi, M.; Martelli, A.; et al. Pharmacological interventions on early functional gastrointestinal disorders. *Ital. J. Pediatr.* **2016**, *42*, 1–8. [CrossRef] [PubMed]
48. Nelson, S.P.; Chen, E.H.; Syniar, G.M. Prevalence of symptoms of gastroesophageal reflux in infancy. *Arch. Pediatr. Adolesc. Med.* **1997**, *151*, 569–572. [CrossRef] [PubMed]
49. Hegar, B.; Dewanti, N.R.; Kadim, M.; Alatas, S.; Firmansyah, A.; Vandenplas, Y. Natural evolution of regurgitation in healthy infants. *Acta Paediatr.* **2009**, *98*, 1189–1193. [CrossRef] [PubMed]
50. Salvatore, S.; Vandenplas, Y. Epidemiology. In *Gastroesophageal Reflux in Children*; Vandenplas, Y., Ed.; Springer: Cham, Switzerland, 2017. [CrossRef]
51. Salvatore, S.; Hauser, B.; Vandemaele, K.; Novario, R.; Vandenplas, Y. Gastroesophageal reflux disease in infants, how much is predictable with questionnaires, pH-metry, endoscopy and histology? *J. Pediatr. Gastroenterol. Nutr.* **2005**, *40*, 210–215. [CrossRef]
52. Orenstein, S.R.; Hassall, E.; Furmaga-Jablonska, W.; Atkinson, S.; Raanan, M. Multicenter, double-blind, randomized, placebo-controlled trial assessing the efficacy and safety of proton pump inhibitor lansoprazole in infants with symptoms of gastroesophageal reflux disease. *J. Pediatr.* **2009**, *154*, 514–520. [CrossRef] [PubMed]
53. Salvatore, S.; Arrigo, S.; Luini, C.; Vandenplas, Y. Esophageal Impedance in Children: Symptom-Based Results. *J. Pediatr.* **2010**, *157*, 949–954.e2. [CrossRef] [PubMed]
54. Vandenplas, Y.; Goyvaerts, H.; Helven, R.; Sacre, L. Gastroesophageal reflux, as measured by 24-hour pH monitoring, in 509 healthy infants screened for risk of sudden infant death syndrome. *Pediatrics* **1991**, *88*, 834–840. [PubMed]
55. Gilger, M.A.; El-Serag, H.B.; Gold, B.D.; Dietrich, C.L.; Tsou, V.; McDuffie, A.; Shub, M.D. Prevalence of Endoscopic Findings of Erosive Esophagitis in Children: A Population-based Study. *J. Pediatr. Gastroenterol. Nutr.* **2008**, *47*, 141–146. [CrossRef] [PubMed]
56. Volonaki, E.; Sebire, N.J.; Borrelli, O.; Lindley, K.J.; Elawad, M.; Thapar, N.; Shah, N. Gastrointestinal endoscopy and mucosal biopsy in the first year of life, indications and outcome. *J. Pediatr. Gastroenterol. Nutr.* **2012**, *55*, 62–65. [CrossRef] [PubMed]
57. Salvatore, S.; Ripepi, A.; Huysentruyt, K.; Van De Maele, K.; Nosetti, L.; Agosti, M.; Salvatoni, A.; Vandenplas, Y. The Effect of Alginate in Gastroesophageal Reflux in Infants. *Pediatr. Drugs* **2018**, *20*, 575–583. [CrossRef] [PubMed]

58. Salvatore, S.; Pagliarin, F.; Huysentruyt, K.; Bosco, A.; Fumagalli, L.; Van De Maele, K.; Agosti, M.; Vandenplas, Y. Distress in Infants and Young Children, Don't Blame Acid Reflux. *J. Pediatr. Gastroenterol. Nutr.* **2020**, *71*, 465–469. [CrossRef]
59. Schoemaker, A.A.; Sprikkelman, A.B.; Grimshaw, K.E.; Roberts, G.; Grabenhenrich, L.; Rosenfeld, L.; Siegert, S.; Dubakiene, R.; Rudzeviciene, O.; Reche, M.; et al. Incidence and natural history of challenge-proven cow's milk allergy in European children—EuroPrevall birth cohort. *Allergy* **2015**, *70*, 963–972. [CrossRef] [PubMed]
60. Saarinen, K.M.; Juntunen-Backman, K.; Järvenpää, A.-L.; Klemetti, P.; Kuitunen, P.; Lope, L.; Renlund, M.; Siivola, M.; Vaarala, O.; Savilahti, E. Breast-Feeding and the Development of Cows' Milk Protein Allergy. *Adv. Exp. Med. Biol.* **2000**, *478*, 121–130. [CrossRef]
61. Meyer, R.; Chebar Lozinsky, A.; Fleischer, D.M.; Vieira, M.C.; Du Toit, G.; Vandenplas, Y.; Dupont, C.; Knibb, R.; Uysal, P.; Cavkaytar, O.; et al. Diagnosis and management of Non-IgE gastrointestinal allergies in breastfed infants-An EAACI Position Paper. *Allergy* **2020**, *75*, 14–32. [CrossRef]
62. Hait, E.J.; McDonald, D.R. Impact of Gastroesophageal Reflux Disease on Mucosal Immunity and Atopic Disorders. *Clin. Rev. Allergy Immunol.* **2019**, *57*, 213–225. [CrossRef] [PubMed]
63. Nowak-Wegrzyn, A.; Chehade, M.; Groetch, M.E.; Spergel, J.M.; Wood, R.A.; Allen, K.; Atkins, D.; Bahna, S.; Barad, A.V.; Berin, C.; et al. International consensus guidelines for the diagnosis and management of food protein-induced enterocolitis syndrome, executive summary-Workgroup Report of the Adverse Reactions to Foods Committee, American Academy of Allergy, Asthma & Immunology. *J. Allergy Clin. Immunol.* **2017**, *139*, 1111–1126. [PubMed]
64. Labrosse, R.; Graham, F.; Caubert, J.C. Non-IgE-mediated gastrointestinal food allergies in children, an update. *Nutrients* **2020**, *12*, 2086. [CrossRef] [PubMed]
65. Gonzalez Ayerbe, J.I.; Hauser, B.; Salvatore, S.; Vandenplas, Y. Diagnosis and Management of Gastroesophageal Reflux Disease in Infants and Children, from Guidelines to Clinical Practice. *Pediatr. Gastroenterol. Hepatol. Nutr.* **2019**, *22*, 107–121. [CrossRef] [PubMed]
66. Salvatore, S.; Savino, F.; Singendonk, M.; Tabbers, M.; Benninga, M.A.; Staiano, A.; Vandenplas, Y. Thickened formula, what to know. *Nutrition* **2018**, *49*, 51–56. [CrossRef]
67. Baldassarre, M.E.; Di Mauro, A.; Pignatelli, M.C.; Fanelli, M.; Salvatore, S.; Di Nardo, G.; Chiaro, A.; Pensabene, L.; Laforgia, N. Magnesium Alginate in Gastro-Esophageal Reflux: A Randomized Multicenter Cross-Over Study in Infants. *Int. J. Environ. Res. Public Health* **2019**, *17*, 83. [CrossRef]
68. Vandenplas, Y. Prevention and Management of Cow's Milk Allergy in Non-Exclusively Breastfed Infants. *Nutrients* **2017**, *9*, 731. [CrossRef]
69. Verduci, E.; D'Elios, S.; Cerrato, L.; Comberiati, P.; Calvani, M.; Palazzo, S.; Martelli, A.; Landi, M.; Trikamjee, T.; Peroni, D. Cow's Milk Substitutes for Children: Nutritional Aspects of Milk from Different Mammalian Species, Special Formula and Plant-Based Beverages. *Nutrients* **2019**, *11*, 1739. [CrossRef]
70. Salvatore, S.; Vandenplas, Y. Hydrolyzed Proteins in Allergy. *Nestle Nutr. Inst. Workshop Ser.* **2016**, *86*, 11–27.
71. Vandenplas, Y.; Dupont, C.; Eigenmann, P.; Host, A.; Kuitunen, M.; Ribes-Koninckx, C.; Shah, N.; Shamir, R.; Staiano, A.; Szajewska, H.; et al. A workshop report on the development of the Cow's Milk-related Symptom Score awareness tool for young children. *Acta Paediatr.* **2015**, *104*, 334–339. [CrossRef]
72. Vandenplas, Y.; Steenhout, P.; Järvi, A.; Garreau, A.-S.; Mukherjee, R. Pooled Analysis of the Cow's Milk-related-Symptom-Score (CoMiSSTM) as a Predictor for Cow's Milk Related Symptoms. *Pediatr. Gastroenterol. Hepatol. Nutr.* **2017**, *20*, 22. [CrossRef] [PubMed]
73. Salvatore, S.; Bertoni, E.; Bogni, F.; Bonaita, V.; Armano, C.; Moretti, A.; Baù, M.; Luini, C.; D'Auria, E.; Marinoni, M.; et al. Testing the Cow's Milk-Related Symptom Score (CoMiSSTM) for the Response to a Cow's Milk-Free Diet in Infants, A Prospective Study. *Nutrients.* **2019**, *11*, 2402. [CrossRef]
74. Vandenplas, Y.; Salvatore, S.; Ribes-Koninckx, C.; Carvajal, E.; Szajewska, H.; Huysentruyt, K. The Cow Milk Symptom Score (CoMiSSTM) in presumed healthy infants. *PLoS ONE* **2018**, *13*, e0200603. [CrossRef] [PubMed]
75. Lucassen, P. Colic in infants. *BMJ Clin. Evid.* **2010**, *2010*, 309.
76. Iacovou, M.; Ralston, R.A.; Muir, J.; Walker, K.Z.; Truby, H. Dietary management of infantile colic, a systematic review. *Matern. Child Health J.* **2012**, *16*, 1319–1331. [CrossRef] [PubMed]
77. Gordon, M.; Biagioli, E.; Sorrenti, M.; Lingua, C.; Moja, L.; Banks, S.S.; Ceratto, S.; Savino, F. Dietary modification for infantile colic. *Cochrane Database Syst. Rev.* **2018**, *10*, CD011029. [CrossRef] [PubMed]
78. Taubman, B. Parental counseling compared with elimination of cow's milk or soy milk protein for the treatment of infant colic syndrome, a randomized trial. *Pediatrics* **1988**, *81*, 756–761.
79. Savino, F.; Palumeri, E.; Castagno, E.; Cresi, F.; Dalmasso, P.; Cavallo, F.; Oggero, R. Reduction of crying episodes owing to infantile colic: A randomized controlled study on the efficacy of a new infant formula. *Eur. J. Clin. Nutr.* **2006**, *60*, 1304–1310. [CrossRef]
80. Orenstein, S.R. Symptoms and reflux in infants, Infant Gastroesophageal Reflux Questionnaire Revised (I-GERQ-R)—Utility for symptom tracking and diagnosis. *Curr. Gastroenterol. Rep.* **2010**, *12*, 431–436. [CrossRef]
81. Gieruszczak-Białek, D.; Konarska, Z.; Skórka, A.; Vandenplas, Y.; Szajewska, H. No Effect of Proton Pump Inhibitors on Crying and Irritability in Infants: Systematic Review of Randomized Controlled Trials. *J. Pediatr.* **2015**, *166*, 767–770.e3. [CrossRef]

82. Lucendo, A.J.; Arias, Á.; Molina-Infante, J. Efficacy of Proton Pump Inhibitor Drugs for Inducing Clinical and Histologic Remission in Patients With Symptomatic Esophageal Eosinophilia, A Systematic Review and Meta-Analysis. *Clin. Gastroenterol. Hepatol.* **2016**, *14*, 13–22. [CrossRef] [PubMed]
83. National Institute for Health and Care Excellence (NICE). Gastro-Oesophageal Reflux Disease, Recognition, Diagnosis and Management in Children and Young People. (*Clinical Guideline 193*). 2015. Available online: http,//www.nice.org.uk/guidance/NG1 (accessed on 30 April 2015).
84. Levy, E.I.; Salvatore, S.; Vandenplas, Y.; De Winter, J.P. Prescription of acid inhibitors in infants: An addiction hard to break. *Eur. J. Pediatr.* **2020**, *179*, 1957–1961. [CrossRef]
85. Mantegazza, C.; Mallardo, S.; Rossano, M.; Meneghin, F.; Ricci, M.; Rossi, P.; Capra, G.; Latorre, P.; Schindler, A.; Isoldi, S.; et al. Laryngeal signs and pH-multichannel intraluminal impedance in infants and children, The missing ring, LPR and MII-pH in children. *Dig. Liver Dis.* **2020**, *52*, 1011–1016. [CrossRef] [PubMed]
86. Kelly, K.J.; Lazenby, A.J.; Rowe, P.C.; Yardley, J.H.; Perman, J.A.; Sampson, H.A. Eosinophilic esophagitis attributed to gastroesophageal reflux: Improvement with an amino acid-based formula. *Gastroenterology* **1995**, *109*, 1503–1512. [CrossRef]
87. Papadopoulou, A.; Koletzko, S.; Heuschkel, R.; Dias, J.; Allen, K.; Murch, S.; Chong, S.; Gottrand, F.; Husby, S.; Lionetti, P.; et al. Management Guidelines of Eosinophilic Esophagitis in Childhood. *J. Pediatr. Gastroenterol. Nutr.* **2014**, *58*, 107–118. [CrossRef] [PubMed]
88. Venter, C.; Fleischer, D.M. Diets for diagnosis and management of food allergy, The role of the dietitian in eosinophilic esophagitis in adults and children. *Ann. Allergy Asthma Immunol.* **2016**, *117*, 468–471. [CrossRef]
89. Atwal, K.; Hubbard, G.P.; Venter, C.; Stratton, R.J. The use of amino acid-based nutritional feeds is effective in the dietary management of pediatric eosinophilic oesophagitis. *Immunity Inflamm. Dis.* **2019**, *7*, 292–303. [CrossRef] [PubMed]
90. Jensen, E.T.; Kuhl, J.T.; Martin, L.J.; Rothenberg, M.E.; Dellon, E.S. Prenatal, intrapartum, and postnatal factors are associated with pediatric eosinophilic esophagitis. *J. Allergy Clin. Immunol.* **2018**, *141*, 214–222. [CrossRef]
91. Votto, M.; Marseglia, G.L.; De Filippo, M.; Brambilla, I.; Caimmi, S.M.E.; Licari, A. Early Life Risk Factors in Pediatric EoE: Could We Prevent This Modern Disease? *Front. Pediatr.* **2020**, *8*, 263. [CrossRef]
92. Cheng, E. Translating New Developments in Eosinophilic Esophagitis Pathogenesis into Clinical Practice. *Curr Treat Options Gastroenterol.* **2015**, *13*, 30–46. [CrossRef]

 *nutrients* MDPI

Article

Identification and Purification of Novel Low-Molecular-Weight Lupine Allergens as Components for Personalized Diagnostics

Uta Jappe [1,2,*], Arabella Karstedt [1], Daniela Warneke [1], Saskia Hellmig [1], Marisa Böttger [1], Friedrich W. Riffelmann [3], Regina Treudler [4], Lars Lange [5], Susanne Abraham [6], Sabine Dölle-Bierke [7], Margitta Worm [7], Nicola Wagner [8], Franziska Ruëff [9], Gerald Reese [10], André C. Knulst [11] and Wolf-Meinhard Becker [1]

1 Division of Clinical and Molecular Allergology, Research Center Borstel, Airway Research Center North (ARCN), German Center for Lung Research, 23845 Borstel, Germany; AKarstedt@web.de (A.K.); dwarneke@fz-borstel.de (D.W.); saskia.hellmig@gmail.com (S.H.); mboettger@fz-borstel.de (M.B.); wolfmeinhard@aol.com (W.-M.B.)
2 Interdisciplinary Allergy Outpatient Clinic, Department of Internal Medicine and Pneumology, University of Lübeck, 23538 Lübeck, Germany
3 Department of Allergology, Fachkrankenhaus Kloster Grafschaft, 57392 Schmallenberg, Germany; F.Riffelmann@fkkg.de
4 Department of Dermatology, Venerology and Allergology, Universität Leipzig, 04103 Leipzig, Germany; Regina.Treudler@medizin.uni-leipzig.de
5 Department of Pediatrics, St. Marien-Hospital, 53115 Bonn, Germany; Lars.Lange@gfo-kliniken-bonn.de
6 Department of Dermatology, University Allergy Center, Medical Faculty Carl Gustav Carus, Technische Universität Dresden, 01307 Dresden, Germany; Susanne.Abraham@uniklinikum-dresden.de
7 Division of Allergy and Immunology, Department of Dermatology, Venerology and Allergology, Charité—Universitätsmedizin Berlin, Corporate Member of Freie Universität Berlin, Humboldt-Universität zu Berlin, and Berlin Institute of Health, 10117 Berlin, Germany; sabine.doelle@charite.de (S.D.-B.); margitta.worm@charite.de (M.W.)
8 Department of Dermatology, University Hospital Erlangen, Friedrich-Alexander-Universität Erlangen-Nürnberg (FAU), 91054 Erlangen, Germany; Nicola.Wagner@uk-erlangen.de
9 Department of Dermatology and Allergology, Klinikum der Universität München, 80337 Munich, Germany; Franziska.Rueff@med.uni-muenchen.de
10 Allergopharma, 21465 Reinbeck, Germany; gerald.reese@allergopharma.com
11 Department of Dermatology/Allergology, University Medical Center Utrecht, Utrecht University, 3508 GA Utrecht, The Netherlands; A.C.Knulst@UmcUtrecht.nl
* Correspondence: ujappe@fz-borstel.de

Citation: Jappe, U.; Karstedt, A.; Warneke, D.; Hellmig, S.; Böttger, M.; Riffelmann, F.W.; Treudler, R.; Lange, L.; Abraham, S.; Dölle-Bierke, S.; et al. Identification and Purification of Novel Low-Molecular-Weight Lupine Allergens as Components for Personalized Diagnostics. *Nutrients* **2021**, *13*, 409. https://doi.org/10.3390/nu13020409

Academic Editor: Sara Manti
Received: 16 December 2020
Accepted: 18 January 2021
Published: 28 January 2021

Publisher's Note: MDPI stays neutral with regard to jurisdictional claims in published maps and institutional affiliations.

Abstract: Lupine flour is a valuable food due to its favorable nutritional properties. In spite of its allergenic potential, its use is increasing. Three lupine species, *Lupinus angustifolius*, *L. luteus*, and *L. albus* are relevant for human nutrition. The aim of this study is to clarify whether the species differ with regard to their allergen composition and whether anaphylaxis marker allergens could be identified in lupine. Patients with the following characteristics were included: lupine allergy, suspected lupine allergy, lupine sensitization only, and peanut allergy. Lupine sensitization was detected via CAP-FEIA (ImmunoCAP) and skin prick test. Protein, DNA and expressed sequence tag (EST) databases were queried for lupine proteins homologous to already known legume allergens. Different extraction methods applied on seeds from all species were examined by SDS-PAGE and screened by immunoblotting for IgE-binding proteins. The extracts underwent different and successive chromatography methods. Low-molecular-weight components were purified and investigated for IgE-reactivity. Proteomics revealed a molecular diversity of the three species, which was confirmed when investigated for IgE-reactivity. Three new allergens, *L. albus* profilin, *L. angustifolius* and *L. luteus* lipid transfer protein (LTP), were identified. LTP as a potential marker allergen for severity is a valuable additional candidate for molecular allergy diagnostic tests.

Keywords: cross-reaction; flour; food allergy; individualized diagnostics; legumes; lupine; lipid transfer protein; peanut; profilin

1. Introduction

Flour from raw lupine seed is used increasingly as a protein source in Australia, New Zealand, The USA and European countries, where lupine serves as a replacement for animal proteins, i.e., milk, egg white, and potentially genetically modified soy products [1]. Since lupine lacks gluten, lupine-containing products are recommended for patients with wheat protein allergy and coeliac disease. Further beneficial effects are increased satiety, reduced energy intake, hypolipidemia, and a decrease in blood glucose concentration [2]. In addition, lupine is an important protein source in the vegan diet, which is experiencing a growing interest. The genus *Lupinus* belongs to the *Papilionaceae* subfamily in the *Leguminosae* family, which among others, also contains peanuts, soybean, beans, peas, chickpeas, lentils, and fenugreek. However, the dietary value of lupine proteins is higher than that of beans or peas, which is mainly due to high concentrations of the essential amino acids lysine, leucine and threonine, which are higher only in soybeans. It has relatively high concentrations of protein and dietary fiber in contrast to digestible carbohydrates and lipids (summarized in [3]). Three lupine species (*Lupinus albus*, *L. luteus*, and *L. angustifolius*) are used as food as well as a food additive to fortify wheat flour, which may contain up to 10 or even 15% of lupine flour.

In spite of its nutritional value, lupine is an upcoming food allergen responsible for severe food allergy. Since the first publication of lupine allergy in a peanut-allergic child [4] and the systematic review of 151 cases in 2010 [3], further cases were reported worldwide. It is, therefore, interesting that detailed knowledge on individual lupine allergens is still sparse and that in vitro allergy diagnostic tests still rely on lupine seed extract only (*Lupinus albus*, ImmunoCAP f335, Thermo Fisher Scientific [5]) [3]. This is in contrast to other relevant allergenic legumes (peanut and soy), for which, besides the whole extract, individual allergens covering relevant protein families and marker allergens are available for routine allergy diagnostic tests. As has been shown for other whole allergen extracts before, it is highly probable that the only available lupine extract lacks some relevant allergens and, in addition, does not satisfactorily address species-specific differences. In addition, most patients who are confronted with the question of food allergy to lupine do not even know that it is a food as well as a food ingredient. Therefore, it can be safely assumed that not all lupine-allergic patients are diagnosed correctly. Most clinical studies on lupine allergy vary with regard to design, population, geographic origin, and endpoints. In general, there are three variants of lupine allergy: primary food allergy, secondary (pollen-associated) food allergy, and occupational inhalant allergy with or without associated intolerance of lupine ingestion [3]. Depending on the selection of the study population, the percentage of clinically relevant (symptoms of allergy after ingestion or inhalation) and non-relevant lupine sensitization (positive IgE-antibody detection only) differs. With regard to the data published so far, cross-reactivity between lupine and soybean, beans, lentils, and peas does not seem to be of much clinical relevance when compared with the cross-reactivity between lupine and peanut (summarized in [3,6]). The study of Moneret-Vautrin and co-authors in 1999 had shown that 28% of peanut-allergic patients also experienced symptoms after lupine ingestion [7], in a study from Peeters and co-authors in 2009, the percentage of lupine-allergy in peanut-allergic patients was 35% [1]. None of these patients had been aware of their allergy to lupine. This unawareness is a considerable problem to the present day [8], indicating that peanut-allergic patients are at risk of reacting to lupine-containing food with severe allergic symptoms. Those patients will definitely avoid peanut, but not necessarily lupine, when they are oblivious to the phenomenon of cross-reactivity. In order to understand cross-reactivity and the fact that raw lupine in the form of seed, flour or dust is known to induce different disease entities (food allergy as well as occupational allergy like baker's asthma) [9], several investigations focus on molecular allergology trying to identify and characterize single lupine allergens (summarized in [3], updated in [6]). Regarding the identification of clinically relevant single allergens in lupine species, the storage proteins were the first to be studied, as they are in other clinically relevant legumes associated with severe allergic reactions. IgE-reactivity was demonstrated for the δ-conglutin (9 kDa and 4 kDa), non-reduced γ-conglutin or α-conglutin with a molecular weight of 43 kDa (glycinin or legumin) (summarized in [3]). The precursor of the β-conglutin of *Lupinus angustifolius* (20 kDa to 80 kDa), which is a vicilin-like storage protein, was the first accepted by the WHO/International Union of Immunological Societies (IUIS) Allergen Nomenclature Subcommittee as allergen Lup an 1 [10]. However, since it is not generally documented by the food producers, which lupine species has been used for flour in foods, and because the phenomenon of clinically relevant cross-reactivity has not been fully understood, it is necessary that investigations for species-specific differences should become possible in the future. Based on a German multicenter study, patients allergic to lupine with and without cross-reactivity to peanut were included, their clinical data recorded, and their sera investigated for sensitization profiles with the three lupine species, revealing species-relevant differences. Subsequently, the sera of these patients were used for the identification of new single lupine allergens, which were purified (lipid transfer protein) and recombinantly produced (profilin) and applied in IgE-detection measures.

2. Materials and Methods

2.1. Study Group

A total of 31 individuals have been included: 5 patients with lupine allergy alone (including one with strongly suspected allergy to lupine-containing foods), 10 patients with peanut allergy and lupine allergy (including four with strongly suspected lupine allergy), 11 patients with peanut allergy and lupine sensitization, two patients with peanut allergy without proven lupine sensitization, two patients with lupine and peanut sensitization only, and one non-allergic individual, whose serum served as negative control were recruited during clinical work in the allergy outpatient clinics of Borstel and Lübeck, Germany, as well as in study centers in Schmallenberg, Leipzig, Bonn, Dresden, Berlin, Erlangen, and Munich, Germany, on an ongoing basis. In addition, the University of Utrecht participated in this investigation. The patients were characterized by standardized questionnaires and specified medical history, and the sensitization to lupine and peanut was investigated via ImmunoCAP (ImmunoCAP, Phadia AB, Freiburg, Germany, and Uppsala, Sweden) (Table 1). In cases without convincing clinical history for anaphylaxis, an open oral food challenge was performed with lupine flour [11] in some centers. Three had undergone double-blind placebo-controlled lupine challenge in Utrecht with positive results and were included in this study [1]. The study was approved by the ethics committee of the University of Lübeck, Germany, with approval numbers 10-124, 10-126, and 13-086. All patients gave informed consent.

2.2. Lupine Extract Production and Protein Identification According to the Molecular Weight

Lupine extracts were produced from dry seeds of *Lupinus angustifolius*, var. Boregine (Saatszucht Steinach, Steinach, Germany), *Lupinus albus*, var. Feodora (Saaten-Union GmbH, Isernhagen, Germany), and *Lupinus luteus*, var. Juno ZS (Feldsaaten Freudenberger GmbH and Co. KG, Krefeld, Germany). For the production of the protein extracts, the lupine seeds were briefly frozen with liquid nitrogen, then ground in a coffee grinder to obtain flour, and afterwards extracted at different pH-levels.

In order to identify as many allergens as possible, we performed two extraction protocols, acidic and alkaline protein extraction, of the three relevant lupine species.

Table 1. Characterization of the patients investigated for IgE-reactivity with lupine extracts and new single allergens.

							BLOT					CAP					
P-Code	Gender	Age	Peanut Allergy	Peanut Sensitization	Lupine Allergy	Lupine Sensitization	*L. luteus* (Whole Extract)	*L. angustifolius* (Whole Extract)	*L. luteus* LTP	*L. angustifolius* LTP	*L. albus* Profilin	Total-IgE	Lupine Seed	Peanut Extract	Ara h 2	Ara h 8	Ara h 9
P 1	M	35		x		x	+	+	+	+	n. d.	>5000	n. d.	84.7	n. d.	n. d.	n. d.
P 2	F	25	x	x	(x)	x	+	+	–	–	n. d.	110	4.75	n. d.	n. d.	n. d.	n. d.
P 3	F	24	x	x	(x)	x	+	+	(+)	(+)	n. d.	n. d.	5.54	n. d.	>100	0	n. d.
P 4	M	76	x	x	?	x	+	+	+	+	n. d.	1053	n. d.	n. d.	n. d.	n. d.	n. d.
P 5	F	27	x	x	?	x	+	+	–	–	n. d.	1828	n. d.	47.5	28.9	11.4	0.06
P 6	F	63	x	x	x	x	+	+	+	+	n. d.	199	26.6	1.27	n. d.	n. d.	5.83
P 7	F	24	x	x	x	x	+	+	+	+	n. d.	n. d.	6.5	>100	n. d.	n. d.	n. d.
P 8	F	42		x	x	x	+	+	+	+	n. d.	>5000	4.9	18	0	50	0
P 9	M	15	x	x	?	x	+	+	+	+	n. d.	705.5	n. d.	21.3	3.21	1.34	3.7
P 10	M	63			x	x					+		0.75	n. d.	n. d.	n. d.	n. d.
P 11	F	30	x	x	?	x	+	+	–	–	+	1696	n. d.	80.7	n. d.	>100	n. d.
P 12	F	57	x	x		x	+	–	–	–	+	357	3.69	4.2	<0.1	6.2	0
P 13	F	27	x	x		x	+	–	–	–	+	1324	0.71	13.7	8.56	4.7	<0.01
P 14	F	50			(x)	x	+	–	–	–	+	n. d.	0.64	n. d.	n. d.	n. d.	n. d.
P 15	F	38	x	x	x		–	–	–	–	+	n. d.	<0.35	0.8	n. d.	n. d.	n. d.

Table 1. *Cont.*

							BLOT					CAP					
P-Code	**Gender**	**Age**	**Peanut Allergy**	**Peanut Sensitization**	**Lupine Allergy**	**Lupine Sensitization**	***L. luteus* (Whole Extract)**	***L. angustifolius* (Whole Extract)**	***L. luteus* LTP**	***L. angustifolius* LTP**	***L. albus* Profilin**	**Total-IgE**	**Lupine Seed**	**Peanut Extract**	**Ara h 2**	**Ara h 8**	**Ara h 9**
P 16	M	24	x	x		x	+	+	n. d.	n. d.	+	208	1.28	39.8	36.4	0.67	0
P 17	M	28	x	x		x	+	+	n. d.	n. d.	+	94.3	0	0.74	0	0	0
P 18	M	24	x	x		x	+	+	–	–	+	1416	11	88.4	34.1	9.93	0.15
P 19	M	33	x	x		x	+	+	n. d.	n. d.	n. d.	595.6	5.55	75.9	51	7.95	0
P 21	M	10	x	x			n. d.	n. d.	+	n. d.	n. d.	531	n. d.	52.1	18.2	0.17	36.9
P 22	M	43		x		x	+	+	n. d.	n. d.	+	83.5	<0.01	0.22	<0.01	1.65	0.30
P 23	F	67		x	x	x	+	+	–	–	–	116	42.9	0	n. d.	0.46	n. d.
P 24	M	32	x	x		x	+	+	n. d.	n. d.	–	n. d.	7.02	37.1	n. d.	n. d.	n. d.
P 25	F	9	x	x	x	x	+	(+)	n. d.	n. d.	–	n. d.	11.2	>100	n. d.	n. d.	n. d.
P 26	F	39	x	x	x	x	+	+	n. d.	n. d.	–	n. d.	1.58	n. d.	n. d.	n. d.	n. d.
P 27	F	34	x	x	(x)		+	+	n. d.	n. d.	–	137	<0.12	2.76	<0.1	0.29	<0.01
P 28	F	32	x		(x)		(+)	(+)	n. d.	n. d.	–	319	16.7	n. d.	67.6	0	n. d.
P 29	F	51			x		–	–	n. d.	n. d.	n. d.	67.3	0	0	0	0	0
P 30	F	26	x	x	x	x	+	+	n. d.	n. d.	–	1023	n.d.	>100	>100	n. d.	<0.01
P 31	M	30	x	x			–	–	n. d.	n. d.	–	936	n. d.	n. d.	0.04	n. d.	n. d.

M: male; F: female; x: convincing history with or without provocation test; (x) strongly suspected lupine allergy; ?: not knowingly consumed; n. d.: not done. Due to the small volumes of some sera, not all parameters could be investigated for every patient.

2.2.1. Alkaline Extraction

Samples of 2 × 4 g lupine flour were dissolved each in 40 mL of 0.2 M ammonium hydrogen carbonate (NH_4HCO_3, pH 8.0) and incubated for 30 min at 37 °C on a shaker. The flour was then centrifuged for 30 min at 13,000× *g*. Afterwards, the dialysis of the supernatant against Milli-Q water was performed with a dialysis tube, cutoff 3 kDa. The solution was again centrifuged at 13,000× *g*, and the supernatant filtered with a 0.45 µm pre-syringe filter (membrane polyethersulfone (PES)).

2.2.2. Acidic Extraction

2 × 4 g lupine flour was dissolved each in 40 mL of 0.1 M ammonium acetate (CH_3COONH_4, pH 5.0) and incubated for 6 h at 4 °C on a shaker. The flour was then centrifuged for 30 min at 13,000× *g*, and the supernatant dialyzed against Milli-Q water. The Milli-Q water was changed twice during this process. The solution was subsequently centrifuged again at 13,000× *g* for 20 min, and the supernatant filtered with a 0.45 µm pre-filter for syringes.

2.2.3. Gel Electrophoresis of Different Lupine Extracts

The separation of proteins of a molecular weight between 14 kDa and 100 kDa was performed via SDS-PAGE according to Laemmli UK (1970) [12].

Depending on the required volume, the samples were taken up 1:2 in 2-fold reducing sample buffer (200 mM Tris/HCl; 2 mM EDTA; 2% SDS; 25% glycerin; 1% dithiothreitol (DTT); 0.02% bromophenol blue, pH 6.8), or 1:5 in 5-fold reducing sample buffer (500 mM Tris/HCl; 5 mM EDTA; 5% SDS; 25% glycerin; 2.5% DTT; 0.02% bromophenol blue (pH 6.8)) and boiled for 5 min at 95 °C. Unless otherwise noted in the results section, 40 µg protein per cm gel were used for SDS-PAGE.

For better separation of proteins in the low molecular range, NuPAGE Bis-Tris 4–12% and NuPAGE Bis-Tris 12% ready-to-use gels (Invitrogen, Carlsbad, CA, USA) were used.

In combination with the 2-(N-morpholino)ethanesulfonic acid (MES) buffer (50 mM MES, 50 mM Tris, 1 mM EDTA, 0.1% SDS), these gels provide a high-resolution separation of the proteins in the low molecular range (<40 kDa).

The proteins in the acidic extracts of *L. angustifolius* and *L. luteus* were further separated by gel filtration (size exclusion chromatography (Superdex 75)) and ion-exchange chromatography (source Q) (see below).

2.2.4. In Silico Analysis

Protein, DNA and EST databases were queried for lupine proteins homologous to already known legume allergens.

IgE-reactive low-molecular-weight (LMW) proteins were further investigated by N-terminal sequencing and mass spectrometric analysis. Homology search in an expressed sequence tag (EST) database revealed a cDNA sequence (FG090100), which was used for expression in *E. coli*. The resulting recombinant protein was used for immunoblot inhibition studies.

2.2.5. 2D Fluorescence Difference Gel Electrophoresis (DIGE)

Protein extracts obtained via alkaline extraction of flour from all three lupine species were studied by 1D- and 2D-SDS-PAGE by means of a Refraction-2D labeling kit (NH DyeAGNOSTICS, 1 × 1, 8 nmol PR08, Halle, Germany). The extracts were solved in Milli-Q water, concentrated in a vacuum concentrator and used in the 2D-labeling kit applying 50 µg of each species and filled up to 10 µL of a compatible buffer. The buffer was produced with Tris (30 mM), urea (7 M), thiourea (2 M), 3-[(3-cholamidopropyl)dimethylammonio]-1-propanesulfonate (CHAPS) (4%) (pH 8.5). The working solution consisted of the different dyes (G-Dye 200 (green color), G-Dye 300 (red color), prepared in 4.5 µL G-Dye solvent. The samples (50 µg each plus buffer, 10 µL) plus 1 µL working solution were briefly stirred and centrifuged. Two samples of *L. angustifolius* extracts and one *L. albus* extract were dyed green, and two *L. luteus* extracts and one *L. albus* extract dyed red. These samples were cooled for 30 min on ice, and the reaction stopped with 1 µL of G-Dye stop solution. These solutions were briefly stirred, put again on ice for 10 min and subsequently investigated or frozen at −80 °C.

The investigation in the first-dimension gel electrophoresis was performed using the labeled lupine extracts as follows: 5 µL of *L. angustifolius* extract, dyed green plus 5 µL of *L. luteus* extract labeled red. The other samples of different lupine species extracts were combined accordingly.

The samples were added to 155 µL of rehydration buffer that consisted of 8 M urea 2% CHAPS, 0.5% ampholyte (Servalyt, pH 3–10, Serva, Heidelberg, Germany), 0.002% of bromophenol blue up to a final concentration of 20 mM DTT to be added directly before use.

The three combinations of lupine extract samples that were to be compared were pipetted into a ZOOM chamber (Invitrogen, Carlsbad, CA, USA), and one Novex™ ZOOM™ IPG-strip pH 3–10 L (Thermo Fisher Scientific, Invitrogen, Carlsbad, CA, USA) each added to the chamber, and incubated overnight at room temperature. The run of proteins in the first dimension was performed at 2 W, 2 mA, 200 V for 20 min, 450 V for 15 min, 750 V for 15 min, 2000 V for 60 min. After discontinuation, the chamber was stored at −80 °C. The second-dimension run was prepared as follows: the ZOOM strips were incubated in an equilibration buffer I and II for 15 min each in a rolling incubator. Equilibration buffer consisted of 0.05 M Tris, 6 M urea, 30% glycerin, 2% SDS, 2% bromophenol blue; 5 mL of equilibration buffer I consisted additionally of 0.05 g DTT; equilibration buffer II (5 mL) consisted additionally of 0.125 g iodoacetamide. After the incubation was finished, agarose (0.5 g agarose in 100 mL MES buffer) and marker proteins were prepared by heating, the strips were added to the gel and sealed with agarose. 10 µL of marker per pocket were added, and the run was started at 50 V for 20 min, then continued at 200 V. Afterwards, the gels (4–12% Zoom Gel, Invitrogen) were conserved with 40% ethanol plus 10% acetic acid. The readout was performed via Chemidoc MP (Bio-Rad, Hercules, CA, USA), Cy 3 adjustment on the Chemidoc for dye 200 (green), Cy 5 adjustment for dye 300 (red).

2.3. *Identification and Purification of New Single Allergens*

2.3.1. Recombinant Production of the *Lupinus albus* Profilin

Based on the sequence information obtained from the EST database (accession number FG090100), the company GeneArt (Regensburg, Germany) was able to identify the gene for the recombinant synthesis of profilin in *E. coli*. The gene was present in the vector pMA-T and thus obtained the antibiotic resistance to ampicillin used for selection. For expression, the ordered gene first had to be cloned. The transformation into living cells was used during cloning to propagate the DNA (plasmid).

The transformation was always performed with calcium-competent *E. coli* cells, first with TOP10F′ cells (Invitrogen, Carlsbad, CA, USA) and then BL21 DE3 cells (Stratagene, La Jolla, CA, USA).

After transformation into chemically competent BL21 DE3 cells of *E. coli* and induction with isopropyl-β-D-thiogalactopyranoside (IPTG), they produced the protein profilin.

10 mL of nutrient medium were inoculated with a clone picked from an agar plate, where *E. coli* cultures were growing, and incubated in an overnight culture at 37 °C on a shaker. The isolation of the plasmid DNA from the bacteria cells was achieved with the GeneJET Plasmid Miniprep Kit according to the manual (Thermo Fisher Scientific, Bremen, Germany) [13]. The restriction of the plasmid DNA from the bacteria cells was made with FastDigest enzymes NdeI and XhoI, according to the manual (Thermo Fisher Scientific, Vilnius, Lithuania). To analyze the restriction, the samples were separated in the agarose gel. Afterwards, the DNA was extracted from the gel according to the manufacturer's instructions with the GeneJET gel extraction kit (Thermo Fisher Scientific, Bremen, Germany) [13].

The pET23b vector (Novagen, Darmstadt, Germany), which is required for expression, contains after the stop codon the information on the synthesis of a His-tag. The His-tag consists of six histidine residues and serves for the later isolation of the protein. For ligation, the pET23b vector was cut with the same restriction enzymes under the same conditions as the insert for ligation. The ligation of the pET23b vector with the profilin insert was made with T4-Ligase according to the manual (Thermo Fisher Scientific, Vilnius, Lithuania). The ligation sample was transformed into new cells, these multiplied in an overnight culture, and the plasmid DNA was isolated by miniprep (above). Using DNA-sequencing (MWG, Biotech AG, Ebersberg, Germany), the DNA sequence could be confirmed. This was followed by the transformation of the Plasmid DNA in BL21 DE3 expression cells.

2.3.2. Expression

10 mL of preculture were infected with a clone picked from the agar plate, incubated overnight at 37 °C and transferred on the following day into 1 L nutrient medium with an additional 1 mL of ampicillin. The culture was incubated to a cell density of OD600 = 0.6–0.9. By adding IPTG (1 mM, final concentration), the protein biosynthesis of the recombinant profilin was induced. This solution was incubated for 3 h at 37 °C on a shaker and centrifuged (4000× *g*, 15 min, 4 °C). The supernatant was removed, and the pellet was processed for purification under native conditions and denaturing conditions.

2.3.3. Purification under Native Conditions

The pellet was dissolved in 27 mL of lysis buffer (native) (300 mM NaCl, 50 mM Na_2HPO_4, 10 mM imidazole pH 8.0), 3 mL of buffer for lysis (bug buster, Merck, Darmstadt, Germany; 100 mg/mL) and 10 µL of benzonase. The bug buster served to break up the bacterial cell walls in order to facilitate the subsequent isolation of the recombinant profilin. The mixture was incubated for 1 h at 4 °C in a rolling incubator and then centrifuged (13,000× *g*, 15 min, 4 °C). This pellet was purified under denaturing conditions (see below). The supernatant was isolated via metal affinity chromatography. The supernatant was mixed with approx. 5 mL of the column material HisPur™ cobalt resin (Thermo Fisher Scientific, Rockford, IL, USA) and incubated for 1 h in a rolling incubator. Meanwhile, the His-tag of the recombinant profilin bound to the cobalt (Co^{2+}) in the column material. Subsequently, the mixture was transferred to a column and the sample re-collected after the column material had settled. After the run, the first washing step was performed with 50 mL of lysis buffer (native, 300 mM NaCl, 50 mM Na_2HPO_4, 10 mM imidazole, pH 8.0), and a second wash step with 20 mL of wash buffer (native, 300 mM NaCl, 50 mM Na_2HPO_4, 40 mM imidazole pH 8.0) was carried out. The recombinant profilin was then eluted in four fractions with 5 mL of elution buffer (300 mM NaCl, 50 mM Na_2HPO_4, 250 mM imidazole each. Through the increased concentration of the imidazole, the elution buffer cleared the recombinant profilin from the column material. All fractions were collected and characterized by means of SDS-PAGE. The fractions, which contained the recombinant profilin, were pooled and dialyzed overnight against Milli-Q water. The protein content was then determined using the Bradford method Pierce™ Coomassie (Bradford) protein Assay Kit, Thermo Fisher Scientific, Waltham, MA, USA) and the profilin further purified via preparative SDS-PAGE (Model 491 Prep Cell, Bio-Rad Laboratories, Inc., Model 200/2.0 power supply; wide mini sub cell; Hercules, CA, USA).

2.3.4. Preparative SDS-PAGE

Using preparative SDS-PAGE (prep cell model 491) from Bio-Rad (Hercules, CA, USA), the recombinant profilin, which was expressed and isolated under native conditions, was separated from further contaminants. The pooled and washed samples from the elution (above) were used in a ratio of 1:2 in a 2-fold-reducing, or 1:5 in a 5-fold-reducing sample buffer and then heated for approx. 5 min at 95 °C. For the preparative SDS-PAGE, an 11% separation gel and a 4% collection gel were used.

The column was constructed and filled with the separating gel. Then the gel was covered with water-saturated n-butanol and left to polymerize overnight. The gel was washed with Milli-Q water, and the collection gel was installed onto the separating gel.

The column with the now polymerized gel was installed in the prep cell, which was filled with running buffer. The sample was applied and separated in the gel at 150 V for 80 min. The voltage was increased to 250 V for the remaining electrophoresis time.

After the bromophenol peak became visible, 95 fractions of 6 mL each were collected. The collected fractions were then investigated for the recombinant profilin.

2.3.5. Separation, Isolation and Purification of Proteins via Different Chromatography Methods

The Superdex 75 column (10/300 GL, GE Healthcare, Uppsala, Sweden) is suitable for optimal separation in the range of 3000–70,000 Da.

The acidic lupine extracts were aliquoted at a protein content of 6 mg/mL each, freeze-dried and stored at −20 °C. For the Superdex 75 procedure, the extract was resuspended in 10 mL Milli-Q water and transferred to the sample loop of the chromatography system (ÄktaPurifier, GE Healthcare, Uppsala, Sweden). A buffer of 0.2 M ammonium hydrogen carbonate was used. At a flow rate of 0.5 mL/min, 25 mL per run were collected as 0.8 mL fractions. The detection of the proteins was performed at 280 nm. The individual fractions were investigated with SDS-PAGE and Coomassie staining and subsequently by immunoblot. The further natural purification of the *L. albus* profilin was not successful, which is why we promoted its recombinant production. Since the purification of the recombinant profilin via further chromatography steps (size exclusion chromatography and ion-exchange chromatography (source 15Q 4.6/100 PE, GE Healthcare, Uppsala, Sweden) were not successful, it was further purified via preparative SDS-PAGE.

2.4. Identification, Isolation and Purification of Natural Lupine Lipid Transfer Proteins

SDS-PAGE and immunoblotting with patients' sera revealed LMW proteins of about 12 kDa in the acidic extracts (see above) of *L. angustifolius* and *L. luteus*. In order to isolate and purify them, 5 mg of whole lupine acidic extract per run underwent size exclusion chromatography (Superdex 75; 10/300 GL, GE Healthcare, Uppsala, Sweden) (above). In total, there were nearly 20 runs, which were performed with ammonium hydrogen carbonate buffer (0.2 M NH_4CO_3).

The obtained fractions underwent SDS-PAGE. Those fractions containing LMW proteins were pooled, re-buffered and concentrated by Amicon Ultra-15, (3 kDa, Millipore) and prepared for purification via ion-exchange chromatography (source 15 S, GE Healthcare, Uppsala, Sweden). The respective sample was first dialyzed three times against Milli-Q water so that the buffer that remained in the Superdex 75 was removed, and then three times against 50 mM sodium acetate buffer (pH 5.5). The run was started using samples in 100% of buffer A (50 mM sodium acetate pH 5.5). It was rinsed until the baseline was reached, then a gradient was generated. This spanned from 100% of buffer A to 100% B (50 mM sodium acetate plus 1 M sodium chloride pH 5.5) within 60 mL. Again, the obtained fractions were concentrated via Amicon Ultra-15, (3 kDa, Millipore), subsequently Superdex-peptide (10/300 GL, GE Healthcare, Uppsala, Sweden) was performed in case the result was not pure enough. SDS-PAGE of the potential *L. angustifolius* (*L. luteus*) lipid transfer protein (LTP) was then performed under reducing and nonreducing conditions. Immunoblotting revealed a single IgE-reactive band at ca. 12 kDa.

N-terminal sequencing (above) and database research was performed on the respective single allergens.

2.5. Immunoblotting

For blotting of the recombinant profilin, the method described above was used. For immunoblot analysis, proteins were transferred to polyvinylidene fluoride (PVDF) membranes (BIORAD, Immuno-Blot PVDF Membranes for protein blotting, BIO-RAD 0.2 µm) by semi-dry blotting for 45 min at 0.8 mA/cm^2 as described previously [14]. Membranes were blocked for 2 h with SynBlock (Bloomington, IN, USA). The subsequently performed blots were incubated in TTBS (100 mM Tris/HCl, 100 mM NaCl, 2.5 mM $MgCl_2$, 0.05% Tween-20, pH 7.4) for 30 min [15].

For immunoblot analysis, the patient sera were applied in a 1:10, 1:20 or 1:40 dilution. (For the different detection methods, see below).

2.6. Immunological Antigen Detection on Blotting Membranes

In our study, two different approaches to immune detections were applied:

Staining with NBT/BCIP (TBS-AP: 100 mM Tris/HCl, 100 mM NaCl, 5 mM $MgCl_2$, pH9.5). The chromogen substrate solution (nitro blue tetrazolium chloride (NBT)/5-bromine-4-chloro-3-indoxyl phosphate (BCIP)) induces an enzymatic reaction, by which the binding of the alkaline phosphatase-conjugated antibody to the protein becomes visible by means of a change in color.

Staining with horseradish peroxidase (HRP): A chemiluminescence reaction is catalyzed by an HRP-conjugated secondary antibody. The HRP catalyzes the oxidation of luminol (Bio-Rad, Hercules, CA, USA) in the presence of hydrogen peroxide (Bio-Rad, Hercules, CA, USA), and its luminescence can be detected with a digital imaging system, the ChemiDoc MP (Bio-Rad, Hercules, CA, USA). The PVDF membranes were first placed in Tris-buffered saline (TBS)-Tween buffer (pH 7.4) and incubated with gentle shaking. TBS-Tween buffer was used to block free binding sites on the blotting membrane to prevent unspecific binding.

The membrane was incubated overnight with the primary antibody, in most cases, patient serum (diluted 1:20 in TBS-Tween buffer pH 7.4). The membrane was washed three times with TBS-Tween buffer (pH 7.4) for 10 min each. Each suitable secondary antibody was incubated for 3 h while softly shaking. For detection with NBT/BCIP and alkaline phosphatase-conjugated (APC) mouse anti-human IgE antibody (dilution 1:10,000 in TBS-Tween buffer pH 7.4) (BD Pharmingen Bioscience, Heidelberg, Germany) was used. As a secondary antibody for the chemiluminescence, an HRP-conjugated mouse anti-human IgE Fc antibody (diluted 1:10,000 in Tris-Tween buffer pH 7.4 (SouthernBiotech, Birmingham, AL, USA)) was used. Free secondary antibody was removed by washing three times with TBS-Tween buffer (pH 7.4). For NBT/BCIP labeling, the membrane was washed a second time in TBS buffer (pH 9.5) to ensure optimal pH conditions for the enzymatic reaction with the chromogen substrate solution. During the washing procedure, the individual solutions, NBT and BCIP, were heated to 37 °C and then mixed with the chromogen substrate solution and applied to the membrane. The solution remained on the membrane until the proteins were clearly visible, then the reaction was stopped by transfer of the membrane into Milli-Q water. For the detection of chemiluminescence, the membrane was incubated for 5 min in 3 mL of luminol/enhancer reagent (Clarity Western ECL Substrat Bio-Rad, Hercules, CA, USA) and with 3 mL of peroxide reagent. The detection and documentation were performed with the ChemiDoc MP (Bio-Rad, Hercules, CA, USA).

2.7. Sequence Alignment

Sequence alignment was performed by use of BLAST (Clustal Omega, Wellcome Genome Campus, Hinxton, Cambridgeshire, CB10 1SD, UK).

2.8. N-Terminal Sequence Analysis

After blotting, the polyvinylidene fluoride (PVDF) membrane was washed with Milli-Q water, stained with 0.1% Coomassie in 50% methanol, destained in 50% methanol and air-dried. Protein bands were excised, and microsequencing was performed on a Procise protein sequencer with an online phenyl thiohydantoin (PTH) amino acid analyzer (PE Biosystems, Weiterstadt, Germany) [15].

2.9. Mass Spectrometry

The molecular masses of the protein fraction were analyzed using a high-resolution electrospray ionization (ESI) Fourier transform ion cyclotron resonance (FT ICR) mass spectrometer (Thermo Fisher Q Exactive Plus serious #387 and Advion NanoMate). For a straightforward interpretation of the heterogeneous samples, the obtained positive ion mass spectra were charge deconvoluted. Mass numbers refer to the monoisotopic mass of the neutral molecules. Tryptic mass fingerprinting was performed as described previously [16]. Briefly, Coomassie-stained protein bands were excised, destained and digested overnight with trypsin (trypsin gold, mass spectrometry grade; Promega, Mannheim, Germany) as described previously [17]. Afterwards, the corresponding tryptic fragments were mixed with 50% ACN/0.1% FA. The samples were analyzed by Thermo Fisher Q Exactive Plus serious #387 and Advion NanoMate with MS/MS. External mass calibration was performed with an appropriate mixture of peptides. Mass spectrometric data were analyzed with XCalibur Software (Thermo Fisher Scientific, Waltham, MA, USA).

3. Results

3.1. Comparison of Different Protein Extractions of Three Lupine Species

Immunoblots of the different extracts from the three lupine species that are of nutritional and commercial relevance for the community revealed differences between the species and the method of extraction (data not shown) with regard to the proteins of certain molecular weight and the respective concentration in the extract (data not shown). Investigations with sera from lupine-allergic patients revealed that seeds of different *Lupinus* species vary quantitatively in their allergen compositions. In order to investigate whether there were also qualitative differences, a 2D-DIGE was performed, including all three species, and in a subsequent experiment, peanut extract as well (Figure 1A–C).

3.1.1. Immunoblot Analyses

Comparison of Different Protein Extractions of Three Lupine Species

The seeds of different *Lupinus* species vary qualitatively and quantitatively in their allergen compositions (Figure 2).

The patients reveal inter-individually different sensitization profiles and react differently to acidic and alkaline extractions of lupine flour. This is evident for the sera from P 3 and P 23 reacting to considerably more proteins in the alkaline *L. angustifolius* extract when compared to the acidic extraction, indicating the necessity to use more than one extraction method when searching for new single allergens. IgE-reactivity showed species-specific differences also for alkaline lupine extracts.

These results confirm on an immunological level the differences between the lupine species as had been expected based on the proteomic analysis (Figure 1B,C).

Whereas our attempt to isolate lupine storage proteins failed insofar as we could enrich only gamma conglutin (43 kDa, data not shown), but not in a sufficient amount for further complex investigations with patients' sera, we aimed at the characterization of those proteins instead, that were identified in the low molecular range. We assumed that in these fractions, proteins of allergen families could be found, which already have some biomarker quality in other relevant food allergen sources like the closely related peanut (defensins, LTP, and profilins).

We, therefore, continued working with the acidic extracts of lupine species and more refined discrimination methods in order to identify, purify, and characterize proteins as potential allergens that were responsible for the binding of IgE in the sera of patients.

We focused on those reactions in immunoblots that were either close to a monosensitization in a patient or that stood out as prominent IgE-reactive protein bands. Since these were to be found in the low molecular range, we suspected LTP or profilin of lupine to be IgE-reactive.

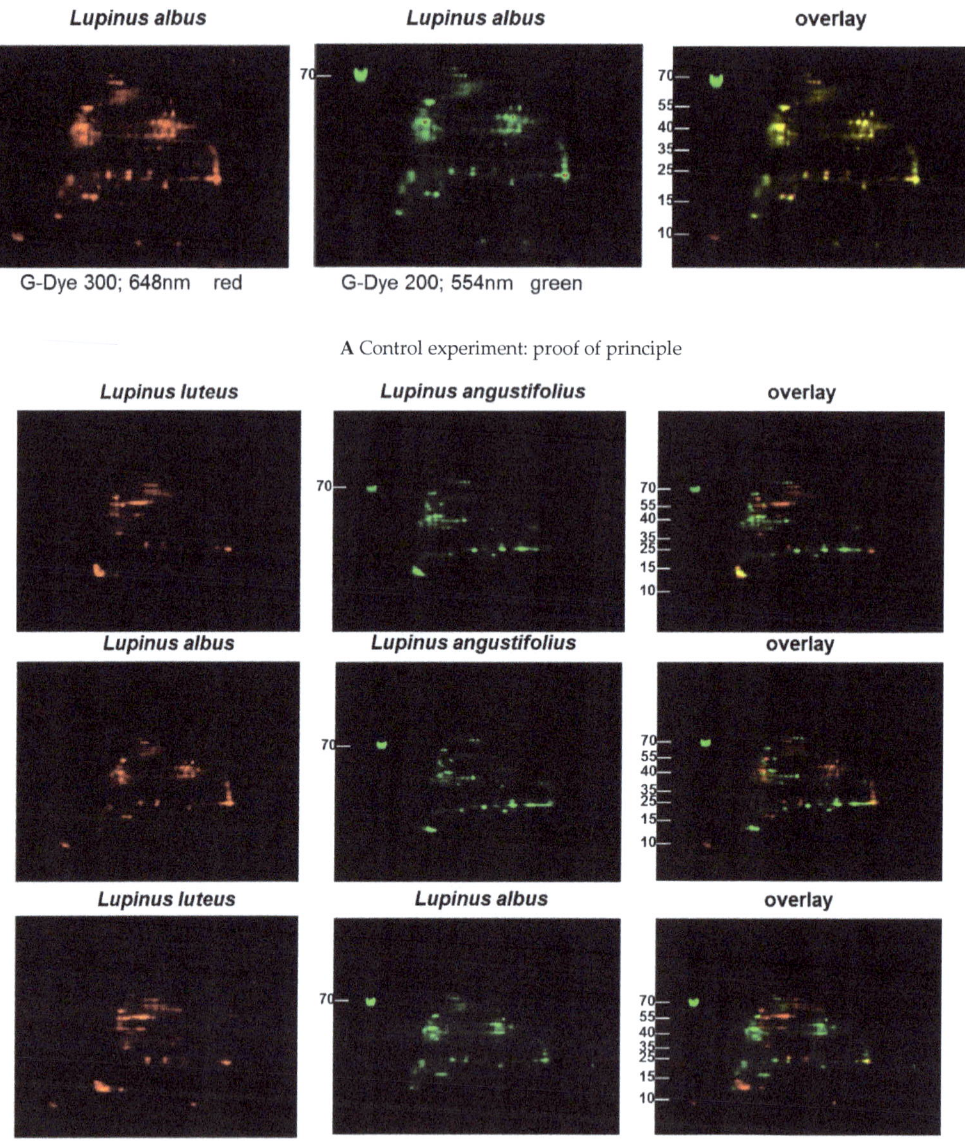

B Comparison of extracts from the three lupine species by 2D-DIGE

Figure 1. *Cont.*

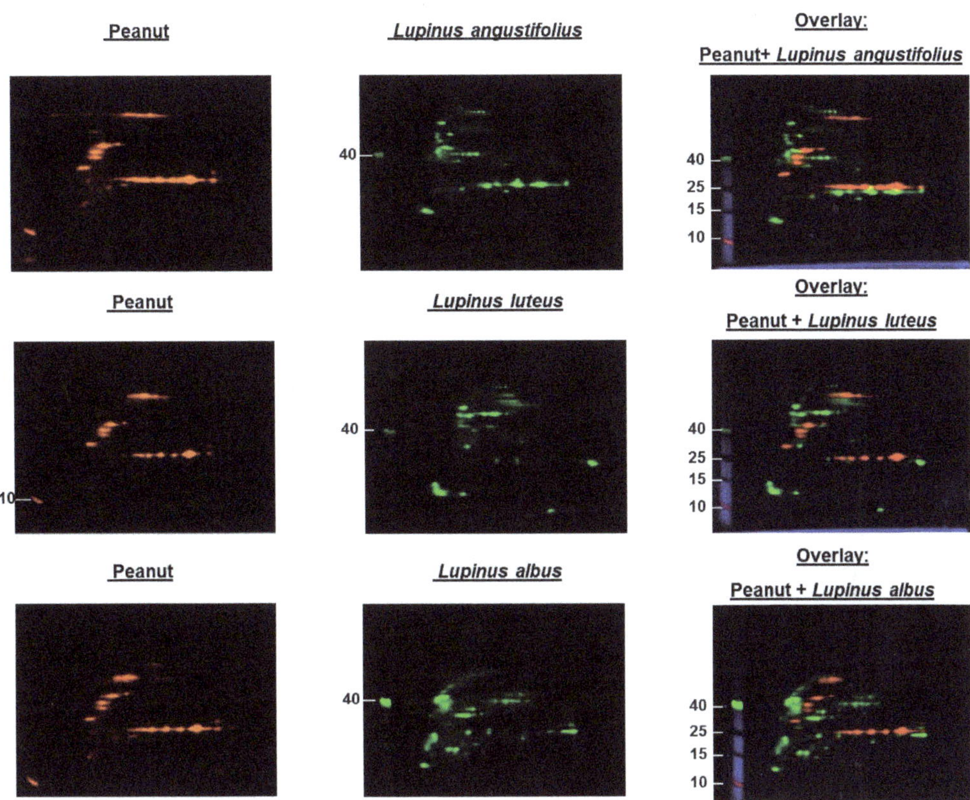

C Comparison of extracts from the three lupine species with peanut extract by 2D-DIGE

Figure 1. (**A–C**) Protein extracts obtained via alkaline extraction of flour from all three lupine species were studied by 2D-DIGE. (**A**) The control experiment shows the complete identity of the differently dyed samples of an alkaline extract of *Lupinus albus* by turning to yellow after having overlaid the two complementary colors, red and green. (**B**) By comparison of extracts from the three lupine species by 2D fluorescence difference gel electrophoresis (2D-DIGE), it becomes evident that there is no strong identity. On the contrary, the dominance of the single colors in the overlay speaks in favor of a broad molecular diversity of the dyed proteins in the different extracts. Whether the diversity is mirrored by immunological diversity was part of the subsequent investigations. (**C**) When comparing peanut extract with the extracts from different lupine species, there are only a few yellow areas. In addition, the differences between the lupine species become evident in this experiment as well, as there are different distributions of proteins colored green (lupine) when overlaid with peanut proteins dyed red.

3.2. Identification of an L. albus Profilin

Searches on sequence information of the lupine species revealed a sequence for a profilin in *L. albus* with the Accession number *FG090100.1*:

SWQTYVDEHLLCDIEGNQLTSAAIIGQDGSVWAQSSSFPQFKPEEITAIVNDFAEPG SLAPTGLYLGGTKYMVIQGEPGAVIRGKKGPGGVTVKKTNQALIIGIYDEPMTPGQCNV VVERLGDYLIDTGL

The sequence was then used to produce the recombinant profilin of *L. albus*, to confirm the sequence after expression and the molecular weight via mass spectrometry (Figure S1A–E, Figure S2A–E) and investigate for its allergenicity (IgE-reactivity) using patients' sera (Figure 3A,B).

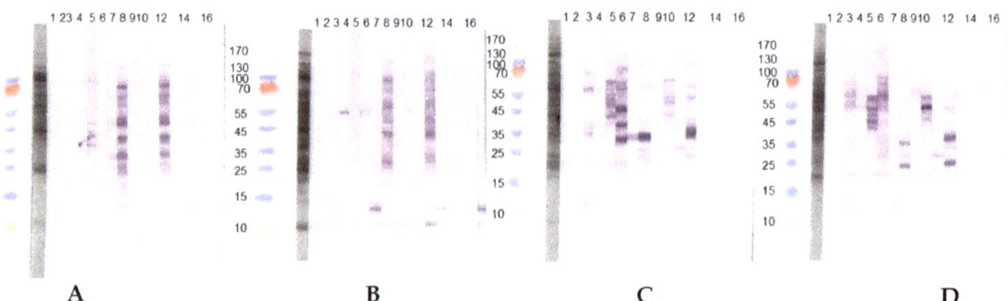

Figure 2. Immunoblots with acidic extracts (0.1 M CH_3COONH_4, pH 5.0), TTBS blocking and dilution of sera 1:10 with two different lupine species, (**A**) *L. angustifolius* and (**B**) *L. luteus* chosen as examples. (1) Tris control; (2) negative control serum; (3) lupine and peanut allergy (P 6); (4) legume allergy, suspected lupine allergy, lupine-sensitized (P 2); (5) peanut allergy, suspected lupine allergy, lupine-sensitized (P 3); (6) lupine allergy (P 23); (7) lupine allergy (P 10); (8) peanut allergy, lupine-sensitized (P 24); (9) lupine and peanut allergy (P 25); (10) lupine and peanut allergy (P 26); (11) peanut allergy, suspected lupine allergy (P 27); (12) peanut allergy, lupine-sensitized (P 4); (13) peanut allergy, lupine-sensitized (P 13); (14) lupine allergy (P 29); (15) peanut allergy and suspected lupine allergy (P 28); (16) suspected lupine allergy, lupine-sensitized (P 14); (17) peanut allergy, lupine-sensitized (P 12). (**C**,**D**) Immunoblots with alkaline extracts (0.2 M NH_4HCO_3, pH 8.0) of (**C**) *L. angustifolius* and (**D**) *L. luteus* with sera from the same patients. For sera from some individuals (P 4, P 10, P 12, P 13, P 14) differences regarding the IgE-reactivity to different lupine species are detectable. Particularly, sera from P 10, P 12, and P 13 showed reactivity to one LMW protein in the *L. luteus* extract, which we decided to work upon further since the reactivity was dominant when compared to weak or missing reactivity to other proteins in the extract. (P-code corresponds with Table 1.).

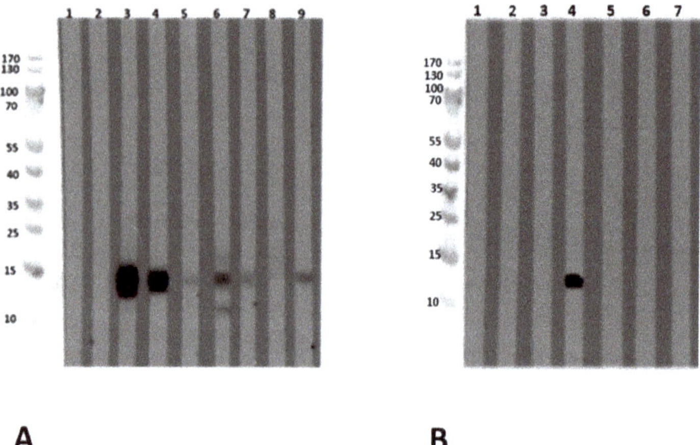

Figure 3. Immunoblot with (**A**) recombinant *L. albus* profilin and (**B**) with natural *L. albus* extract. (**A**) (1) Tris (buffer negative control) (2) negative serum; (3) peanut allergy, lupine-sensitized (P 11); (4) peanut allergy, lupine-sensitized (P 12); (5) suspected lupine allergy (P 14); (6) peanut allergy, lupine-sensitized (P 13); (7) peanut and lupine allergy (P 15); (8) peanut allergy, lupine-sensitized (P 16); (9) peanut allergy, lupine-sensitized (P 17). (**B**) The immunoblot was designed as an inhibition assay using natural *L. albus* extract and sera pre-incubated with different concentrations of recombinant *L. albus* profilin using one profilin-reactive serum (4). (1) Tris, (2) negative serum, (3) negative serum + 50 µg rProfilin, (4) serum 4 (P 12), (5) serum 4 + 5 µg rProfilin, (6) serum 4 + 50 µg rProfilin, (7) serum 4 + 500 µg lupine extract. Blocking was performed with Synblock; serum dilution was 1:20, dilution of the HRP-conjugated mouse-anti-human IgE Fc-antibody was 1:5000 (SouthernBiotech, Birmingham, AL, USA). Immunologically, the recombinant profilin performed similarly to the natural allergen. P-code refers to Table 1.

3.2.1. Allergen Identification via IgE-Reactivity with Sera from Lupine-Sensitized and Lupine-Allergic Patients

Sera from lupine- and/or peanut-allergic patients showed an IgE-binding to an LMW protein of 15 kDa in an acidic extract of *L. albus*. Mass spectrometric analysis of the corresponding LMW compound provided the first evidence for the presence of *L. albus* profilin. EST database search revealed a cDNA sequence of *L. albus*, which showed more than 87% amino acid homology to the peanut (*Arachis hypogaea*) profilin Ara h 5. Since the natural profilin could not be purified in a sufficient amount via different serial purification steps, it was subsequently produced as a recombinant protein. For this, the *L. albus* cDNA sequence was used for the expression of a recombinant *L. albus* profilin. Nucleotide and protein sequences were taken from the EST sequence using the data (Figure S1A–E) in the suppository.

Sera from lupine- and/or peanut-allergic patients who showed IgE-reactivity against the natural LMW compound also reacted with the recombinant *L. albus* profilin. This was confirmed by the experiment, where the recombinant *L. albus* profilin was able to inhibit the IgE-binding to the natural LMW compound in immunoblot analysis with the acidic *L. albus* extracts (Figure 3A,B).

3.2.2. Sequence Alignment

The sequence alignment of the lupine profilin with profilin sequences from other plants with allergy relevance revealed a 92% identity with Ara h 5 (*Arachis hypogaea*, peanut profilin) and a 78% identity with Bet v 2, the profilin from birch pollen (*Betula verrucosa*). The newly identified profilin, therefore, is another component that could be responsible for cross-reactivity with peanut. In addition, the fact that patient sera are IgE reactive to lupine profilin provides some further evidence for a pollen-induced lupine sensitization.

The recombinantly produced *L. albus* profilin was consequently submitted by us to and accepted by the WHO/IUIS allergen nomenclature subcommittee as Lup a 5, isoallergen No. 0101. The molecular weight was 13.849 kDa (as deduced from the sequence), under reducing conditions (and still with the His-tag), it was 15 kDa (Table 2). An investigation for glycosylation was not performed. However, no sequence pattern for a putative N-glycosylation (NXT/S) was found in the sequence.

Table 2. Synopsis of single allergens in lupine species compared to peanut allergens.

Plant Food Allergens (Protein Families)	Peanut Allergen (*Arachis hypogaea*) Ara h x	MW [kDa]	Lupine Allergen (*Lupinus angustifolius*): Lup an x (*Lupinus albus*): Lup a x	MW [kDa]
Vicilin-type storage protein; 7S globulin	Ara h 1 (IUIS)	64	Lup an 1 β-conglutin (IUIS) *Lup a 1*	55–61
Conglutin-like storage protein; 2S albumin	Ara h 2 (IUIS)	17	δ-conglutin *Lup a δ-conglutin* *Lup an δ-conglutin*	2 4
Legumin-type; 11S globulin	Ara h 3 (IUIS)	60	α-conglutin Lup a α-conglutin Lup an α-conglutin	43
Profilin	Ara h 5 (IUIS)	15	Lup a 5 (IUIS)	15
Conglutin; 2S albumin	Ara h 6 (IUIS)	15		n.a.
Conglutin; 2S albumin	Ara h 7 (IUIS)	15	*Lup a γ-conglutin?* *Lup an γ-conglutin?*	n.a
PR-10 Bet v 1-super family	Ara h 8 (IUIS)	17	*Lup a 4* *Lup l 4*	16.5 17
Non-specific lipid transfer protein	Ara h 9 (IUIS) Ara h 16 (IUIS) Ara h 17 (IUIS)	9.8 8.5 11	Lup an 3 (IUIS) *L. luteus*	11 11

[3], modified; [18]. Italics: new, but not documented/accepted by the WHO/IUIS Allergen Nomenclature Subcommittee [19]; MW for IUIS accepted allergens are given as MW (SDS-PAGE); n.a.: not available.

By tryptic mass fingerprinting and MS/MS analysis from the reduced protein sample (apparent molecular mass from SDS-PAGE: 15 kDa), the masses given in Figure S1A–E were obtained and correspond to the calculated masses.

3.2.3. *Lupinus albus* Profilin Allergenicity

L. albus profilin is an allergen of 15 kDa and was shown to be IgE-reactive. Nine subjects with a case history of food allergy either to lupine (two, one of them proven by a food challenge, and one additional individual with strongly suspected lupine allergy) and/or to peanut with a suspected lupine allergy/lupine sensitization (seven out of nine), and one individual sensitized to both, lupine and peanut (Figure 3A, Table 1) were tested IgE-positive by immunoblot analysis with the recombinant profilin.

3.3. *Isolation and Purification of a Non-Specific Lipid Transfer Protein from L. angustifolius Seeds*

After the detection of an LMW protein of nearly 10 kDa in immunoblot with an acidic extract of *L. angustifolius* and *L. luteus*, this protein was isolated and further purified via size exclusion chromatography and ion-exchange chromatography from *L. angustifolius* extract and investigated for IgE-reactivity (Figure 4A,B). N-terminal sequencing and BLAST revealed that it was an LTP. The database was queried in January 2017 after the LTP sequence was published in December 2016. The LTP was purified and, after MS/MS analysis, identified as such. However, it was shown that the LTP was unintentionally co-purified with a cysteine proteinase inhibitor, both at approximately 11 kDa, and, therefore, indistinguishable from each other. This could not be seen in the SDS-PAGE under reducing conditions. After the existence of cysteine proteinase inhibitor had become known, SDS-PAGE was performed under non-reducing conditions, where both proteins could be differentiated (Figure 4C, Figure S3A–D). Under reducing conditions, the patient IgE was directed against the 11 kDa "protein band", whereas the immunoblot under non-reducing conditions shows solely IgE-reactivity to the LTP (16 kDa band) and not to the cysteine proteinase inhibitor at 11 kDa (Figure 4D).

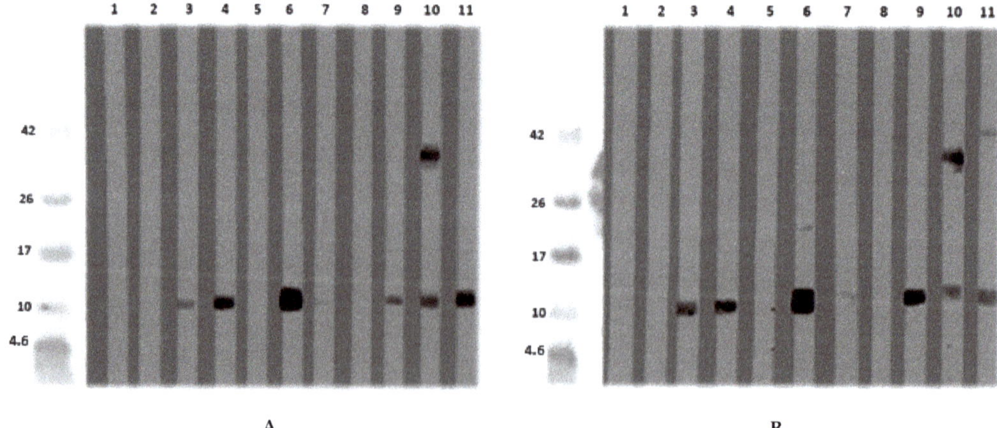

Figure 4. *Cont.*

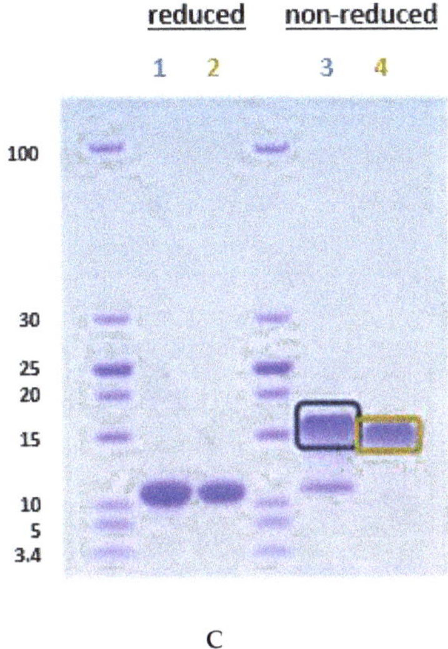

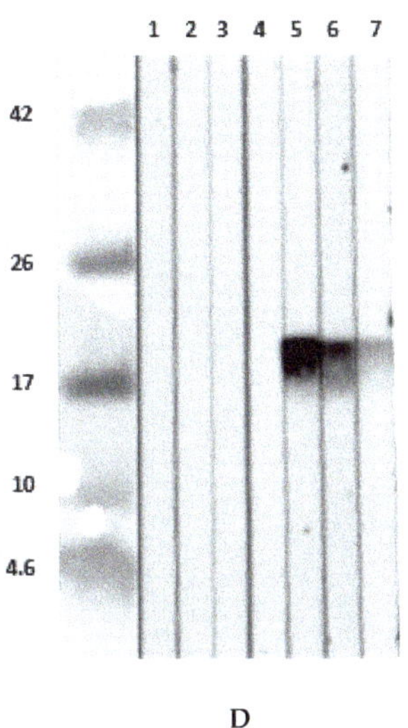

C D

Figure 4. (**A**) Immunoblot with naturally purified *L. luteus* LTP; (**B**) with naturally purified *L. angustifolius* LTP, using in a first step Synblock, in a second step anti-human-IgG 1:500. The sera were diluted 1:10. (1) Tris control; (2) negative control serum; (3) lupine and peanut allergy (P 6); (4) peanut allergy, lupine-sensitized (P 9); (5) peanut allergy, lupine-sensitized (P 5); (6) peanut allergy, lupine-sensitized (P 4); (7) peanut allergy and suspected lupine allergy (P 3); (8) peanut and suspected lupine allergy (P 2); (9) peanut and lupine sensitization (P 1); (10) lupine allergy and peanut sensitization (P 8); (11) lupine and peanut allergy (P 7). (**C**) SDS-PAGE for MS-analysis: *L. angustifolius* LTP (1) and *L. luteus* LTP (2) under reducing conditions. *L. angustifolius* LTP (3) and *L. luteus* LTP (4) under non-reducing conditions. (**D**) Immunoblot with natural *L. angustifolius* LTP (non-reducing conditions) (1) Tris (2nd antibody control) (2) negative serum I (3) negative serum II (4) negative serum III (5) peanut allergy and lupine-sensitized (P 4); (6) peanut allergy and lupine-sensitized (P 9); (7) lupine and peanut allergy (P 6). (P-codes refer to Table 1).

MS-analysis revealed the following sequences (similarity calculated in %). XP_0194 46786.1:

Lane 3: ITCGQVTANLAQCLNYLRSGGAVPAPCCNGIKNILNLAKTTPDRRTACNCLKAAAANTPGLNPSNAGSLPGKCGVNIPYKISTSTNCASIK: 93.4%

Lane 4: ITCGQVTANLAQCLNYLRSGGAVPAPCCNGIKNILNLAKTTPDRRTACNCLKAAAANTPGLNPSNAGSLPGKCGVNIPYKISTSTNCASIK: 33.3%

Investigation for glycosylation of the LTP from *L. angustifolius* was not performed in detail. However, one potential glycosylation at position 87 (NPSN, analyzed by http://www.cbs.dtu.dk/services/NetNGlyc/) exists. This was not verified by mass spectrometric analysis (meaning AAAANTPGLNPSNAGSLPGK was identified without glycosylation). Protein sequence (complete) was: MAGIVKLACAVLICMVVVSAPLTKAITCGQVTANLAQCLNYLRSGGAVPAPCCNGIKNILNLAKTTPDRR TACNCLKAAAANTPGLNPSNAGSLPGKCGVNIPYKISTSTNCASIK (XP_019446786.1). The N-terminal sequencing revealed the following N-terminus ITXGQVTANLAQ, which was confirmed by LC–MS/MS, and the sequence coverage of the expected full-length protein was 100% (see Supplementary Figure S3).

Lupinus angustifolius and *L. luteus* LTP Allergenicity

The investigation for the allergenicity of the newly found LTPs in immunoblot analysis revealed that eight out of 17 patients with lupine allergy and/or peanut allergy and a polysensitized individual were IgE-positive for lupine LTPs.

Three of these individuals had undergone an oral provocation test with lupine flour and developed symptoms [1], and all three were IgE-positive to *Lupinus angustifolius*-LTP.

The polysensitized individual was also LTP-IgE-positive. The *L. angustifolius* LTP was submitted to and accepted by the WHO/IUIS allergen nomenclature subcommittee in 2019 as Lup an 3 (Table 2).

4. Discussion

Molecular allergology based on many individual allergens belonging to only a few concise protein families has revolutionized the understanding of the pathomechanism of allergies and allergy diagnostic procedures considerably. However, there are still important gaps in diagnostic test sensitivity and specificity [20,21]. Several examples outline the importance of the inclusion of single allergens of clinically relevant allergen sources into diagnostic tests either as component-resolved diagnostics or as additional ingredients in the whole extracts used for diagnostic tests ("spiking" of the extracts with single allergens). Many single allergens have already gone into routine allergy diagnostic tests; however, lupine species are not among them. In vitro tests for lupine allergy presently are based only on lupine seed extract of one species. This is critical because lupine is an upcoming relevant food, and allergy to lupine can be severe, which is why allergists should be prepared and diagnostic procedures updated considerably.

The study presented here is, to the best of our knowledge, the first experimental comparative investigation on potential allergens of all three relevant *Lupinus* species and shows qualitative and quantitative species-associated differences in the protein content. In addition, we achieved for the first time the detection of three single LMW lupine proteins as new allergens, one pan allergen, the profilin of *L. albus*, and two LTPs, and thereby potential marker allergens for severe reactions, a non-specific (ns) LTP of *L. angustifolius* and an ns-LTP of *L. luteus*. The latter could not be submitted as a new allergen to the WHO/IUIS allergen nomenclature subcommittee as yet because our results could not be confirmed with a published sequence at that time and are presently being re-evaluated.

Having single allergens available for routine diagnostic tests allows for the identification of the primary sensitizing food (primary or pollen-associated lupine allergy), the detection of potential cross-reactivity, an increase of in vitro test sensitivity in cases where the extract-based diagnostic test lacks single allergens, and the identification of patients at risk to suffer from severe reactions.

The reactivity of patients to lupine profilin provides evidence for a pollen-induced sensitization to lupine. In general, IgE to profilins are associated with mild symptoms in pollen allergy but can be severe in pollen-associated food allergy [22].

In contrast, severe reactions are mostly associated with storage proteins—which is also hypothesized for lupine—but there are increasing cases not only in the Mediterranean but also in Central and Northern Europe, where severe reactions are associated with ns-LTPs [23]. LTPs are small, lipophilic proteins (91 to 95 amino acids), eight cysteines forming disulfide bridges, basic isoelectric point, α-helical structure [24], an altogether stable structure that is resistant to heat and digestion. They are ubiquitous in the plant kingdom, and investigations regarding potential cross-reactivity revealed that some show a strong structural similarity even when part of plants with a distant taxonomic relationship [25]. This is particularly relevant for a clinical phenomenon called the LTP syndrome, where patients characteristically react to LTPs from phylogenetically different plant food sources [26,27]. Although there is still a clinically silent sensitization to be considered, there are observations that whenever a food-allergic patient has IgE against more than five LTPs from different food sources, there is a risk of developing severe reactions [28]. In addition, there are data on endoluminal food allergy (gastrointestinal symptoms only) as a sequel to ingestion of LTP-containing food [28–30]. As can be imagined, these isolated gastrointestinal symptoms are very often not classified as allergic, except an experienced allergist elucidates this connection and supports it with plausible results of allergy diagnostic tests. These are important clinical observations that strongly speak in favor of a broadening of allergen panels to investigate for sensitization patterns that allow the correct phenotyping and—in case marker allergens for the severity of a reaction are involved—an adequate risk evaluation and management. The first LTP, fully characterized as an allergen, was Pru p 3 [31]. Presently, this allergen, which is already part of routine allergy tests, is being used as representative LTP in case a clinician suspects LTP-association of a severe allergic reaction.

The lupine proteins that have been documented so far by the WHO/IUIS allergen nomenclature subcommittee as allergens [19] belong to different lupine species. Single allergens belonging to one protein family have not been isolated from all three lupine species in parallel yet. It is the same in our study. According to our own experience, this is due to methodical difficulties in protein isolation and purification, maybe even due to species-specific differences. In our comparison experiments, we detected a high degree of molecular diversity between the three lupine species, which partly was mirrored by an immunological diversity in so far as patients had different IgE-reactivity to different lupine species. Unfortunately, the 2D-PAGE experiment could not be performed with patient sera due to the lack of sera volumes. Therefore, the question remains as to whether a patient allergic to one lupine species may tolerate another.

After we had purified two lupine allergens, we tested more patients and used these allergens to identify the culprit food in patients with no unambiguous evidence regarding the cause of their food-associated (severe) symptoms. One source of potentially severe lupine allergy is a pre-existing peanut allergy. In our study, we also included peanut-allergic individuals who mostly showed lupine sensitization, but some also suffered from lupine allergy.

All in all, not many lupine-allergic patients were admitted to our outpatient clinic, and—although this is a multicenter study—not many could be included in this investigation. Most of them were oblivious to their lupine allergy and did not even know that lupine is a food, although it must be declared on ingredient lists [32]. Therefore, we believe that there is a huge number of unrecorded allergic and even anaphylactic cases based on lupine, which are most probably documented as idiopathic anaphylaxis.

We think that only after the allergen profiles of foods with high anaphylactic potential have been elucidated, a correct diagnosis accompanied by correctly proposed prophylactic measures can be made. Apart from Lup an 1, ours are the only two other new lupine allergens accepted by the WHO/IUIS allergen nomenclature subcommittee presently as Lup an 3 and Lup a 5. Lup a 4 and Lup l 4 (the Bet v 1-homolog) have already been described and documented in Allergome only so that, in general, some of the most relevant allergen families represented in lupine could be detected in case these allergens went into routine diagnostics. Some research still must be done on the purification of lupine storage proteins, as they are most probably associated with severe reactions and also must be included in routine diagnostic measures. Although building slowly, there will be component-resolved diagnostics for lupine allergy and anaphylaxis in the near future.

Supplementary Materials: The following are available online at https://www.mdpi.com/2072-6643/13/2/409/s1, Figure S1: Identification of the sequence information on *L. albus* (A) available from the EST data base, (B) sequence used for recombinant expression of the authentic protein, and (C–E) preparation for mass spectrometry analysis, Figure S2: Synopsis of the mass spectrometry analyses of *L. albus* recombinant profilin and the *L. albus* profilin purified from the natural source, Figure S3: *Lupinus angustifolius* lipid transfer protein (LTP): (A) database protein sequence, (B) molecular weight determination of the mature protein (under reducing and non-reducing conditions), (C) N-terminal sequencing, and (D) mass spectrometry.

Author Contributions: U.J. designed the project, obtained positive votes from the local ethics committee, recruited and characterized the patients, designed and wrote the manuscript, designed Tables 1 and 2, discussed the experimental setup and results. S.A., S.D.-B., A.C.K., L.L., N.W., F.W.R., R.T., M.W., F.R. provided sera from well-characterized patients; A.K., S.H., D.W., and M.B. performed the experiments on allergen identification and purification the immunoblots, and the 2D-DIGE (D.W.). G.R. performed the EST analysis, W.-M.B. discussed the design of the project, the experimental setup and the results. All authors have read and agreed to the published version of the manuscript.

Funding: This research received no external funding.

Institutional Review Board Statement: The study was conducted according to the guidelines of the Declaration of Helsinki and approved by the Ethics Committee of the University of Lübeck, Germany, with approval numbers 10-124, 10-126, and 13-086.

Informed Consent Statement: Informed consent was obtained from all subjects involved in the study.

Data Availability Statement: Data is contained within the article and the supplementary material.

Acknowledgments: The two master students (Saskia Hellmig 2011, Arabella Karstedt 2016), whose excellent work in parts is being presented here, Skadi Kull for performing mass spectrometry and her most valuable support regarding the allergen-submission to the WHO/IUIS allergen nomenclature subcommittee, and Arnd Petersen (in my research group until 02/2014) for his support in establishing the early experimental investigations on lupine protein extraction and LTP identification, are gratefully acknowledged. We also want to thank Brigitte Kunz and Dominik Schwudke for providing the MS-services and technical support.

Conflicts of Interest: The authors declare no conflict of interest.

References

1. Peeters, K.A.; Koppelman, S.J.; Penninks, A.H.; Lebens, A.; Bruijnzeel-Koomen, C.A.; Hefle, S.L.; Taylor, S.L.; van Hoffen, E.; Knulst, A.C. Clinical relevance of sensitization to lupine in peanut-sensitized adults. *Allergy* **2009**, *64*, 549–555. [CrossRef] [PubMed]
2. Lee, Y.P.; Mori, T.A.; Sipsas, S.; Barden, A.; Puddey, I.B.; Burke, V.; Hall, R.S.; Hodgson, J.M. Lupin-enriched bread increases satiety and reduces energy intake acutely. *Am. J. Clin. Nutr.* **2006**, *84*, 975–980. [CrossRef] [PubMed]
3. Jappe, U.; Vieths, S. Lupine, a source of new as well as hidden food allergens. *Mol. Nutr. Food Res.* **2010**, *54*, 1–14. [CrossRef] [PubMed]
4. Hefle, S.L.; Lemanske, R.F., Jr.; Bush, R.K. Adverse reaction to lupine-fortified pasta. *J. Allergy Clin. Immunol.* **1994**, *94*, 167–172. [CrossRef] [PubMed]
5. Thermo Fisher Scientific. *Lupinus albus*, ImmunoCAP f335. Available online: https://www.thermofisher.com/diagnostic-education/hcp/de/de/resource-center/allergen-encyclopedia/whole-allergens.html?key=f335 (accessed on 1 December 2020).
6. Cabanillas, B.; Jappe, U.; Novak, N. Allergy to peanut, soy bean and other legumes: Recent advances on allergen characterization, stability to processing, and IgE cross-reactivity. *Mol. Nutr. Food Res.* **2018**, *62*, 1–9. [CrossRef]
7. Moneret-Vautrin, D.A.; Guerin, L.; Kanny, G.; Flabbee, J.; Frémont, S.; Morisset, M. Cross-allergenicity of peanut and lupine: The risk of lupine allergy in patients allergic to peanuts. *J. Allergy Clin. Immunol.* **1999**, *104*, 883–888. [CrossRef]
8. Jappe, U.; Kull, S.; Opitz, A.; Zabel, P. Anaphylaxis to vanilla ice cream: A near fatal cross reactivity phenomenon. Letter to the Editor. *J. Eur. Acad. Dermatol. Venereol.* **2018**, *32*, e22–e23. [CrossRef]
9. Campbell, C.P.; Jackson, A.S.; Johnson, A.R.; Thomas, P.S.; Yates, D.H. Occupational sensitzation to lupin in the workplace: Occupational asthma, rhinitis, and work-aggravated asthma. *J. Allergy Clin. Immunol.* **2007**, *119*, 1133–1139. [CrossRef]
10. Goggin, D.E.; Mir, G.; Smith, W.; Stuckey, M.; Smith, P.M.C. Proteomic analysis of lupin seed proteins to identify conglutin ß as an allergen, Lup an 1. *J. Agric. Food Chem.* **2008**, *56*, 6370–6377. [CrossRef]
11. Niggemann, B.; Beyer, K.; Erdmann, S.; Fuchs, T.; Kleine-Tebbe, J.; Lepp, U.; Raithel, M.; Reese, I.; Saloga, J.; Schäfer, C.; et al. Standardisierung von oralen Provokationstests bei Verdacht auf Nahrungsmittelallergie. *Allergo J.* **2011**, *20*, 149–160. [CrossRef]
12. Laemmli, U.K. Cleavage of structural proteins during the assembly of the head of bacteriophage T4. *Nature* **1970**, *227*, 680–685. [CrossRef] [PubMed]
13. Thermo Fisher Scientific. User Guide: GeneJET Gel Extraktion Kit. Available online: https://tools.thermofisher.com/content/sfs/manuals/MAN0012661_GeneJET_Gel_Extraction_UG.pdf (accessed on 15 December 2016).
14. Kyhse-Andersen, J. Electroblotting of multiple gels: A simple apparatus without buffer tank for rapid transfer of proteins from polyacrylamide to nitrocellulose. *J. Biochem. Biophys. Methods* **1984**, *10*, 203–209. [CrossRef]
15. Petersen, A.; Kull, S.; Rennert, S.; Becker, W.M.; Krause, S.; Ernst, M.; Gutsmann, T.; Bauer, J.; Lindner, B.; Jappe, U. Peanut defensins: Novel allergens isolated from lipophilic peanut extract. *J. Allergy Clin. Immunol.* **2015**, *136*, 1295–1301. [CrossRef] [PubMed]
16. Petersen, A.; Suck, R.; Lindner, B.; Georgieva, D.; Ernst, M.; Notbohm, H.; Becker, W.M. Phl p 3: Structural and immunological characterization of a major allergen of timothy grass pollen. *Clin. Exp. Allergy* **2006**, *36*, 840–849. [CrossRef] [PubMed]
17. Shevchenko, A.; Wilm, M.; Vorm, O.; Mann, M. Mass spectrometric sequencing of proteins silver-stained polyacrylamide gels. *Anal. Chem.* **1996**, *68*, 850–858. [CrossRef] [PubMed]
18. Available online: www.allergome.org (accessed on 20 January 2021).
19. Available online: www.allergen.org (accessed on 20 January 2021).

20. Jappe, U.; Nikolic, J.; Opitz, A.; Homann, A.; Zabel, P.; Gavrovic-Jankulovic, M. Apparent IgE-negative anaphylactic reaction to banana combined with kiwi-allergy-Complementary diagnostic value of purified single banana allergens. *J. Eur. Acad. Dermatol. Venereol.* **2016**, *30*, 1220–1222. [CrossRef]
21. Jappe, U.; Schwager, C. Relevance of lipophilic allergens in food allergy diagnosis. *Curr. Allergy Asthma Rep.* **2017**, *17*, 61. [CrossRef]
22. Alvarado, M.I.; Jimeno, L.; De La Torre, F.; Boissy, P.; Rivas, B.; Lázaro, M.J.; Barber, D. Profilin as a severe food allergen in allergic patients overexposed to grass pollen. *Allergy* **2014**, *69*, 1610–1616. [CrossRef]
23. Guelsen, A.; Jappe, U. Lipid transfer protein sensitization in an apple-allergic patient: A case report from Northern Europe. *Eur. Ann. Allergy Clin. Immunol.* **2019**, *51*, 80–83. [CrossRef]
24. Schocker, F.; Lüttkopf, D.; Scheurer, S.; Petersen, A.; Cisteró-Bahima, A.; Enrique, E.; San Miguel-Moncín, M.; Akkerdaas, J.; van Ree, R.; Vieths, S.; et al. Recombinant lipid transfer protein Cor a 8 from hazelnut: A new tool for in vitro diagnosis of potentially severe hazelnut allergy. *J. Allergy Clin. Immunol.* **2004**, *113*, 141–147. [CrossRef]
25. Asero, R.; Mistrello, G.; Roncarolo, D.; Amato, S.; Caldironi, G.; Barocci, F.; van Ree, R. Immunological cross-reactivity between lipid transfer proteins from botanically unrelated plant-derived foods: A clinical study. *Allergy* **2002**, *57*, 900–906. [CrossRef] [PubMed]
26. Pastorello, E.A.; Robino, A.M. Clinical role of lipid transfer proteins in food allergy. *Mol. Nutr. Food Res.* **2004**, *48*, 356–362. [CrossRef] [PubMed]
27. Van Winkle, R.C.; Chang, C. The biochemical basis and clinical evidence of food allergy due to lipid transfer proteins: A comprehensive review. *Clin. Rev. Allergy Immunol.* **2014**, *46*, 211–224. [CrossRef] [PubMed]
28. Scala, E.; Till, S.J.; Asero, R.; Abeni, D.; Guerra, E.C.; Pirrotta, L.; Paganelli, R.; Pomponi, D.; Giani, M.; De Pità, O.; et al. Lipid transfer protein sensitization: Reactivity profiles and clinical risk assessment in an Italian cohort. *Allergy* **2015**, *70*, 933–943. [CrossRef] [PubMed]
29. Pastorello, E.A.; Farioli, L.; Pravettoni, V.; Scibilia, J.; Mascheri, A.; Borgonovo, L.; Piantanida, M.; Primavesi, L.; Stafylaraki, C.; Pasqualetti, S.; et al. Pru p 3-sensitised Italian peach-allergic patients are less likely to develop severe symptoms when also presenting IgE antibodies to Pru p 1 and Pru p 4. *Int. Arch. Allergy Immunol.* **2011**, *156*, 362–372. [CrossRef]
30. Gomez, F.; Aranda, A.; Campo, P.; Diaz-Perales, A.; Blanca-Lopez, N.; Perkins, J.; Garrido, M.; Blanca, M.; Mayorga, C.; Torres, M.J. High prevalence of lipid transfer protein sensitization in apple allergic patients with systemic symptoms. *PLoS ONE* **2014**, *9*, e107304. [CrossRef]
31. Pastorello, E.A.; Farioli, L.; Pravettoni, V.; Ortolani, C.; Ispano, M.; Monza, M.; Baroglio, C.; Scibola, E.; Ansaloni, R.; Incorvaia, C.; et al. The major allergen of peach (Prunus persica) is a lipid transfer protein. *J. Allergy Clin. Immunol.* **1999**, *103 Pt 1*, 520–526. [CrossRef]
32. Commission Directive 2006/142/EC of 22 December 2006 amending Directive 2000/13/EC listing the ingredients which must under all circumstances appear on the labelling of foodstuffs. *Off. J. Union* **2006**, *L368*, 110.

 *nutrients* 

Review

Component-Resolved Diagnosis of Hazelnut Allergy in Children

Carlo Caffarelli [1,*], Carla Mastrorilli [2], Angelica Santoro [1], Massimo Criscione [3] and Michela Procaccianti [1]

1 Clinica Pediatrica, Dipartimento di Medicina e Chirurgia, Università di Parma, Azienda Ospedaliero-Universitaria di Parma, 43126 Parma, Italy; angelica.santoro204@gmail.com (A.S.); michela.procaccianti@outlook.it (M.P.)
2 UO Pediatria e Pronto Soccorso, Azienda Ospedaliero-Universitaria Consorziale Policlinico, Ospedale Pediatrico Giovanni XXIII, 70126 Bari, Italy; carla.mastrorilli@icloud.com
3 Dipartimento di Medicina e Chirurgia, Università di Parma, Azienda Ospedaliero-Universitaria di Parma, 43126 Parma, Italy; massimo.criscione@studenti.unipr.it
* Correspondence: carlo.caffarelli@unipr.it; Tel.: +39-0521-7022-07

Abstract: Hazelnuts commonly elicit allergic reactions starting from childhood and adolescence, with a rare resolution over time. The definite diagnosis of a hazelnut allergy relies on an oral food challenge. The role of component resolved diagnostics in reducing the need for oral food challenges in the diagnosis of hazelnut allergies is still debated. Therefore, three electronic databases were systematically searched for studies on the diagnostic accuracy of specific-IgE (sIgE) on hazelnut proteins for identifying children with a hazelnut allergy. Studies regarding IgE testing on at least one hazelnut allergen component in children whose final diagnosis was determined by oral food challenges or a suggestive history of serious symptoms due to a hazelnut allergy were included. Study quality was assessed by the Quality Assessment of Diagnostic Accuracy Studies-2 tool. Eight studies enrolling 757 children, were identified. Overall, sensitivity, specificity, area under the curve and diagnostic odd ratio of Cor a 1 sIgE were lower than those of Cor a 9 and Cor a 14 sIge. When the test results were positive, the post-test probability of a hazelnut allergy was 34% for Cor a 1 sIgE, 60% for Cor a9 sIgE and 73% for Cor a 14 sIgE. When the test results were negative, the post-test probability of a hazelnut allergy was 55% for Cor a 1 sIgE, 16% for Cor a9 sIgE and 14% for Cor a 14 sIgE. Measurement of IgE levels to Cor a 9 and Cor a 14 might have the potential to improve specificity in detecting clinically tolerant children among hazelnut-sensitized ones, reducing the need to perform oral food challenges.

Keywords: hazelnut; hypersensitivity; component-resolved diagnostics; children; food allergy; Cor a 1; Cor a 14; Cor a 9; IgE

Citation: Caffarelli, C.; Mastrorilli, C.; Santoro, A.; Criscione, M.; Procaccianti, M. Component-Resolved Diagnosis of Hazelnut Allergy in Children. *Nutrients* **2021**, *13*, 640. https://doi.org/10.3390/nu13020640

Academic Editor: Giacomo Caio
Received: 24 December 2020
Accepted: 10 February 2021
Published: 16 February 2021

Publisher's Note: MDPI stays neutral with regard to jurisdictional claims in published maps and institutional affiliations.

1. Introduction

Corylus avellana belongs to the same tree family of alders and birches (Betulaceae). Hazelnut is recognized as a common nut triggering allergic reactions from childhood and adolescence, and its prevalence varies by region. The self-reported prevalence of hazelnut allergies is approximately 0.2% in children [1] and up to 4.5% among adults from birch-endemic areas [2]. Resolution of a hazelnut allergy is rare (9% of cases), and children tend to have the disease for their whole life [3]. Clinical presentation differs from age [4]. Hazelnut allergies are associated with severe reactions in childhood and are one of the most common causes of anaphylactic death in adolescents and young adults [5]. On the contrary, adults mainly experience localized oral symptoms due to cross-reactions with pollens, in particular birch and alder.

Current management in childhood is based on a strict elimination diet, along with education of patients, families, and caregivers on managing allergic reactions caused by accidental ingestion [6], which is frequent in allergic individuals [7,8]. Since hazelnut is widespread in many processed foods and bakery products (especially pastries and chocolates), the dietary restriction and the constant fear of a severe reaction significantly worsen the quality of life of affected patients and their families. Therefore, it is essential to correctly diagnose hazelnut allergy with the aim of avoiding unnecessary therapeutic measures that limit the patient's quality of life. Hazelnut allergies are diagnosed with a combination of a convincing clinical history, serum-specific IgE (sIgE), skin prick testing (SPT), and oral food challenges (OFCs) [6,9–12]. Extract-based hazelnut tests (SPT and sIgE) have high sensitivity but low specificity (6-28% for SPT, using respectively natural and commercial extracts; 17–77% for sIgE, depending on the cutoff) [13] due to cross-sensitization with pollen or other food allergens that present high homology with the allergen tested. IgE sensitization to hazelnut extract is common, especially in birch endemic areas, it can occur whether patients react to hazelnut with severe or mild symptoms or even if no reaction occurs, and it often requires an OFC to assess the clinical significance [4,14–17]. Raising the cut-off values does not increase the sensitivity of SPT and sIgE [18]. OFC is considered the gold standard for diagnosis [19], even if it is an expensive, time-consuming test, with the risk for the patient of potentially life-threatening allergic reactions.

Component-resolved diagnostics (CRDs) has been introduced in clinical practice to more accurately discriminate patients who are not only sensitized to hazelnut but also allergic, and it is becoming an essential tool able to improve diagnostic accuracy [20].

The allergens of hazelnut belong to the families of seed storage proteins, pathogenesis-related proteins (PR-10), lipid transfer proteins (LTP), profillin and oleosine.

Genuine allergies to hazelnuts are generally due to sensitization to storage proteins or LPTs in children. Storage proteins are heat-stable and resistant to gastric digestion. They are well represented in hazelnuts and may account for more severe reactions, in particular Cor a 9, 11S globulin; and Cor a 14, 2s albumin. Also, LTPs (Cor a 8) are resistant to heat and digestion and are correlated with more serious symptoms. [15,17,21]. The pathogenesis-related class 10 proteins (PR-10) belong to one of the 11 subfamilies of the Bet v 1 family. Cross-reactive allergy to hazelnuts develops in birch pollen allergic individuals sensitized to Bet v 1 (PR-10).

Hazelnut contains Cor a 1, a PR-10 labile to heat and digestion that is a highly cross-reactive allergen shared with the main birch protein, Bet v 1. Cor a 1 was introduced in 2007 by one of the commercial producers of a serum hazelnut-specific sIgE test to improve the test's sensitivity for birch-related reactions to hazelnut, but it resulted in positive tests without clinical relevance [20]. Patients with an allergy to pollen birch generally develop mild to moderate symptoms. Clinical relevance of other hazelnut components, such as the 7S-vicilinlike protein Cor a 11 and two oleosins (Cor a 12 and Cor a 13), has not been confirmed [22]. The aim of this systematic review was to assess the diagnostic accuracy of sIgE on individual hazelnut proteins in the diagnostic work-up of hazelnut allergies in children.

2. Materials and Methods

We systematically searched three key electronic databases: MEDLINE (Pubmed), EMBASE (Ovid) and the Cochrane library. The databases were searched from 2010 to February 28th, 2020, using the search terms: "IgE", "prick", "SPT", "diagnosis", "challenge", "allergy", "DBPCFC", "OFC", "patch", "CRD", "component resolved diagnosis", "hazelnut", "corylus avellana", and "tree nut" in the title and abstract, aged 0-18. The Preferred Reporting Items for Systematic Reviews and Meta-Analyses (PRISMA) checklist was used to report this systematic review [23]. We include clinical trials, case-control, and cross-sectional studies. Reviews, discussion papers, editorials, qualitative studies, case reports, case series, conference abstracts and animal studies were excluded. We included studies that presented sufficient data to calculate sensitivity and specificity for at least one allergen component (Cor a 1, Cor a 8, Cor a 9, Cor a 14) (index test). Index tests were sIgE to hazelnut components. All studies were required to have a defined study population, limited to paediatric patients (0–18 years) who were suspected of hazelnut allergies. The reference standard was OFC, open or single-blind or double-blind placebo-controlled food challenge (DBPCFC). Alternatively, the reference standard was a suggestive history of anaphylaxis or serious symptoms due to hazelnut allergy confirmed by an allergist. At least 50% of patients must have performed an oral food challenge.

2.1. Study Selection and Data Collection

Two reviewers (M.C. and M.P.) independently screened titles and abstracts and then reviewed the full texts of studies that were considered to potentially meet the inclusion criteria, to identify eligible studies. If a study missed some information necessary to meet the inclusion criteria, authors were contacted. Where we received no response, we used data previously provided by the authors to other reviewers [24]. No language restrictions for included studies were applied: literature in languages other than English has been translated. Data of the following information were extracted: first author, date of publication, country, type of study, sample size, age (0–18 years), gender, and diagnostic tests (sIgE, OFC). Two reviewers (C.C. and M.P.) assessed the quality of the included studies using the Quality Assessment of Diagnostic Accuracy Studies-2 (QUADAS-2) tool [25]. Discrepancies were resolved by discussion and consensus. Measures of diagnostic test accuracy (DTA) and 95% confidence intervals were calculated if they were not reported in the papers.

2.2. Statistical Analyses

For each analysis, the cut off threshold was 0.35 kilounits of antibodies per litre (kUA/L). Given the significant heterogeneity found among the results of the included studies, quantified by Chi2, a random-effect meta-analysis model using the DerSimonian-Laird method was run to estimate the pooled test results. The random-effects model was utilized because it considers the risk of significant heterogeneity among studies and gives larger confidence intervals (CIs) than fixed-effect models [26]. Diagnostic odds ratios (DOR), positive and negative likelihood ratios (LR+/LR−), and area under the curve (AUC) were calculated. Fagan nomograms, which consider the LR+ and LR- obtained from the meta-analysis, were also used to estimate the clinical value of the index test [27]. Calculation of post-test probabilities was performed by assuming a pre-test probability that was equal to the prevalence of hazelnut allergies reported in the selected studies. The results were obtained as follows: pretest odds = prevalence/1-prevalence; post-test odds = pretest odds x LR- (LR+); and post-test probability = post-test odds/1+post-test odds. Positive predictive values (PPV) and negative predictive values (NPV) were computed. Publication bias was assessed with the funnel plot proposed by Egger [28]. Statistical analyses were performed using StatsDirect Statistical Software(StatsDirect statistical software. http://www.statsdirect.com. England: StatsDirect Ltd.) and Meta-DiSc Software (Meta-analysis of studies of evaluations of Diagnostic and Screening tests. http://www.hrc.es/investigacion/metadisc_en.htm. Spain: Unit of Clinical Biostatistics team of the Ramón y Cajal Hospital in Madrid.).

3. Results

The literature search found 1609 articles. After removing 199 duplicates, 1410 articles were reviewed based on their title and abstract. Among them, 24 full texts were assessed for inclusion, while 1386 articles were excluded based on their title and abstract. Eight studies (Figure 1) that met the research criteria were identified and included in the analysis. All the studies recruited pediatric patients only.

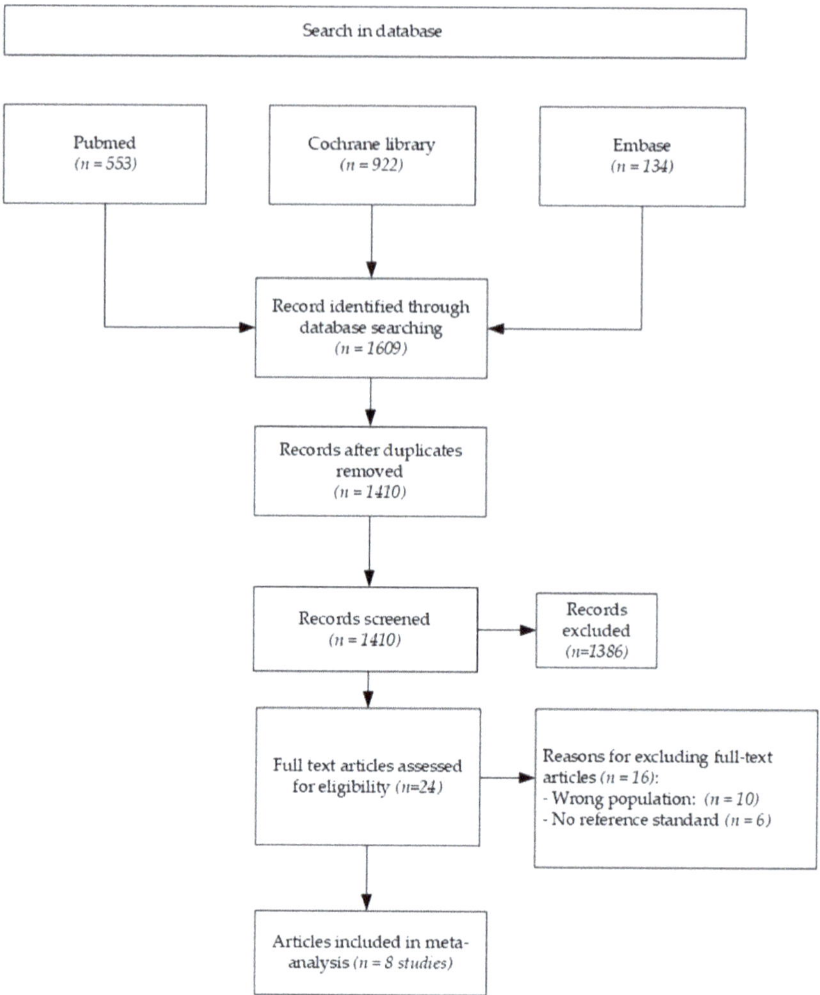

Figure 1. Flow diagram of included studies.

3.1. Risk of Bias

Results of the QUADAS-2 tool are reported in Figure 2. One study was found to have high risk of bias (ROB) in the patient selection domain because of case-control design [29]. The remaining studies had an unclear ROB because they failed to meet at least one of the criteria of the domain, mostly the sampling methodology [30–35]. There was no concern of ROB for applicability in this domain. In the index test domain, two studies were rated as having high ROB because a pre-specified threshold was not used [31,33]. The remaining studies were ranked as having unclear ROB because it was undetermined whether index test results were interpreted without knowledge of OFC results [29,30,32,34,35]. There was no concern of ROB for applicability. Regarding their reference standard, all studies were scored as having low ROB [29–35]. There was no concern of ROB for applicability except in one study with unclear ROB in this domain [29]. In flow and timing domains, a study had high ROB because it did not use OFC in all patients and did not include all patients in the analysis [34]. Six studies had unclear ROB because they did not meet a criterion [16,29,30,32,33,35]. One study had low ROB in this domain [31].

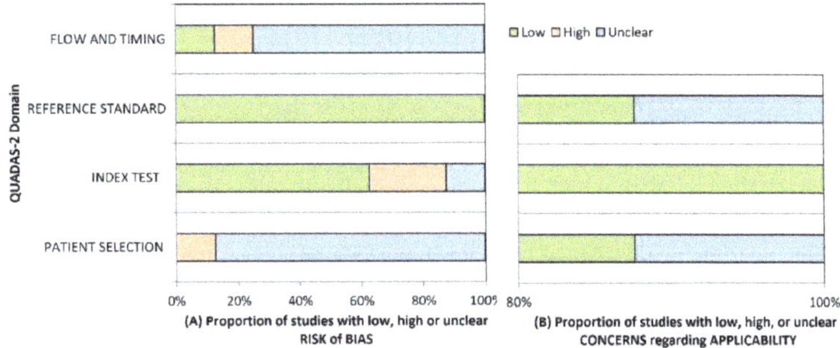

Figure 2. QUADAS-2 Domain. (**A**) Risk of bias; (**B**) Concerns regarding applicability.

3.2. Study Characteristics

Extracted data were summarized in Table 1.

Table 1. Summary of the included studies. In all studies, specific IgE to Cor a 1, Cor a 8, Cor a 9, and Cor a 14 was measured. DBPCFCF = double-blind placebo-controlled food challenge.

Summary of the Included Studies					
				Oral Food Challenge	
Author, Country, Year	**Study Population**	**Age (years) M/F**	**Lower limit IgE Positive Test (kUA/L)**	**Participants (%)**	**Number Positive (%)**
Beyer, Germany, 2015 [30]	143 children with suspected hazelnut allergy	Age (median, quartile) Tolerant 4.7 (2.1–8.1) Allergic 4.3 (2.2–6.1) –98/45	0.10	143 (100%) of which 46/143 (32%) DBPCFCF	99 (69%)
Brandström, Sweden, 2015 [31]	40 children referred for oral challenge for hazelnut allergy suspicion	Age (median, range) 11 (6–18) 23/17	0.10	40 (100%) DBPCFC	8 (20%)
Buyuktiryaki, Turkey, 2016 [32]	64 children with hazelnut allergy to determine resolution of hazelnut allergy	Age (median, interquartile) 3.4 (2.1–7.2) 45/19	0.10	56 (87.5%) DBPCFC 8 not performed because of anaphylaxis within the last 12 months	24 (42%)
Eller, Denmark, 2016 [33]	155 children with suspected hazelnut allergy	Age 5.1 (0.7–15.5) 100/55	0.35	140 (90%) open challenge 15 DBPCFC	65 (41%)
Grabenhenrich, Germany, 2016 [16]	142 children with suspected hazelnut allergy	Age (median, interquartile) 4.5 (2.1–7.6) 97/45	>0	142 (100%) open, single blind, double blind challenge.	44 (31%)
Inoue, Japan, 2019 [35]	91 children sensitized to hazelnut	Age (median, interquartile) 7.3 (5.9–10.5) 63/28	0.35	91 (100%) open food challenge	9 (9%)
Kattan, US, 2014 [34]	33 children with clinical impression of hazelnut allergy 9 children with history of objective symptoms with hazelnut ingestion	-	0.10	33 (78%) open challenge 9 not performed because of a history of objective symptoms with ingestion of hazelnut	4 (12%)
Masthoff, Netherlands, 2013 [29]	81 children Retrospective equally powered groups with positive/ negative challenge	Age (median, interquartile) 8 (7–12) 54/27	0.35	81 (100%) DBPCFC	40 (49%)

Studies were conducted in Europe (n = 7), Japan (n = 1), and the United States (n = 1). We found a total of 757 pediatric cases of suspected hazelnut allergy. All studies measured levels of sIgE to Cor a 1, Cor a 8, Cor a 9, and Cor a 14, using the same assay (ImmunoCAP, ThermoFisher, Uppsala, Sweden). Studies varied at the lower detection limit of hazelnut components between > 0 and 0.35 kilounits of antibody per litre (kUA/L) (Table 1). Regarding inclusion criteria, some studies enrolled children based on clinical history of a suspected hazelnut allergy [16,30,31,33], while others selected children with a clinical impression or convincing history of a hazelnut allergy [34]. A trial investigated children with hazelnut sensitization [35]. Other studies selected children based on the outcome of a food challenge [29] or to determine whether children had reached tolerance to hazelnuts [32]. All studies except one [34] reported the age of children, which ranged from 0.7 to 18 years. Median age varied from a low of 3.4 to 11 years. All papers but one [34] described the gender of recruited children. There were 480 (67%) males. The reference standard was an oral challenge using hazelnuts in all studies. However, OFCs were conducted with different protocols, including open, single blind or double blind. When the OFC was blinded, hazelnut was masked in chocolate products including mousse [16,30], balls [31], pudding [32], bars [33]) or Nutella [34] (Ferrero U.S.A., Inc., Somerset, NJ). In one study [29] the challenge was performed with defatted hazelnut flour for the first 9 doses (blinded) and a portion of 10 hazelnuts for the last dose (unblinded). In one study [35], roasted hazelnuts were used. The outcome of 741 hazelnut challenges was positive in 293 (39%) instances. In 16 patients, the challenge was not performed, and diagnosis was based on clinical history or recent anaphylactic reactions to hazelnuts.

3.3. *Diagnostic Accuracy*

There was variability in the diagnostic accuracy of sIgE to hazelnut components among studies (Tables 2–4).

Table 2. Diagnostic accuracy of sIgE to Cor a1, Cor a9, Cor a14. DOR = diagnostic odd ratio.

Author	Sensitivity (%)	(95%CI)	Specificity (%)	(95%CI)	DOR	(95%CI)
Cor a 1						
Brandström [31]	50	(5.7–84.4)	12.5	(3.5–29.0)	0.14	(0.03–0.81)
Eller [33]	49.2	(36.6–61.9)	58.9	(48.0–69.2)	1.39	(0.73–2.64)
Masthoff [29]	70.0	(53.5–83.4)	9.8	(2.7–23.1)	0.25	(0.07–0.87)
Pooled	56.6	(47.0–65.9)	37.4	(30.0–45.3)	0.42	(0.09–1.89)
Heterogeneity, Chi2	4.60 p = 0.100		43.3 p = 0.000		9.9 p = 0.007	
Cor a 9						
Brandstrom [31]	100	(63.1–100)	56.3	(37.7–73.6)	21.69	(1.15–407.76)
Eller [33]	74.2	(61.5–84.0)	67.9	(57.1–77.3)	5.94	(2.93–12.06)
Kattan [34]	84.6	(54.6–98.1)	65.5	(45.7–82.1)	10.45	(1.93–56.64)
Masthoff [29]	83.0	(67.2–92.7)	80.0	(65.1–91.2)	19.43	(6.32–59.75)
Pooled	79.5	(71.5–86.2)	68.1	(60.9–74.6)	9.45	(4.92–18.13)
Heterogeneity, Chi2	5.4 p = 0.145		4.9 p = 0.180		3.5 p = 0.320	
Cor a 14						
Beyer [30]	84.1	(69.9–93.4)	80.8	(71.7–88.0)	22.26	(8.61–57.56)
Brandstrom [31]	100	(63.1–100)	84.6	(67.2–94.7)	85.00	(4.25–1699.61)
Buyuktiryaki [32]	84.6	(65.1–95.6)	88.0	(68.8–97.5)	49.00	(11.14–215.60)
Eller [33]	80	(68.2–88.9)	84.4	(75.3–91.2)	21.71	(9.44–49.96)
Kattan [34]	69.2	(38.6–90.9)	82.8	(64.2–94.2)	10.80	(2.36–49.46)
Masthoff [29]	70	(53.85–83.4)	75.6	(59.7–87.6)	7.23	(2.71–19.32)
Pooled	80.2	(74.0–85.5)	82.4	(77.7–86.4)	18.27	(10.24–32.59)
Heterogeneity, Chi2	8.4 p = 0.135		2.35 p = 0.799		6.92 p = 0.227	

Table 3. Area under the curve (AUC) of sIgE to hazelnut components.

Author	AUC	95%CI
Cor a1		
Masthoff [29]	0.43	0.3–0.55
Beyer [30]	0.56	0.46–0.66
Grabenhenrich [16]	0.55	0.46–0.65
Inoue [35]	0.72	0.55–0.9
Pooled	0.55	0.46–0.64
Cor a8		
Masthoff [29]	0.51	0.39–0.64
Beyer [30]	0.63	0.53–0.73
Grabenhenrich [16]	0.62	0.52–0.72
Inoue [35]	0.58	0.39–0.78
Pooled	0.59	0.54–0.65
Cor a9		
Masthoff [29]	0.87	0.79–0.96
Beyer [30]	0.8	0.72–0.88
Eller [33]	0.78	0.7–0.85
Grabenhenrich [16]	0.8	0.72–0.88
Inoue [35]	0.71	0.52–0.89
Pooled	0.81	0.77–0.84
Cor a14		
Masthoff [29]	0.8	0.7–0.9
Beyer [30]	0.89	0.83–0.95
Eller [33]	0.85	0.77–0.94
Grabenhenrich [16]	0.89	0.83–0.95
Buyuktiryaki [32]	0.93	0.85–1
Inoue [35]	0.65	0.44–0.86
Pooled	0.87	0.82–0.92

Table 4. Positive predictive value (PPV), negative predictive value (NPV), likelihood ratio (LR) of IgE to Cor a1, Cor a9, Cor a14.

	PPV (%)	(95%CI)	NPV (%)	(95%CI)	LR+	(95%CI)	LR−	(95%CI)
Cor a 1								
Brandström [31]	12.5	(1–24)	50.0	(15.4–84.6)	0.57	(0.28–1.16)	4.0	(1.27–2.62)
Eller [33]	46.4	(34.6–58.1)	61.6	(51.4–71.9)	1.2	(0.84–1.7)	0.86	(0.64–1.16)
Masthoff [26]	43.1	(31–55.1)	25	(3.8–46.2)	0.78	(0.62–0.97)	3.08	(1.08–8.74)
Pooled					0.85	(0.58–1.26)	1.99	(0.63–6.21)
Heterogeneity, Chi2					6.0 $p = 0.050$		11.5 $p = 0.003$	
Cor a 9								
Brandstrom [31]	36.4	(16.3–56.5)	100	(100–100)	2.15	(1.42–3.26)	0.1	(0.07–1.5)
Eller [33]	62.3	(51.5–73.2)	78.2	(69–84.4)	2.29	(1.64–3.2)	0.39	(0.25–0.60)
Kattan [34]	52.4	(31–73.7)	90.5	(77.9–100)	2.45	(1.41–4.26)	0.24	(0.06–0.86)
Masthoff [29]	80.5	(68.4–2.6)	82.5	(70.7–94.3)	4.16	(2.20–7.83)	0.21	(0.11–0.43)
Pooled					2.47	(1.93–3.17)	0.31	(0.21–0.45)
Heterogeneity, Chi2					3.6 $p = 0.309$		3.1 $p = 0.377$	
Cor a 14								
Beyer [30]	66.1	(53.7–78.5)	92	(86.2–97.7)	4.38	(2.87–6.70)	0.20	(0.1–0.39)
Brandstrom [31]	61.5	(35.1–88)	100	(100–100)	5.67	(2.60–12.35)	0,07	(0–1)
Buyuktiryaki [32]	88	(75.3–100)	84.6	(70.7–98.5)	7.00	(2.77–17.67)	0.14	(0.06–0.36)
Eller [33]	78.8	(68.9–88.7)	85.4	(78.1–92.7)	5.14	(3.13–8.45)	0.24	(0.14–0.39)
Kattan [34]	64.3	(39.2–89.4)	85.7	(72.8–98.7)	4.02	(1.67–9.64)	0.37	(0.16–0.85)
Masthoff [29]	73.7	(59.7–87.7)	72.1	(58.7–85.5)	2.87	(1.61–5.1)	0.40	(0.24–0.66)
Pooled					4.44	(3.48–5.67)	0.26	(0.18–0.37)
Heterogeneity, Chi2					3.9 $p = 0.560$		7.4 $p = 0.196$	

Overall, both the sensitivity and specificity of IgE to hazelnut components were low (Table 2). The sensitivity and specificity of Cor a 1 sIgE were significantly lower than those of sIgE to both Cor a 9 and to Cor a 14, since a 95% confidence interval (CI) did not overlap. There is only one study by Masthoff et al [29] on Cor a 8 sIgE in children. They found that sIgE to Cor a 8 had a significantly lower sensitivity (5.0 (CI 95%, 0.6–16.9)) than other hazelnut components. The specificity of Cor a 9 sIgE [29] was also significantly lower than that of both Cor a 14 sIgE and Cor a 8 sIgE (95.1 (CI 95%, 83.5–99.4)).

AUC (Table 3) showed that the chance to be able to distinguish between positive and negative Cor a 1 sIgE was only 55%. For Cor a 8 sIgE, it was 58%. The AUCs of Cor a 9 and Cor a 14 were higher, 81% and 87%, respectively, and significantly different from those of Cor a 1 and Cor a 8, as shown by no overlapping 95% CIs. There was no difference between the AUC of Cor a 9 and the AUC of Cor a 14.

Regarding index test predictivity (Table 4), the LR+ of Cor a 1 sIgE was significantly lower than those of sIgE to both Cor a 9 and Cor a 14. Cor a 1 sIgE did not increase the probability of a hazelnut allergy, while sIgE to both Cor a 9 and Cor a 14 slightly increased it. The LR- of Cor a 1 sIgE slightly decreased the probability of having a hazelnut allergy, while both Cor a 9 sIgE and Cor a 14 sIgE moderately decreased it. According to the Fagan nomogram, we fixed the pre-test probability to 39% for hazelnut allergies, which was estimated by the number of children who reacted to hazelnuts in the selected studies. If the test result was positive, the post-test probability of a hazelnut allergy was 34% for Cor a 1 sIgE, 60% for Cor a9 sIgE and 73% for Cor a 14 sIgE. On the other hand, if the test result was negative, the post-test probability of a hazelnut allergy was 55% for Cor a 1 sIgE, 16% for Cor a9 sIgE and 14% for Cor a 14 sIgE.

The DOR (Table 2) of sIgE to Cor a1 was 0.42—lower than that of sIgE to both Cor a 9 and Cor a 14. DOR of Cor a 14 sIgE was not significantly higher than that of Cor a 9 sIgE (18.27 vs. 9.45). Positive predictive value of Cor a 1 sIgE varied from 12% to 46%, Cor a 9 sIgE from 36% to 80% and Cor a 14 sIgE from 61% to 88%. Negative predictive values of sIgE to Cor a 1, Cor a 9 and Cor a 14, respectively, ranged from 25% to 61%, 82% to 100%, and 72% to 100%.

Only one study [33] assessed whether IgE to hazelnut components were associated with the severity of objective symptoms in the hazelnut challenge. They found no correlation between IgE to Cor a 9 and Cor a 14 and the grade of allergic reaction.

Four of the selected studies considered the diagnostic value of combined IgE to hazelnut components. Beyer [30], Eller [33] and Inoue [35] did not find that the performance of diagnostic tests was improved by combining different components. In contrast, Masthoff [29] reported that combined positive IgE to Cor a 9 and Cor a 14 had a sensitivity that was similar to that of single molecules and a specificity of 98% that was higher than those of Cor a 9 and Cor a 14.

4. Discussion

This study has provided an assessment of the diagnostic accuracy of sIgE on individual hazelnut proteins in the diagnostic work-up of hazelnut allergies in children. Available hazelnut component tests include storage proteins Cor a 9 and Cor a 14, PR-10 Cor a 1, and LPT Cor a 8 [22,36,37].

The studies included in the present research had sensitivity and specificity varying from 50% to 70% and 9% to 58%, respectively, for Cor a 1 sIgE; from 74% to 100% and from 56% to 80%, respectively, for Cor a 9 sIgE; and from 69% to 100% and 75% to 88%, respectively, for Cor a 14 sIgE. When we performed an overall estimate of sensitivity and specificity, Cor a 9 sIgE and Cor a 14 sIgE were superior to Cor a 1 sIgE.

AUC that was unaffected by the prevalence of disease, since it was based on combined sensitivity and specificity, showed a moderate diagnostic accuracy for Cor a 1 sIgE and Cor a 8 sIgE. The AUCs of Cor a 9 sIgE and Cor a 14 sIgE were significantly more elevated than those of Cor a 1 sIgE and Cor a 8 sIgE. The AUC of Cor a 9 sIgE was similar to that of Cor a 14 sIgE. It is unclear why Cor a 1 sIgE had a lower sensitivity/specificity. Several hypotheses may be offered. Hazelnut sensitization can be genuine or due to IgE-mediated cross-reactivity to Bet v 1. Children who were primarily sensitized to PR-10 from birch or birch-related tree pollen [38] can have positive Cor a 1-sIgE as the result of a cross-reaction, which may be asymptomatic [39]. Since Cor a 1 is sensitive to gastric digestion and heat-labile, children who are only sensitized to this component often do not develop allergic symptoms. However, the severity of positive OFC on hazelnuts was not associated with positive results for any IgE to hazelnut components. The studies we selected did not allow us to separately assess the diagnostic accuracy of IgE to hazelnut components in children with or without Bet v 1 sensitization. Only three of the selected studies considered hazelnut allergies in children in relation to birch pollen allergies. Buyuktiryaki [32] found that only 4 (7,8%) children with hazelnut allergies were sensitized to tree pollen, and it was not stated how many children were allergic to birch. Eller found that sensitization to Bet v 1 was not associated with hazelnut allergies. Masthoff found that most children with subjective hazelnut allergies were sensitized to birch pollen and Bet v 1. Neither Buyuktiryaki [32], Eller [33] or Masthoff [29] reported levels of IgE to hazelnut allergen components (Cor a1, Cor a8, Cor a9 and Cor a14) in children with birch pollen sensitization compared with those in children who were not sensitized to birch pollen. Other explanations may be the smaller amount of Cor a 1 available compared with other components in fruit with reduced recognition, or less induction of IgE production by Cor a 1. These speculations require confirmation by further studies. There is not sufficient data to consider the specificity and sensitivity of sIgE to Cor a 8 since we have found only one study. Cor a 8 is more difficult to evaluate due to its great variability depending on the geographic area considered. Sensitization to LTP is more common in Mediterranean areas, but its clinical relevance is still debated. Another question is whether combining the results of studies addressing IgE to hazelnut components may improve diagnostic accuracy. Since there are contrasting data on this issue, further studies are necessary.

In clinical practice, it is recommended that children should not avoid hazelnuts without a clear diagnosis. On the other hand, children with hazelnut allergies should be carefully identified since serious reactions may develop following hazelnut ingestion. The gold standard for diagnosing hazelnut allergies is the OFC. However, extensive use of OFC is not economically sustainable. It requires a large amount of healthcare resources, and it is a stressful event both for patients and their caregivers. Moreover, OFC is a diagnostic procedure that involves some risks and requires an appropriate setting with personnel able to manage severe reactions such as anaphylaxis. Therefore, we have assessed whether component-resolved diagnosis for hazelnuts might predict hazelnut allergies and reduce the number of patients who need an OFC. This is of greater importance in the SARS-CoV-2 pandemic context, in which it is necessary to limit hospital tests as much as possible [40,41].

The prevalence of hazelnut allergies varied from 9% to 69% in the populations of the selected studies, and it is higher than in the general population of children who reported a hazelnut allergy in 3% of cases [42]. Higher prevalence of the disease increases PPV and decreases NPV. So, it is better to consider likelihood ratios that are not affected by the prevalence of the disease in the studied population. We calculated the post-test probability by using LRs and Fagan nomograms. We fixed the pre-test probability of a hazelnut allergy to 39%, which corresponds to the number of children with confirmed hazelnut allergies in the selected studies. We have determined that the post-test probability of a positive result was 34% for Cor a 1 sIgE, 60% for Cor a9 sIgE and 73% for Cor a 14 sIgE. Therefore, a positive result of sIgE to hazelnut components is not able to correctly identify children with hazelnut allergies. Negative hazelnut component sIgEs are more able to predict tolerance to hazelnuts. However, the post-test probability of negative result is too high for Cor a9 sIgE (16%), Cor a 14 sIgE (14%), and especially for Cor a 1 sIgE (55%) to reach a distinct diagnosis.

The results of DOR, which may vary from zero to infinity, are along the same lines. Higher values of DOR indicate a greater chance of a positive result for the index test in a person with a hazelnut allergy, compared with children who tolerate hazelnuts. The DOR of sIgE to Cor a1 was low, while the DORs of sIgE to Cor a 9 and Cor a 14 were significant. The DOR of sIgE to Cor a 1 was lower than that of sIgE to both Cor a 9 and Cor a 14. Children with positive Cor a 14 sIgE were at higher risk of having hazelnut allergies than those with positive Cor a 9 sIgE, but the difference was not significant. To our knowledge, only one systematic review with metanalysis about hazelnut allergy testing has been published until now (24). In agreement with our results, it found that the diagnostic accuracy of sIgE to Cor a 1 was lower than that of Cor a9 sIgE and Cor a 14 sIgE.

The strength of this study is a highly sensitive research strategy performed without language limitations, using various databases that permitted a complete literature review. All the studies included in the metanalysis performed an OFC to reach the diagnosis of a hazelnut allergy in more than 50% of the patients enrolled. A wide number of children were included. The strength of the findings can be limited by differences in inclusion criteria among the selected studies. It was not possible to analyze the diagnostic value of IgE to hazelnut components in studies with similar inclusion criteria since data were insufficient. However, we think that the criteria of our study are large enough to comprise the diversity of studies and the conditions in which the test is used, which also being satisfactorily narrow to obtain important answers when studies are considered together. Another relevant limitation is that if children with a genuine allergy to hazelnuts and those with a birch allergy and cross-reactive allergy to hazelnut are mixed up, the sensitivity and specificity of IgE components are diluted. As a result, the strength of Cor a9 and Cor a14 assessment in genuine hazelnut allergies in children is lost. A weakness may be represented by the limited number of studies retrieved. There is especially a paucity of studies on Cor a 8 sIgE. Another limitation is that there was heterogeneity across the studies on Cor a 1 sIgE. Finally, data were not divided on the basis of other variables, including sex or age, since information was lacking in the included studies.

5. Conclusions

Our analysis has shown that an OFC will still, in many cases, be necessary to prove clinical manifestations of hazelnut allergies. Measurement of IgE levels to Cor a 9 and Cor a 14 might have the potential to improve specificity in detecting clinically tolerant children among hazelnut-sensitized ones. This may lead to a reduction in the number of OFCs. Studies on the general population are warranted to elucidate this issue.

Author Contributions: Conceptualization, C.C.; validation, C.C., M.P., C.M. and A.S.; methodology, formal analysis, M.C.; investigation, M.C., M.P., C.M. and A.S.; data curation, M.C.; writing—original draft preparation, M.P. and A.S.; writing—review and editing, C.C. and C.M.; visualization, M.P. All authors have read and agreed to the published version of the manuscript.

Funding: This research received no external funding.

Institutional Review Board Statement: Not applicable

Informed Consent Statement: Not applicable.

Data Availability Statement: Data sharing is not applicable to this article.

Acknowledgments: We thank Giuseppe Pedrazzi for help in the statistical analysis.

Conflicts of Interest: The authors declare no conflict of interest.

References

1. Sicherer, S.H.; Muñoz-Furlong, A.; Godbold, J.H.; Sampson, H.A. US Prevalence of Self-Reported Peanut, Tree Nut, and Sesame Allergy: 11-Year Follow-Up. *J. Allergy Clin. Immunol.* **2010**, *12*, 1322–1326. [CrossRef] [PubMed]
2. McWilliam, V.; Koplin, J.; Lodge, C.; Tang, M.; Dharmage, S.; Allen, K. The Prevalence of Tree Nut Allergy: A Systematic Review. *Curr. Allergy Asthma Rep.* **2015**, *15*, 54. [CrossRef]
3. Fleischer, D.M.; Conover-Walker, M.K.; Matsui, E.C.; Wood, R.A. The Natural History of Tree Nut Allergy. *J. Allergy Clin. Immunol.* **2005**, *116*, 1087–1093. [CrossRef] [PubMed]
4. De Knop, K.J.; Verweij, M.M.; Grimmelikhuijsen, M.; Philipse, E.; Hagendorens, M.M.; Bridts, C.H.; De Clerck, L.S.; Stevens, W.J.; Ebo, D.G. Age-Related Sensitization Profiles for Hazelnut (Corylus Avellana) in a Birch-Endemic Region. *Pediatr. Allergy Immunol.* **2011**, *22*, e139–e149. [CrossRef]
5. Turner, P.J.; Gowland, M.H.; Sharma, V.; Ierodiakonou, D.; Harper, N.; Garcez, T.; Pumphrey, R.; Boyle, R.J. Increase in Anaphylaxis-Related Hospitalizations but No Increase in Fatalities: An Analysis of United Kingdom National Anaphylaxis Data, 1992–2012. *J. Allergy Clin. Immunol.* **2015**, *135*, 956–963. [CrossRef] [PubMed]

6. Stiefel, G.; Anagnostou, K.; Boyle, R.J.; Brathwaite, N.; Ewan, P.; Fox, A.T.; Huber, P.; Luyt, D.; Till, S.J.; Venter, C.; et al. BSACI Guideline for the Diagnosis and Management of Peanut and Tree Nut Allergy. *Clin. Exp. Allergy.* **2017**, *47*, 719–739. [CrossRef]
7. Sicherer, S.H.; Muñoz-Furlong, A.; Sampson, H.A. Prevalence of Peanut and Tree Nut Allergy in the United States Determined by Means of a Random Digit Dial Telephone Survey: A 5-Year Follow-up Study. *J. Allergy Clin. Immunol.* **2003**, *112*, 1203–1207. [CrossRef]
8. Ewan, P.W.; Clark, A.T. Long-Term Prospective Observational Study of Patients with Peanut and Nut Allergy after Participation in a Management Plan. *Lancet* **2001**, *357*, 111–115. [CrossRef]
9. Sampson, H.A.; Aceves, S.; Bock, S.A.; James, J.; Jones, S.; Lang, D.; Nadeau, K.; Nowak-Wegrzyn, A.; Oppenheimer, J.; Perry, T.T.; et al. Food Allergy: A Practice Parameter Update—2014. *J. Allergy Clin. Immunol.* **2014**, *134*, 1016–1025. [CrossRef]
10. Sicherer, S.H.; Sampson, H.A. Food Allergy: A Review and Update on Epidemiology, Pathogenesis, Diagnosis, Prevention, and Management. *J. Allergy Clin. Immunol.* **2018**, *141*, 41–58. [CrossRef]
11. Caffarelli, C.; Garrubba, M.; Greco, C.; Mastrorilli, C.; Dascola, C.P. Asthma and Food Allergy in Children: Is There a Connection or Interaction? *Front. Pediatr.* **2016**, *4*, 34. [CrossRef] [PubMed]
12. Mastrorilli, C.; Caffarelli, C.; Hoffmann-Sommergruber, K. Food Allergy and Atopic Dermatitis: Prediction, Progression, and Prevention. *Pediatr. Allergy Immunol.* **2017**, *28*, 831–840. [CrossRef]
13. Ortolani, C.; Ballmer-Weber, B.K.; Hansen, K.S.; Ispano, M.; Wüthrich, B.; Bindslev-Jensen, C.; Ansaloni, R.; Vannucci, L.; Pravettoni, V.; Scibilia, J.; et al. Hazelnut allergy: A double-blind, placebo-controlled food challenge multicenter study. *J. Allergy Clin Immunol.* **2000**, *105*, 577–581. [CrossRef]
14. Hansen, K.S.; Ballmer-Weber, B.K.; Sastre, J.; Lidholm, J.; Andersson, K.; Oberhofer, H.; Lluch-Bernal, M.; Östling, J.; Mattsson, L.; Schocker, F.; et al. Component-Resolved In Vitro Diagnosis of Hazelnut Allergy in Europe. *J. Allergy Clin. Immunol.* **2009**, *123*, 1134–1141. [CrossRef]
15. Schocker, F.; Lüttkopf, D.; Scheurer, S.; Petersen, A.; Cisteró-Bahima, A.; Enrique, E.; San Miguel-Moncín, M.; Akkerdaas, J.; Van Ree, R.; Vieths, S.; et al. Recombinant Lipid Transfer Protein Cor a 8 from Hazelnut: A New Tool for in Vitro Diagnosis of Potentially Severe Hazelnut Allergy. *J. Allergy Clin. Immunol.* **2004**, *113*, 141–147. [CrossRef]
16. Grabenhenrich, L.; Lange, L.; Härtl, M.; Kalb, B.; Ziegert, M.; Finger, A.; Harandi, N.; Schlags, R.; Gappa, M.; Puzzo, L.; et al. The Component-Specific to Total IgE Ratios Do Not Improve Peanut and Hazelnut Allergy Diagnoses. *J. Allergy Clin. Immunol.* **2016**, *137*, 1751–1760. [CrossRef] [PubMed]
17. Beyer, K.; Grishina, G.; Bardina, L.; Grishin, A.; Sampson, H.A. Identification of an 11S Globulin as a Major Hazelnut Food Allergen in Hazelnut-Induced Systemic Reactions. *J. Allergy Clin. Immunol.* **2002**, *110*, 517–523. [CrossRef]
18. Clark, A.T.; Ewan, P.W. Interpretation of Tests for Nut Allergy in One Thousand Patients, in Relation to Allergy or Tolerance. *Clin. Exp. Allergy* **2003**, *33*, 1041–1045. [CrossRef]
19. Caffarelli, C.; Ricò, S.; Rinaldi, L.; Povesi Dascola, C.; Terzi, C.; Bernasconi, S. Blood Pressure Monitoring in Children Undergoing Food Challenge: Association With Anaphylaxis. *Ann. Allergy Asthma Immunol.* **2012**, *108*, 285–286. [CrossRef]
20. Uotila, R.; Röntynen, P.; Pelkonen, A.S.; Voutilainen, H.; Kaarina Kukkonen, A.; Mäkelä, M.J. For Hazelnut Allergy, Component Testing of Cor a 9 and Cor a 14 Is Relevant Also in Birch-Endemic Areas. *Allergy* **2020**, *75*, 2977–2980. [CrossRef] [PubMed]
21. Matricardi, P.M.; Kleine-Tebbe, J.; Hoffmann, H.J.; Valenta, R.; Hilger, C.; Hofmaier, S.; Aalberse, R.C.; Agache, I.; Asero, R.; Ballmer-Weber, B.; et al. EAACI Molecular Allergology User's Guide. *Pediatr. Allergy Immunol.* **2016**, *27*, 1–250. [CrossRef] [PubMed]
22. Costa, J.; Mafra, I.; Carrapatoso, I.; Oliveira, M.B.P.P. Hazelnut Allergens: Molecular Characterization, Detection, and Clinical Relevance. *Crit. Rev. Food Sci. Nutr.* **2016**, *56*, 2579–2605. [CrossRef]
23. Moher, D.; Liberati, A.; Tetzlaff, J.; Altman, D.G. Preferred Reporting Items for Systematic Reviews and Meta-Analyses: The PRISMA Statement. *PLoS Med.* **2009**, *6*, e1000097. [CrossRef] [PubMed]
24. Nilsson, C.; Berthold, M.; Mascialino, B.; Orme, M.; Sjölander, S.; Hamilton, R. Allergen Components in Diagnosing Childhood Hazelnut Allergy: Systematic Literature Review and Meta-Analysis. *Pediatr. Allergy Immunol.* **2020**, *31*, 186–196. [CrossRef]
25. Whiting, P.F.; Rutjes, A.W.S.; Westwood, M.E.; Mallett, S.; Deeks, J.J.; Reitsma, J.B.; Leeflang, M.M.G.; Sterne, J.A.C.; Bossuyt, P.M.M. Quadas-2: A Revised Tool for the Quality Assessment of Diagnostic Accuracy Studies. *Ann. Intern. Med.* **2011**, *155*, 529–536. [CrossRef] [PubMed]
26. DerSimonian, R.; Laird, N. Meta-Analysis in Clinical Trials. *Control. Clin. Trials* **1986**, *7*, 177–188. [CrossRef]
27. Fagan, T.J.F. Nomogram for Bayes's Theorem. *N. Engl. J. Med.* **1975**, *293*, 257. [CrossRef]
28. Egger, M.; Smith, G.D.; Schneider, M.; Minder, C. Bias in Meta-Analysis Detected by a Simple, Graphical Test. *Br. Med. J.* **1997**, *315*, 629–634. [CrossRef]
29. Masthoff, L.J.N.; Mattsson, L.; Zuidmeer-Jongejan, L.; Lidholm, J.; Andersson, K.; Akkerdaas, J.H.; Versteeg, S.A.; Garino, C.; Meijer, Y.; Kentie, P.; et al. Sensitization to Cor a 9 and Cor a 14 Is Highly Specific for a Hazelnut Allergy with Objective Symptoms in Dutch Children and Adults. *J. Allergy Clin. Immunol.* **2013**, *132*, 393–399. [CrossRef]
30. Beyer, K.; Grabenhenrich, L.; Härtl, M.; Beder, A.; Kalb, B.; Ziegert, M.; Finger, A.; Harandi, N.; Schlags, R.; Gappa, M.; et al. Predictive Values of Component-Specific IgE for the Outcome of Peanut and Hazelnut Food Challenges in Children. *Allergy* **2015**, *70*, 90–98. [CrossRef]

31. Brandström, J.; Nopp, A.; Johansson, S.G.O.; Lilja, G.; Sundqvist, A.C.; Borres, M.P.; Nilsson, C. Basophil Allergen Threshold Sensitivity and Component-Resolved Diagnostics Improve Hazelnut Allergy Diagnosis. *Clin. Exp. Allergy* **2015**, *45*, 1412–1418. [CrossRef] [PubMed]
32. Buyuktiryaki, B.; Cavkaytar, O.; Sahiner, U.M.; Yilmaz, E.A.; Yavuz, S.T.; Soyer, O.; Sekerel, B.E.; Tuncer, A.; Sackesen, C. Cor a 14, Hazelnut-Specific IgE, and SPT as a Reliable Tool in Hazelnut Allergy Diagnosis in Eastern Mediterranean Children. *J. Allergy Clin. Immunol. Pract.* **2016**, *4*, 265–272. [CrossRef] [PubMed]
33. Eller, E.; Mortz, C.G.; Bindslev-Jensen, C. Cor a 14 Is the Superior Serological Marker for Hazelnut Allergy in Children, Independent of Concomitant Peanut Allergy. *Allergy* **2016**, *71*, 556–562. [CrossRef]
34. Kattan, J.D.; Sicherer, S.H.; Sampson, H.A. Clinical Reactivity to Hazelnut May Be Better Identified by Component Testing than Traditional Testing Methods. *J. Allergy Clin. Immunol. Pract.* **2014**, *2*, 633–634. [CrossRef] [PubMed]
35. Inoue, Y.; Sato, S.; Takahashi, K.; Yanagida, N.; Yamamoto, H.; Shimizu, N.; Ebisawa, M. Component-Resolved Diagnostics Can Be Useful for Identifying Hazelnut Allergy in Japanese Children. *Allergol. Int.* **2020**, *69*, 239–245. [CrossRef]
36. Weinberger, T.; Sicherer, S. Current Perspectives on Tree Nut Allergy: A Review. *J. Asthma Allergy* **2018**, *11*, 41–51. [CrossRef] [PubMed]
37. Ebo, D.G.; Verweij, M.M.; Sabato, V.; Hagendorens, M.M.; Bridts, C.H.; De Clerck, L.S. Hazelnut Allergy: A Multi-Faced Condition with Demographic and Geographic Characteristics. *Acta Clin. Belg.* **2012**, *67*, 317–321. [CrossRef] [PubMed]
38. Cipriani, F.; Mastrorilli, C.; Tripodi, S.; Ricci, G.; Perna, S.; Panetta, V.; Asero, R.; Dondi, A.; Bianchi, A.; Maiello, N.; et al. Diagnostic Relevance of IgE Sensitization Profiles to Eight Recombinant Phleum Pratense Molecules. *Allergy Eur. J. Allergy Clin. Immunol.* **2018**, *73*, 673–682. [CrossRef]
39. Mastrorilli, C.; Cardinale, F.; Giannetti, A.; Caffarelli, C. Pollen-Food Allergy Syndrome: A Not so Rare Disease in Childhood. *Med.* **2019**, *55*, 641. [CrossRef]
40. Cardinale, F.; Ciprandi, G.; Barberi, S.; Bernardini, R.; Caffarelli, C.; Calvani, M.; Cavagni, G.; Galli, E.; Minasi, D.; Del Giudice, M.M.; et al. Consensus Statement of the Italian Society of Pediatric Allergy and Immunology for the Pragmatic Management of Children and Adolescents with Allergic or Immunological Diseases during the COVID-19 Pandemic. *Ital. J. Pediatr.* **2020**, *46*, 84. [CrossRef]
41. D'Auria, E.; Anania, C.; Cuomo, B.; Decimo, F.; Cosimo Indirli, G.; Mastrorilli, V.; Santoro, A.; Sartorio, M.U.A.; Veronelli, E.; Caffarelli, C.; et al. COVID-19 and Food Allergy in Children. *Acta Biomed.* **2020**, *91*, 204–206. [CrossRef] [PubMed]
42. Caffarelli, C.; Coscia, A.; Ridolo, E.; Povesi Dascola, C.; Gelmett, C.; Raggi, V.; Volta, E.; Vanell, M.; Dall'Aglio, P.P. Parents' Estimate of Food Allergy Prevalence and Management in Italian School-Aged Children. *Pediatr. Int.* **2011**, *53*, 505–510. [CrossRef] [PubMed]

MDPI

Review

Primary Prevention of Food Allergy—Environmental Protection beyond Diet

Hanna Sikorska-Szaflik * and Barbara Sozańska

Department and Clinic of Paediatrics, Allergology and Cardiology, Wroclaw Medical University, ul. Chałubińskiego 2a, 50-368 Wrocław, Poland; bsoz@go2.pl
* Correspondence: hanna.sikorska-szaflik@umed.wroc.pl; Tel.: +48-71-770-3093

Abstract: A food allergy is a potentially life-threatening disease with a genetic and environmental background. As its prevalence has increased significantly in recent years, the need for its effective prevention has been emphasized. The role of diet modifications and nutrients in food allergy reduction has been extensively studied. Much less is known about the role of other environmental factors, which can influence the incidence of this disease. Changes in neonates gut microbiome by delivery mode, animal contact, inhalant allergens, oral and then cutaneous allergen exposure, air pollution, smoking, infections and vaccinations can be the potential modifiers of food allergy development. There is some data about their role as the risk or preventive factors, but yet the results are not entirely consistent. In this paper we present the current knowledge about their possible role in primary prevention of food allergies. We discuss the mechanisms of action, difficulties in designing accurate studies about food allergy and the potential biases in interpreting the connection between environmental factors and food allergy prevention. A better understanding of the role of environmental factors in food allergies development may help in implementing practical solutions for food allergy primary prevention in the future.

Keywords: allergy; food allergy; environmental factors; primary prevention

Citation: Sikorska-Szaflik, H.; Sozańska, B. Primary Prevention of Food Allergy—Environmental Protection beyond Diet. *Nutrients* **2021**, *13*, 2025. https://doi.org/10.3390/nu13062025

Academic Editors: Sara Manti, Gian Luigi Marseglia and Salvatore Leonardi

Received: 24 April 2021
Accepted: 10 June 2021
Published: 12 June 2021

1. Introduction

Food allergy, a potentially life-threatening condition, is defined as an adverse reaction occurring reproducibly on exposure to a specific food. It can be IgE mediated (an immediate onset of symptoms caused by mediator release induced by IgE binding antibodies, mast cells and basophils), non-IgE mediated (delayed onset of symptoms connected with T-cell inflammatory responses) and also mixed IgE and non-IgE mediated food allergy [1].

Food allergies are diagnosed in up to 10% of the population (mostly children) and the prevalence has been increasing in the last decades [2]. The disease may affect children's growth, patients' quality of life and can be potentially life-threatening, even after exposure to a very small amount of allergen. The gold standard for food allergy diagnosis is the double-blind, placebo-controlled food challenge (DBPCFC) [3]. Despite significant progress in understanding food allergy epidemiology and involved mechanisms, treatment remains mainly based on exclusion of the harmful food from the diet. There are also many questions to be solved about food allergy pathogenesis. Genetic predisposition, epigenetic modifications and environmental exposures may be the risk factors. However, the possibility of some environmental interventions seems to be an opportunity for its primary prevention.

The influence of diet and nutrients on food allergy risk has been extensively studied. Early introduction of peanuts in the diet of high-risk infants reduced the peanut allergy rate in the following years [4]. Unfortunately, similar intervention with other food types gave less certain results [5,6]. The role of vitamin D, antioxidants, vitamins, pre- and probiotics and other diet ingredients were also extensively studied [7]. In the 'dual-allergen exposure' hypothesis the role of cutaneous contact with allergens rather than oral early in life was proposed to increase the harmful effect [8]. Much less is known other than dietary environmental factors that affect the risk of food allergy and may potentially be the strategic tool for its avoidance [9]. Modification of microbiome by delivery mode, inhalant allergens exposure, air pollution, smoking, infections and vaccinations may influence the immunological system and its reactivity to different allergens. It is known that these factors can act on different ages (pregnancy, infancy, childhood, adulthood) and modify the risk of allergic diseases. What is their role in food allergies?

In this review we discuss the possibilities of primary prevention of food allergies given by environmental factors other than dietary and nutrients (Figure 1). We present the newest research on potential mechanisms of action and perspectives of implementing their findings in the practical reduction of food allergy risk.

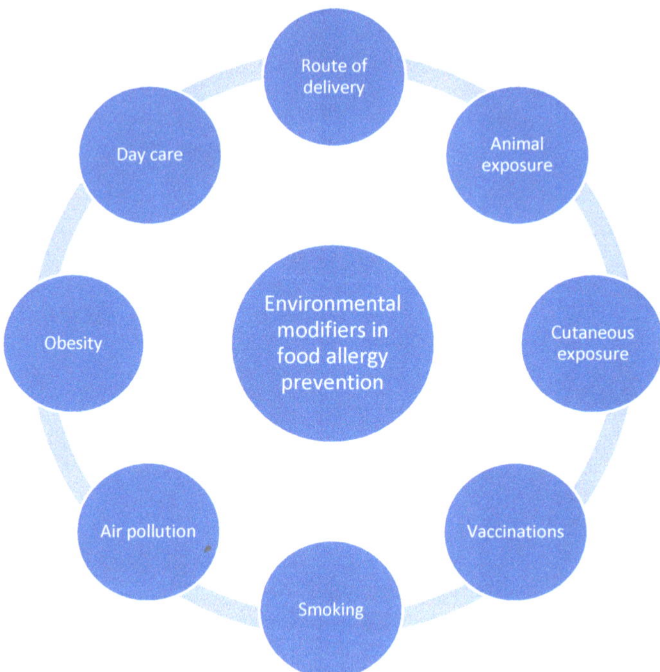

Figure 1. Primary prevention of food allergy—potential environmental modifiers.

2. Route of Delivery

The role of the gut microbiota in many processes of the human organism has been thoroughly investigated. Microbiota is believed to influence the development and maturation of the intestinal lymphoid tissue, strengthen and maintain the integrity of the mucosa and activate the intestinal immune defense [10]. Studies published recently suggest an important role of the intestinal flora in the development of food allergy. The disruption of the original microbiota configuration would have been related to higher risk of this disease [11,12]. Gut microbiota is supposed to have a regulatory role in the manifestation of food allergy but the exact mechanism of this influence is still under investigation. Presumably, it influences the immune system by altering the host's metabolism and altering adaptive immunity [13]. Caesarean section is among the factors potentially responsible for dysbiosis, defined as a change in the microbiota composition, function and imbalance of the gut microbiota homeostasis. Therefore, caesarean section is being considered as having the potential to increase the risk of food allergy [9,14]. Infants who are born through the birth canal are directly exposed to their maternal gastrointestinal flora by mouth, whereas caesarean born infants do not have such a high direct exposure, allowing lower levels of colonization from maternal skin microbiota and areas of low biomass to colonize the neonatal gastrointestinal tract. Compared to infants born naturally, babies born by caesarean section have lower levels of Bacteroides, E-shigella and E. coli [15] and higher levels of Clostridium difficile [16] in the gut microbiota. However, Dominguez-Bello et al. showed that the microbiota of infants' skin and mouth acquired a more adult-like configuration after the first week of life and this was independent from the delivery mode. As the difference in microbiota composition is relatively short-lived, considering its influence on food allergy development should be explored cautiously [17]. Omission of the maternal flora is suspected of predisposing to a proallergic Th2 phenotype and to food allergy simultaneously [17]. These observations are in line with the biodiversity hypothesis assumption that microbial exposure in early life regulates the immune response and has a role in the prevention of allergic diseases [18].

A Swedish longitudinal cohort study conducted on more than one million children reported an increased risk of food allergy in children born by means of elective or emergency caesarean delivery compared to those delivered vaginally. Moreover, infants born large for gestational age or with low Apgar score also had higher risk of developing food allergy. In contrast, children born <32 week of pregnancy were less likely to be diagnosed with food allergy in adolescence [19]. In another study, caesarean delivery was connected to an increased risk of cow's milk allergy in children, regardless of whether it was an elective or emergency caesarean section [20]. In the population-based birth cohort, the Pollution and Asthma Risk: an Infant Study (PARIS) sensitization to food allergens also tended to be more frequent in children born by caesarean section [21]. Increased risk of food allergy, but only in predisposed children with a family history of allergies, was shown by Eggesbø et al. Children whose mothers were allergic and who were delivered by means of caesarean section had a seven-fold increased risk of parentally perceived reactions to egg, fish or nuts and a four-fold increased risk of confirmed egg allergy. In contrast, children with no family history of allergy born by caesarean section did not show a significantly increased risk of food allergy [22]. Similarly, in a Greek cohort study it was shown that caesarean section predisposed food allergy development and that risk was more pronounced in children with allergic family history [23].

In the study of Norwegian children, 6.8% of the study group developed symptoms of food allergy during the first two years of life. In this group of food allergic children, 7.6% were delivered by caesarean section and 6.5% were born vaginally. As the difference was not significant, the authors stated that there was no increased risk of food allergy in the first two years of life in children born by means of caesarean section compared to those delivered vaginally [24]. In the Urban Environment and Childhood Asthma birth cohort study, the children were followed through the age of five years and the influence of many factors on food allergy to milk, egg and peanut were checked. Type of delivery appeared to have no effect on the occurrence of food sensitization or food allergy in children [25]. Pyrhönen et al. did not find an association between the mode of delivery and allergy manifestation, including food allergy during the first four years of life. However, the authors admit that due to the wide confidence intervals in their study, it is not possible to completely exclude the existence of a relationship between caesarean section and the development of allergic symptoms in children [26].

The discrepancies in the studies' results may be partially explained by the differences in the study design and enrolled populations (for example, different family histories), methods of diagnosis and time of observations. Despite the fact that DBPCFC is widely considered as the "gold standard" for the diagnosis of food allergy [2], the diagnostic procedures varied among studies. Assessing specific immunoglobulin E (sIgE) or conducting skin prick tests were used, but sometimes the lack of an objective way to measure the outcomes was applied. The cutoff points when food allergy was diagnosed, or variability in the food allergen chosen in the diagnostic process, can also be a shortcoming and can make it difficult to compare different study results. Some studies are also vulnerable to bias due to missing data from study participants or from conducting the trial using specific groups of a population. Women giving birth by cesarean section have easy access to health care providers, so in the case of allergy symptoms in their child they found a specialist consultation quickly. On the contrary, people with minimal contact with medical care may not seek advice from an allergy specialist or some symptoms may be simply ignored. Interestingly, in the EAT study where the DBPCFC was used to diagnose food allergy, despite the differences in the composition of infants' gut microbiota, no difference in the risk of food allergy was found in children [27].

Considering that most of the studies on the relationship between caesarean section and food allergy have not reach statistical significance, it is not possible to unequivocally state the influence of this type of delivery on the occurrence of food allergies in children. However, as the percentage of caesarean sections (including those based on maternal request) is increasing nowadays, further studies should be conducted to establish the role of caesarean section in food allergy development.

3. Animal Exposure

Another factor conceivably related to the onset of food allergy is animal exposure. Available studies suggest that contact with animals, both pets and farm animals, has a possible influence on the occurrence of atopic disease. This is based on the concept, similar in assumptions to the aforementioned biodiversity hypothesis, that early exposure to animal allergens reduces the organism's susceptibility to allergic diseases. Keeping a pet increases exposure to endotoxin and might therefore contribute to a lower risk of atopy [28]. Most studies focus on the effect of exposure to animals on the development of atopy, asthma and allergic rhinitis. Numerous studies confirmed the lower risk of atopy in children living on the farm [29–31]. However, the rural environmental protection may be connected not only with farm animal contact but also with some dietary habits, like drinking unpasteurized milk [32].

The role of exposure to pets and food allergies and other allergic diseases has been studied but the results are conflicting [33,34]. Regarding the interventional Enquiring About Tolerance (EAT) study participants, Marrs et al. showed that owning a dog is a potential protective factor against food allergy in children. Researchers enrolled three-months old infants and then examined the children at one and three years of age, establishing if the allergy to egg, peanut, milk, wheat, cod and sesame developed in connection to pet ownership. Having a dog appeared to reduce the odds for food allergy by 90%. Moreover, the protective effect increased with the number of dogs owned by the child. Keeping a cat had an effect on reducing the risk of food allergy only if a dog was also present in the household [35]. In the HealthNuts study population, the relationship between pet ownership and egg allergy was studied. Keeping a pet appeared to be a protective factor on egg allergy also in children without a family history of allergic disease [36].

On the contrary, in a study conducted in the United Kingdom, no association between pets at home and the development of food allergy in children up to two years of age was shown [37]. Levin et al. revealed that exposure to farm animals was related to a decrease of food sensitization and food allergy. The authors underlined the difference between urban and rural environments regarding risk and protective factors of food allergy [38].

A protective effect of animal exposure on developing food allergy has been found in some studies, whereas others do not show this association. Distinguishing children with positive allergy family history from those without is important when final conclusions are made about the pet effects on allergy. The potential role of animal exposure in prevention of food allergy is still inconclusive. The inaccuracy may be a result of the variety of study designs. Some analyzed infants, while others focused on older children. Furthermore, the diagnosis of food allergy in the studies was made on the basis of diverse methodology. The gold standard to diagnose food allergy is DBPCFC [2], but this was used in just a few studies, such as in the EAT study [35]. In many studies, the diagnosis was based on parental reports or specific IgE testing or skin prick testing, which may lead to overdiagnosis of the disease. Not using validated gold standard diagnosis is a weakness of these studies and makes the results unreliable and not comparable with the results of other well-designed surveys.

What is more, distinguishing children with positive allergy family history from those without is important when final conclusions are made about the pet effects on allergy. The potential role of animal exposure in prevention of food allergy is still inconclusive.

4. Cutaneous Exposure—Dual-Allergen Exposure Hypothesis

The dual-allergen exposure hypothesis proposed by Lack suggests that low-dose cutaneous exposure to food allergens is a risk factor of food allergy, while early intake of an allergen induces oral tolerance [8]. Du Toit stated that this theory is a precise illustration of one of the mechanisms of developing a food allergy [39]. When the skin barrier is impaired, such as in atopic dermatitis, this skin exposure might be even greater, which may also partly explain the frequent coexistence of atopic dermatitis and food allergy. In a systematic review, a strong association between atopic dermatitis, food sensitization and food allergy was found. In most cases atopic dermatitis arises before the development of food sensitization, which supports the theory of a causal relationship [40]. In the aforementioned population of the HealthNuts study, Martin et al. showed that atopic dermatitis in infancy, especially with early and severe onset, is a strong risk factor for IgE-mediated food allergy, and that eczema increases the odds of food allergy nearly five times compared to children with healthy skin [41]. Similar risk factors of food allergy—early onset of atopic dermatitis and its severity—were shown in children at three months of age [42]. Studies demonstrated the risk of sensitization to peanuts or peanut allergy after skin exposure to peanut allergen in household dust [43]. To determine the route by which infants become sensitized, the authors observed peanut protein levels in the child's environment, household peanut consumption and the development of peanut allergy in children. They made a conclusion that infant's environmental exposure to peanut antigens in the dust through an impaired skin barrier is a probable route for peanut sensitization and allergy [43]. Another study confirmed the increased risk of peanut food allergy after environmental exposure in infancy but also stated that consumption of peanuts in these infants in the first year of life was a protective factor against peanut allergy [44]. Peanut proteins were also detectable in the house dust of families that restricted peanuts and peanut products at home [45]. Taking into consideration environmental exposure of other food allergens, Trendelenburg et al. found that hen's egg allergen was also detectable in house dust samples. They concluded that high environmental exposure to hen's egg allergens may also be a risk factor for cutaneous sensitization, especially in small children with atopic dermatitis [46]. As a defect in the skin barrier is proposed to be a risk factor of development of peanut allergy, attention is also paid to the filaggrin-filament binding protein in the stratum corneum of the skin. Filaggrin mutations lead to increased skin permeability and reduced skin barrier function [47]. To investigate the association between filaggrin loss-of-function mutations and peanut allergy, a case-control study was conducted and showed a strong and significant association with peanut allergy in the food challenge-positive patients [48].

The dual-allergen exposure hypothesis suggests that somehow if the skin barrier can be improved then the risk of food allergy can be reduced. Proper treatment of atopic dermatitis may lower the risk of food allergy. What is more, actions taken to reduce the risk of eczema may lead to decreasing the food allergy occurrence, as was shown in case of reducing bathing frequency and protection against eczema in infants [49]. Therefore, for primary prevention of food allergy the appropriate skin care should also be recommended. However, as Perkin et al. found, moisturizing infants' skin can also promote food allergy. The authors explained that moisturizers can make it easier for food allergens to break the skin barrier or damage the skin barrier and enable the penetration of the food allergen [50]. Beyond proper skin care, a second important aspect of dual-allergen exposure theory is the importance of introducing foods orally at the right time to stimulate tolerance. As it is stated, a window of opportunity exists during the child's first year of life within which there is a possibility to influence a tolerogenic response [39].

5. Vaccinations

As a Bacillus Calmette–Guerin (BCG) vaccination has the potential to stimulate Th1 action involving cytokine response patterns and simultaneously inhibiting the Th2 immunologic response [51], the association between vaccination and the risk of atopic disorders has been studied. In the review by Arnoldussen et al., two original studies considering the association between BCG vaccination and a risk of food allergy were discussed [52]. The results of these studies were too heterogeneous to be pooled as they did not show statistically significant evidence that BCG vaccination has the potential to reduce the risk of food allergy or food sensitization [53,54]. However, in one of these studies Steenhuis et al. presumed that there could be a smaller beneficial effect of BCG vaccination, as in vaccinated children less eczema and significantly less use of medication for eczema was shown. The authors paid attention to the timing of when children got a vaccine. It is likely that BCG vaccination may have a protective immunomodulatory effect when it is administered in the early neonatal period, but not later [54]. The effect of BCG on the risk of atopic diseases, time of vaccine administration and also the impact of neonatal vitamin A supplementation were aims of the study conducted by Kiraly [55]. There is a possibility of BCG vaccination influencing food allergy occurrence in children under certain conditions, however, the results of the study are inconclusive [55].

There is little evidence that BCG vaccination may have a potential to reduce a risk of food allergy. More studies are needed to verify the association between atopic sensitization and administration of BCG. Therefore the latest guidelines of the European Academy of Allergy and Clinical Immunology (EAACI) are against recognizing the BCG vaccination as a method of food allergies prevention [56]. Pertussis vaccine was also considered as having a positive effect on reducing the risk of allergies, but Venter et al. found no such association in investigated children [57].

It is worth underlining that the studies conducted on large groups of patients showed no increased risk of atopic diseases including food allergy, confirming the safety of vaccinations [58,59]. Parents concerned about the possibility of developing atopic diseases in their children should be reassured about the safety of vaccines given early in life.

6. Smoking

Exposure to tobacco smoke is another factor possibly contributing to the development of food allergy [60]. The mechanism of action is still unclear, but there are some assumptions. It is possible that breathing air polluted by tobacco smoke in the first months of life may disrupt the skin barrier and lead to exposition of food allergens via this way. As it was described earlier in this article, skin exposition to food allergens may promote food allergy [6]. Moreover, cigarette smoke reduces the activity of Th1 lymphocytes and thus may contribute to the development of allergic diseases [61].

The role of exposure to tobacco smoke in utero and postnatally for IgE sensitization to allergens in children at four years of age was studied. The results were different depending on the time of exposure. Infants exposed to smoke only during pregnancy did not present a higher risk of sensitization to food (cow's milk, hen's egg, peanut, soy, wheat, cod fish) and indoor inhalant allergens. However, small children whose parents were smoking during the first months of the children's life demonstrated higher risk of sensitization and the effect was dose-dependent [62]. While examining the same group of patients, however, conducting a longer follow-up of children up to 16 years of age, Feldman stated that exposure to tobacco smoke in the second month of life increased the risk of food allergy. It remained significant for egg and peanuts (OR 1.79 and 1.50, respectively) [63]. Similar results were presented in a meta-analysis of studies—passive smoking during childhood but not in utero was associated with an increased risk for food allergy [64].

In contrast to the results mentioned above, a population-based study conducted in Sweden revealed a nonsignificant relation between exposure to tobacco smoke and allergies in children. Only children exposed to secondhand smoke and simultaneously having a family history of atopic diseases showed an increased risk of developing an allergy. Researchers emphasized the synergistic effect of inheritance and the influence of passive tobacco smoke exposure [65].

Smoking is a known risk factor for the development of respiratory tract diseases in children, and also allergic diseases. Of course, research on the influence of smoking on the occurrence of allergic diseases has its limitations—the effect of accompanying factors cannot be ruled out. Moreover, these studies are often based on data collected from an interview the parent does not always want, or they do not always remember factual information. Nevertheless, considering that almost every second child is exposed to tobacco smoke [66] and about 14% of children are exposed to smoking during pregnancy [67], educational information about the harmfulness of smoking and the effects of passive smoking on children should be an important part of medical appointments.

7. Air Pollution

Poor air quality indoors may result not only from exposure to tobacco smoke but also to the quality of the outside air. Traffic-related pollutants, including particles and nitrogen oxides (NOx), are said to induce inflammation of the airway and may increase airway responsiveness [68]. Air pollution is also a possible environmental factor playing a role in food allergy development. Diesel exhaust particles may act as adjuvants to allergen and in that way escalate the sensitization response. What is more, NOx are associated with suppressing the Th1 response and promoting the Th2 proallergic response [69]. The effect of gene–environment interactions in explaining the impact of air pollution on allergic diseases has been proposed. Melén et al. studied the impact of air pollution on a single nucleotide polymorphism. They revealed that variants in the glutathione S-transferase P1 and tumor necrosis factor (TNF) genes modify the result of early long-term exposure to air regarding the sensitization to allergens in children [70]. Canadian children who participated in a prospective longitudinal national birth cohort were observed for over one year and NO2 concentrations in the place of their location were measured. Then the occurrence of allergy to inhalant and food (milk, eggs, peanuts, soy) allergens was determined. The authors showed that exposure to air pollution during the first year of life was correlated with positive results of skin prick tests with the mentioned food allergens at the age of one year. Similar to tobacco smoke studies, the allergic effect was not seen when the exposure to traffic took place during pregnancy [71]. Other prospective birth cohort studies analyzed the impact of air pollution in children from birth to four years of age [72]. The findings confirmed the positive, however nonsignificant, association between sensitization to food allergens and pollution exposure.

As the exposure to air of poor quality can be harmful for children's health, it is important to prevent this exposure as much as possible. Limiting going outside on days when the air quality is very poor may even be needed. Actions leading to the improvement of air quality are also important—limiting car traffic where possible, using public transport and changing fuel to limit pollution.

8. Obesity

The incidence of both atopic diseases and obesity in the pediatric population is steadily increasing. Obesity can be described as a chronic systemic inflammation resulting from the interaction between adipocytes and macrophages recruited to adipose tissue in obesity. TNF-α, leptin and adiponectin are involved in this inflammation, and an increase in gene expression of proteins related to inflammation in obese people was shown—including genes of TNF-α, chemokines, IL8, MCP-1, complement proteins and other acute phase proteins [73,74]. This inflammatory state connected to obesity may be associated with an increased risk of atopic diseases, including food allergy. Another possibility is the influence of leptin—a hormone derived from adipose tissue. Its concentration is significantly increased in obese people [75]. It affects lymphocytes and the production of Th1-specific cytokines, and at the same time inhibits Treg proliferation, which may promote allergic diseases [76].

In a National Health and Nutrition Examination Survey, a relationship between obesity and allergic disease including food allergy was shown. Researchers examined children aged 2–19 years old and tested them for total IgE and specific IgE to i.e., peanut, milk, egg, and, in the case of six-year-old children or older, also to shrimp. It appeared that increased weight was associated with higher allergic predisposition and the effect was more pronounced in obese children compared to those who were overweight. However, the authors underline that this result cannot be understood unequivocally that obesity is a certain risk factor of food allergy. There is a possibility of the mutual influence of both these conditions, and perhaps there are additional issues influencing this relationship [77].

Irei et al. checked the association between overweight and food allergy in a population of 2027 children aged 9–13 years old from Japan, Taiwan and Vietnam. They showed inconsistent results in children from different regions—there was an association between being obese and having food allergy but only in part of the studied population [78]. However, even the authors admit that their study assumed only a questionnaire declaration of the presence of allergies, which may lead to a distortion of the result. Difficulties when interpreting the research results may include the different age of the respondents, methods of assessing the occurrence of food allergy (information from the parent/from the patient itself/objective sIgE assessment) and differentiated classification of BMI values as overweight and obesity. Therefore there is no strong evidence to consider obesity as a risk factor of food allergy.

9. Daycare

Increased risk of food allergy among children attending daycare facilities is also plausible. In Sweden, over 10,000 children were studied and revealed an association between daycare attendance and allergy symptoms (including allergic reactions to food). Children cared at daycare centers were more likely to present food allergy reactions compared with children cared for at home. It was more pronounced in children who started daycare before the age of one than in children who started attending daycare after the age of two [79]. In contrast, a recent study of 5517 participants aged 1 to 18 years did not show an increased risk of atopic diseases (including food allergy) in children attending daycare during the first year of life [80]. Similarly, Koplin et al. revealed lower risk of challenge-proven egg, sesame and peanut allergy in children at daycare in the first six months of life compared to those cared for at home [81]. This result may be related to the microbial exposure influence [9].

It could be assumed that attending a nursery or kindergarten could have a protective effect against allergies in a similar way as some other environmental factors—increasing contact with infectious agents would reduce the risk of allergies. However, since attending daycare does not have a clear antiallergic protective effect, the relationship is not unambiguously confirmed. It is also possible that the relationship between daycare and food allergy is not a causative nor protective one. It may be related to the fact that children at daycare centers are more often offered food that they would not get or try at home. What is more, they consume the meals because other children do. Sometimes this may lead to the occurrence of allergy symptoms. Possibly children taken care of at home who refuse some potential allergens somehow protect themselves against food allergy symptoms. Certainly, many factors have an influence, as a study of atopic wheezing showed that attending daycare can have a different effect (protective or not) depending on the varied expression of TLR2 genes in a child [82].

10. Conclusions

The knowledge about the role of environmental factors other than food in the development of food allergy is still surprisingly scarce. A better understanding of how external exposures interfere with the immunological system in food tolerance or food sensitization development is crucial for implementation of prevention methods. In the newest guidelines from the EAACI on food allergy prevention in infants and young children, none of the discussed possibilities reached a sufficient level of evidence to be recommended [55]. This may reflect the difficulties in designing accurate studies about food allergy; the dangers of elicitation bias, selection bias and losing participants to follow up; the importance of double-blind food challenges and the importance of ensuring representatives access across different populations. Therefore, there is an urgent need to conduct further studies organized on the basis of available guidelines for food allergy diagnosis and with objective measurements of environmental exposures.

Funding: This research received no external funding.

Conflicts of Interest: The authors declare no conflict of interest.

References

1. Calvani, M.; Anania, C.; Caffarelli, C.; Martelli, A.; Miraglia Del Giudice, M.; Cravidi, C.; Duse, M.; Manti, S.; Tosca, M.A.; Cardinale, F.; et al. Food allergy: An updated review on pathogenesis, diagnosis, prevention and management. *Acta Biomed.* **2020**, *15*, 91.
2. Sicherer, S.H.; Sampson, H.A. Food allergy: A review and update on epidemiology, pathogenesis, diagnosis, prevention, and management. *J. Allergy Clin. Immunol.* **2018**, *141*, 41–58. [CrossRef]
3. Sampson, H.A.; Gerth van Wijk, R.; Bindslev-Jensen, C.; Sicherer, S.; Teuber, S.S.; Burks, A.W.; Dubois, A.E.; Beyer, K.; Eigenmann, P.A.; Spergel, J.M.; et al. Standardizing double-blind, placebo-controlled oral food challenges: American Academy of Allergy, Asthma & Immunology-European Academy of Allergy and Clinical Immunology PRACTALL consensus report. *J. Allergy Clin. Immunol.* **2012**, *130*, 1260–1274.
4. Du Toit, G.; Roberts, G.; Sayre, P.H.; Bahnson, H.T.; Radulovic, S.; Santos, A.F.; Brough, H.A.; Phippard, D.; Basting, M.; Feeney, M.; et al. Randomized trial of peanut consumption in infants at risk for peanut allergy. *N. Engl. J. Med.* **2015**, *372*, 803–813. [CrossRef] [PubMed]
5. Perkin, M.; Logan, K.; Tseng, A.; Raji, B.; Ayis, S.; Peacock, J.; Brough, H.; Marrs, T.; Radulovic, S.; Craven, J.; et al. Randomized trial of introduction of allergenic foods in breast-fed infants. *N. Engl. J. Med.* **2016**, *374*, 1733–1743. [CrossRef] [PubMed]
6. Ierodiakonou, D.; Garcia-Larsen, V.; Logan, A.; Groome, A.; Cunha, S.; Chivinge, J.; Robinson, Z.; Geoghegan, N.; Jarrold, K.; Reeves, T.; et al. Timing of allergenic food introduction to the infant diet and risk of allergic or autoimmune disease: A systematic review and meta-analysis. *JAMA* **2016**, *316*, 1181–1892. [CrossRef] [PubMed]
7. Venter, C.; Agostoni, C.; Arshad, S.H.; Ben-Abdallah, M.; Du Toit, G.; Fleischer, D.M.; Greenhawt, M.; Glueck, D.H.; Groetch, M.; Lunjani, N.; et al. Dietary factors during pregnancy and atopic outcomes in childhood: A systematic review from the European Academy of Allergy and Clinical Immunology. *Pediatr. Allergy Immunol.* **2020**, *31*, 889–912. [CrossRef]
8. Lack, G. Epidemiologic risks for food allergy. *J. Allergy Clin. Immunol.* **2008**, *121*, 1331–1336. [CrossRef]
9. Marrs, T.; Bruce, K.D.; Logan, K.; Rivett, D.W.; Perkin, M.R.; Lack, G.; Flohr, C. Is there an association between microbial exposure and food allergy? A systematic review. *Pediatr. Allergy Immunol.* **2013**, *24*, 311–320. [CrossRef] [PubMed]
10. Young, V.B. The role of the microbiome in human health and disease: An introduction for clinicians. *BMJ* **2017**, *356*, j831. [CrossRef]
11. Shu, S.A.; Yuen, A.W.T.; Woo, E.; Chu, K.H.; Kwan, H.S.; Yang, G.X.; Yang, Y.; Leung, P.S.C. Microbiota and Food Allergy. *Clin. Rev. Allergy Immunol.* **2019**, *57*, 83–97. [CrossRef] [PubMed]
12. Chinthrajah, R.S.; Hernandez, J.D.; Boyd, S.D.; Galli, S.J.; Nadeau, K.C. Molecular and cellular mechanisms of food allergy and food tolerance. *J. Allergy Clin. Immunol.* **2016**, *137*, 984–997. [CrossRef] [PubMed]
13. Lee, K.H.; Song, Y.; Wu, W.; Yu, K.; Zhang, G. The gut microbiota, environmental factors, and links to the development of food allergy. *Clin. Mol. Allergy.* **2020**, *18*, 5. [CrossRef] [PubMed]
14. Molloy, J.; Allen, K.; Collier, F.; Tang, M.L.K.; Ward, A.C.; Vuillermin, P. The potential link between gut microbiota and IgE-mediated food allergy in early life. *Int J. Environ. Res. Public Health* **2013**, *10*, 7235–7256. [CrossRef] [PubMed]
15. Azad, M.B.; Konya, T.; Maughan, H.; Guttman, D.S.; Field, C.J.; Chari, R.S.; Sears, M.R.; Becker, A.B.; Scott, J.A.; Kozyrskyj, A.L.; et al. Gut microbiota of healthy canadian infants: Profiles by mode of delivery and infant diet at 4 months. *CMAJ* **2013**, *185*, 385–394. [CrossRef]
16. Adlerberth, I.; Wold, A.E. Establishment of the gut microbiota in western infants. *Acta Paediat.* **2009**, *98*, 229–238. [CrossRef]
17. Dominguez-Bello, M.G.; De Jesus-Laboy, K.M.; Shen, N.; Cox, L.M.; Amir, A.; Gonzalez, A.; Bokulich, N.A.; Song, S.J.; Hoashi, M.; Rivera-Vinas, J.I.; et al. Partial restoration of the microbiota of cesarean-born infants via vaginal microbial transfer. *Nat. Med.* **2016**, *22*, 250–253. [CrossRef] [PubMed]

18. Haahtela, T. A biodiversity hypothesis. *Allergy* **2019**, *74*, 1445–1456. [CrossRef]
19. Mitselou, N.; Hallberg, J.; Stephansson, O.; Almqvist, C.; Melén, E.; Ludvigsson, J.F. Cesarean delivery, preterm birth, and risk of food allergy: Nationwide Swedish cohort study of more than 1 million children. *J. Allergy Clin. Immunol.* **2018**, *142*, 1510–1514.e2. [CrossRef]
20. Metsala, J.; Lundqvist, A.; Kaila, M.; Gissler, M.; Klaukka, T.; Virtanen, S.M. Maternal and Perinatal Characteristics and the Risk of Cow's Milk Allergy in Infants up to 2 Years of Age: A Case-Control Study Nested in the Finnish Population. *Am. J. Epidemiol.* **2010**, *171*, 1310–1316. [CrossRef]
21. Gabet, S.; Just, J.; Couderc, R.; Seta, N.; Momas, I. Allergic sensitisation in early childhood: Patterns and related factors in PARIS birth cohort. *Int. J. Hyg. Environ. Health* **2016**, *219*, 792–800. [CrossRef]
22. Eggesbø, M.; Botten, G.; Stigum, H.; Nafstad, P.; Magnus, P. Is delivery by cesarean section a risk factor for food allergy? *J. Allergy Clin. Immunol.* **2003**, *112*, 420–426. [CrossRef]
23. Papathoma, E.; Triga, M.; Fouzas, S.; Dimitriou, G. Cesarean section delivery and development of food allergy and atopic dermatitis in early childhood. *Pediatr. Allergy Immunol.* **2016**, *27*, 419–424. [CrossRef]
24. Kvenshagen, B.; Halvorsen, R.; Jacobsen, M. Is there an increased frequency of food allergy in children delivered by caesarean section compared to those delivered vaginally? *Acta Paediatr.* **2009**, *98*, 324–327. [CrossRef] [PubMed]
25. McGowan, E.C.; Bloomberg, G.R.; Gergen, P.J.; Visness, C.M.; Jaffee, K.F.; Sandel, M.; O'Connor, G.; Kattan, M.; Gern, J.; Wood, R.A. Influence of early-life exposures on food sensitization and food allergy in an inner-city birth cohort. *J. Allergy Clin. Immunol.* **2015**, *135*, 171–178. [CrossRef] [PubMed]
26. Pyrhönen, K.; Näyhä, S.; Hiltunen, L.; Läärä, E. Caesarean section and allergic manifestations: Insufficient evidence of association found in population-based study of children aged 1 to 4 years. *Acta Paediatr.* **2013**, *102*, 982–989. [CrossRef] [PubMed]
27. Marrs, T.; Jo, J.H.; Perkin, M.R.; Rivett, D.W.; Witney, A.A.; Bruce, K.D.; Logan, K.; Craven, J.; Radulovic, S.; Versteeg, S.A.; et al. Gut microbiota development during infancy: Impact of introducing allergenic foods. *J. Allergy Clin. Immunol.* **2021**, *147*, 613–621. [CrossRef]
28. Heinrich, J.; Gehring, U.; Douwes, J.; Koch, A.; Fahlbusch, B.; Bischof, W.; Wichmann, H.E.; INGA-Study Group. Pets and vermin are associated with high endotoxin levels in house dust. *Clin. Exp. Allergy* **2001**, *31*, 1839–1845. [CrossRef]
29. Brunekreef, B.; von Mutius, E.; Wong, G.K.; Odhiambo, J.A.; Clayton, T.O.; ISAAC Phase Three Study Group. Early life exposure to farm animals and symptoms of asthma, rhinoconjunctivitis and eczema: An ISAAC Phase Three Study. *Int. J. Epidemiol.* **2012**, *41*, 753–761. [CrossRef]
30. Portengen, L.; Sigsgaard, T.; Omland, Ø.; Hjort, C.; Heederik, D.; Doekes, G. Low prevalence of atopy in young Danish farmers and farming students born and raised on a farm. *Clin. Exp. Allergy* **2002**, *32*, 247–253. [CrossRef]
31. von Mutius, E. Asthma and allergies in rural areas of Europe. *Proc. Am. Thorac. Soc.* **2007**, *4*, 212–216. [CrossRef]
32. Sozańska, B.; Pearce, N.; Dudek, K.; Cullinan, P. Consumption of unpasteurized milk and its effects on atopy and asthma in children and adult inhabitants in rural Poland. *Allergy* **2013**, *68*, 644–650. [CrossRef]
33. Chen, C.M.; Rzehak, P.; Zutavern, A.; Fahlbusch, B.; Bischof, W.; Herbarth, O.; Borte, M.; Lehmann, I.; Behrendt, H.; Krämer, U.; et al. Longitudinal study on cat allergen exposure and the development of allergy in young children. *J. Allergy Clin. Immunol.* **2007**, *119*, 1148–1155. [CrossRef]
34. Mandhane, P.J.; Sears, M.R.; Poulton, R.; Greene, J.M.; Lou, W.Y.; Taylor, D.R.; Hancox, R.J. Cats and dogs and the risk of atopy in childhood and adulthood. *J. Allergy Clin. Immunol.* **2009**, *124*, 745–750. [CrossRef]
35. Marrs, T.; Logan, K.; Craven, J.; Radulovic, S.; McLean, W.H.A.I.; Lack, G.; Flohr, C.; Perkin, M.R.; EAT Study Team. Dog ownership at three months of age is associated with protection against food allergy. *Allergy* **2019**, *74*, 2212–2219. [CrossRef]
36. Koplin, J.J.; Dharmage, S.C.; Ponsonby, A.L.; Tang, M.L.; Lowe, A.J.; Gurrin, L.C.; Osborne, N.J.; Martin, P.E.; Robinson, M.N.; Wake, M.; et al. Environmental and demographic risk factors for egg allergy in a population-based study of infants. *Allergy* **2012**, *67*, 1415–1422. [CrossRef] [PubMed]
37. Grimshaw, K.E.; Maskell, J.; Oliver, E.M.; Morris, R.C.; Foote, K.D.; Mills, E.N.; Margetts, B.M.; Roberts, G. Diet and food allergy development during infancy: Birth cohort study findings using prospective food diary data. *J. Allergy Clin. Immunol.* **2014**, *133*, 511–519. [CrossRef] [PubMed]
38. Levin, M.E.; Botha, M.; Basera, W.; Facey-Thomas, H.E.; Gaunt, B.; Gray, C.L.; Kiragu, W.; Ramjith, J.; Watkins, A.; Genuneit, J. Environmental factors associated with allergy in urban and rural children from the South African Food Allergy (SAFFA) cohort. *J. Allergy Clin. Immunol.* **2020**, *145*, 415–426. [CrossRef]
39. Du Toit, G.; Sampson, H.A.; Plaut, M.; Burks, A.W.; Akdis, C.A.; Lack, G. Food allergy: Update on prevention and tolerance. *J. Allergy Clin. Immunol.* **2018**, *141*, 30–40. [CrossRef] [PubMed]
40. Tsakok, T.; Marrs, T.; Mohsin, M.; Baron, S.; du Toit, G.; Till, S.; Flohr, C. Does atopic dermatitis cause food allergy? A systematic review. *J. Allergy Clin. Immunol.* **2016**, *137*, 1071–1078. [CrossRef]
41. Martin, P.E.; Eckert, J.K.; Koplin, J.J.; Lowe, A.J.; Gurrin, L.C.; Dharmage, S.C.; Vuillermin, P.; Tang, M.L.; Ponsonby, A.L.; Matheson, M.; et al. Which infants with eczema are at risk of food allergy? Results from a population-based cohort. *Clin. Exp. Allergy* **2015**, *45*, 255–264. [CrossRef]
42. Flohr, C.; Perkin, M.; Logan, K.; Marrs, T.; Radulovic, S.; Campbell, L.E.; MacCallum, S.F.; McLean, W.H.I.; Lack, G. Atopic dermatitis and disease severity are the main risk factors for food sensitization in exclusively breastfed infants. *J. Investig. Dermatol.* **2014**, *134*, 345–350. [CrossRef] [PubMed]

43. Brough, H.A.; Santos, A.F.; Makinson, K.; Penagos, M.; Stephens, A.C.; Douiri, A.; Fox, A.T.; Du Toit, G.; Turcanu, V.; Lack, G. Peanut protein in household dust is related to household peanut consumption and is biologically active. *J. Allergy Clin. Immunol.* **2013**, *132*, 630–638. [CrossRef] [PubMed]
44. Fox, A.T.; Sasieni, P.; du Toit, G.; Syed, H.; Lack, G. Household peanut consumption as a risk factor for the development of peanut allergy. *J. Allergy Clin. Immunol.* **2009**, *123*, 417–423. [CrossRef] [PubMed]
45. Shroba, J.; Barnes, C.; Nanda, M.; Dinakar, C.; Ciaccio, C. Ara h2 levels in dust from homes of individuals with peanut allergy and individuals with peanut tolerance. *Allergy Asthma Proc.* **2017**, *38*, 192–196. [CrossRef] [PubMed]
46. Trendelenburg, V.; Tschirner, S.; Niggemann, B.; Beyer, K. Hen's egg allergen in house and bed dust is significantly increased after hen's egg consumption-A pilot study. *Allergy* **2018**, *73*, 261–264. [CrossRef]
47. Kelleher, M.M.; Tran, L.; Boyle, R.J. Prevention of food allergy–skin barrier interventions. *Allergol. Int.* **2020**, *69*, 3–10. [CrossRef]
48. Brown, S.; Asai, Y.; Cordell, H.; Campbell, L.; Zhao, Y.; Liao, H.; Northstone, K.; Henderson, J.; Alizadehfar, R.; Ben-Shoshan, M. Loss-of-function variants in the filaggrin gene are a significant risk factor for peanut allergy. *J. Allergy Clin. Immunol.* **2011**, *127*, 661–667. [CrossRef]
49. Marrs, T.; Perkin, M.R.; Logan, K.; Craven, J.; Radulovic, S.; McLean, W.H.I.; Versteeg, S.A.; van Ree, R.; Lack, G.; Flohr, C.; et al. Bathing frequency is associated with skin barrier dysfunction and atopic dermatitis at three months of age. *J. Allergy Clin. Immunol Pract.* **2020**, *8*, 2820–2822. [CrossRef]
50. Perkin, M.R.; Logan, K.; Marrs, T.; Radulovic, S.; Craven, J.; Boyle, R.J.; Chalmers, J.R.; Williams, H.C.; Versteeg, S.A.; van Ree, R.; et al. EAT Study Team. Association of frequent moisturizer use in early infancy with the development of food allergy. *J. Allergy Clin. Immunol.* **2021**, *147*, 967–976.e1. [CrossRef]
51. Rousseau, M.C.; Parent, M.E.; St-Pierre, Y. Potential health effects from non-specific stimulation of the immune function in early age: The example of BCG vaccination. *Pediatr. Allergy Immunol.* **2008**, *19*, 438–448. [CrossRef] [PubMed]
52. Arnoldussen, D.L.; Linehan, M.; Sheikh, A. BCG vaccination and allergy: A systematic review and meta-analysis. *J. Allergy Clin. Immunol.* **2011**, *127*, 246–253. [CrossRef] [PubMed]
53. Alm, J.S.; Lilja, G.; Pershagen, G.; Scheynius, A. Early BCG vaccination and development of atopy. *Lancet* **1997**, *350*, 400–403. [CrossRef]
54. Steenhuis, T.J.; van Aalderen, W.M.; Bloksma, N.; Nijkamp, F.P.; van der Laag, J.; van Loveren, H.; Rijkers, G.T.; Kuis, W.; Hoekstra, M.O. Bacille-Calmette-Guerin vaccination and the development of allergic disease in children: A randomized, prospective, single-blind study. *Clin. Exp. Allergy* **2008**, *38*, 79–85. [CrossRef] [PubMed]
55. Kiraly, N.; Benn, C.S.; Biering-Sřrensen, S.; Rodrigues, A.; Jensen, K.J.; Ravn, H.; Allen, K.J.; Aaby, P. Vitamin A supplementation and BCG vaccination at birth may affect atopy in childhood: Long-term follow-up of a randomized controlled trial. *Allergy* **2013**, *68*, 1168–1176. [CrossRef]
56. Halken, S.; Muraro, A.; de Silva, D.; Khaleva, E.; Angier, E.; Arasi, S.; Arshad, H.; Bahnson, H.T.; Beyer, K.; Boyle, R.; et al. EAACI guideline: Preventing the development of food allergy in infants and young children (2020 update). *Pediatr. Allergy Immunol.* **2021**. Online ahead of print. [CrossRef]
57. Venter, C.; Stowe, J.; Andrews, N.J.; Miller, E.; Turner, P.J. No association between atopic outcomes and type of pertussis vaccine given in children born on the Isle of Wight 2001–2002. *J. Allergy Clin. Immunol. Pract.* **2016**, *4*, 1248–1250. [CrossRef]
58. Grüber, C.; Warner, J.; Hill, D.; Bauchau, V.; EPAAC Study Group. Early atopic disease and early childhood immunization—Is there a link? *Allergy* **2008**, *63*, 1464–1472. [CrossRef]
59. Matheson, M.C.; Haydn Walters, E.; Burgess, J.A.; Jenkins, M.A.; Giles, G.G.; Hopper, J.L.; Abramson, M.J.; Dharmage, S.C. Childhood immunization and atopic disease into middle-age—A prospective cohort study. *Pediatr. Allergy Immunol.* **2010**, *21*, 301–306. [CrossRef]
60. di Mauro, G.; Bernardini, R.; Barberi, S.; Capuano, A.; Correra, A.; De' Angelis, G.L.; Iacono, I.D.; de Martino, M.; Ghiglioni, D.; Di Mauro, D.; et al. Prevention of food and airway allergy: Consensus of the Italian Society of Preventive and Social Paediatrics, the Italian Society of Paediatric Allergy and Immunology, and Italian Society of Pediatrics. *World Allergy Organ. J.* **2016**, *18*, 9–28. [CrossRef]
61. Yu, M.; Zheng, X.; Peake, J.; Joad, J.P.; Pinkerton, K.E. Perinatal environmental tobacco smoke exposure alters the immune response and airway innervation in infant primates. *J. Allergy Clin. Immunol.* **2008**, *122*, 640–647. [CrossRef] [PubMed]
62. Lannero, E.; Wickman, M.; van Hage, M.; Bergstrom, A.; Pershagen, G.; Nordvall, L. Exposure to environmental tobacco smoke and sensitisation in children. *Thorax* **2008**, *63*, 172–176. [CrossRef] [PubMed]
63. Feldman, L.Y.; Thacher, J.D.; van Hage, M.; Kull, I.; Melén, E.; Pershagen, G.; Wickman, M.; To, T.; Protudjer, J.L.; Bergström, A. Early-life secondhand smoke exposure and food hypersensitivity through adolescence. *Allergy* **2018**, *73*, 1558–1561. [CrossRef]
64. Saulyte, J.; Regueira, C.; Montes-Martínez, A.; Khudyakov, P.; Takkouche, B. Active or passive exposure to tobacco smoking and allergic rhinitis, allergic dermatitis, and food allergy in adults and children: A systematic review and meta-analysis [published correction appears in PLoS Med. 2016 Feb;13:e1001939]. *PLoS Med* **2014**, *11*, 1001611.
65. Hansen, K.; Mangrio, E.; Lindström, M.; Rosvall, M. Early exposure to secondhand tobacco smoke and the development of allergic diseases in 4 year old children in Malmö, Sweden. *BMC Pediatr.* **2010**, *10*, 61. [CrossRef]
66. WHO Website. Available online: https://www.who.int/health-topics/tobacco#tab=tab_1 (accessed on 21 April 2021).
67. Murin, S.; Rafii, R.; Bilello, K. Smoking and smoking cessation in pregnancy. *Clin. Chest. Med.* **2011**, *32*, 75–91. [CrossRef]

68. Gilmour, M.I.; Jaakkola, M.S.; London, S.J.; Nel, A.E.; Rogers, C.A. How exposure to environmental tobacco smoke, outdoor air pollutants, and increased pollen burdens influences the incidence of asthma. *Environ. Health Perspect.* **2006**, *114*, 627–633. [CrossRef] [PubMed]
69. Sydbom, A.; Blomberg, A.; Parnia, S.; Stenfors, N.; Sandström, T.; Dahlén, S.E. Health effects of diesel exhaust emissions. *Eur. Respir. J.* **2001**, *17*, 733–746. [CrossRef]
70. Melén, E.; Nyberg, F.; Lindgren, C.M.; Berglind, N.; Zucchelli, M.; Nordling, E.; Hallberg, J.; Svartengren, M.; Morgenstern, R.; Kere, J. Interactions between glutathione S-transferase P1, tumor necrosis factor, and traffic-related air pollution for development of childhood allergic disease. *Environ. Health Perspect.* **2008**, *116*, 1077–1084. [CrossRef] [PubMed]
71. Sbihi, H.; Allen, R.W.; Becker, A.; Brook, J.R.; Mandhane, P.; Scott, J.A.; Sears, M.R.; Subbarao, P.; Takaro, T.K.; Turvey, S.E. Perinatal Exposure to Traffic-Related Air Pollution and Atopy at 1 Year of Age in a Multi-Center Canadian Birth Cohort Study. *Environ. Health Perspect.* **2015**, *123*, 902–908. [CrossRef]
72. Brauer, M.; Hoek, G.; Smit, H.A.; de Jongste, J.C.; Gerritsen, J.; Postma, D.S.; Kerkhof, M.; Brunekreef, B. Air pollution and development of asthma, allergy and infections in a birth cohort. *Eur. Respir. J.* **2007**, *29*, 879–888. [CrossRef] [PubMed]
73. Shore, S. Obesity, airway hyperresponsiveness, and inflammation. *J. Appl. Physiol.* **2010**, *108*, 735–743. [CrossRef]
74. Guo, X.; Cheng, L.; Yang, S.; Che, H. Pro-inflammatory immunological effects of adipose tissue and risk of food allergy in obesity: Focus on immunological mechanisms. *Allergol. Immunopathol. (Madr.)* **2020**, *48*, 306–312. [CrossRef] [PubMed]
75. Johnston, R.A.; Theman, T.A.; Lu, F.L.; Terry, R.D.; Williams, E.S.; Shore, S.A. Diet-induced obesity causes innate airway hyperresponsiveness to methacholine and enhances ozone- induced pulmonary inflammation. *J. Appl. Physiol.* **2008**, *104*, 1727–1735. [CrossRef] [PubMed]
76. Fantuzzi, G. Adipose tissue, adipokines, inflammation. *J. Allergy Clin. Immunol.* **2005**, *115*, 911–919. [CrossRef] [PubMed]
77. Visness, C.M.; London, S.J.; Daniels, J.L.; Kaufman, J.S.; Yeatts, K.B.; Siega-Riz, A.M.; Liu, A.H.; Calatroni, A.; Zeldin, D.C. Association of obesity with IgE levels and allergy symptoms in children and adolescents: Results from the National Health and Nutrition Examination Survey 2005–2006. *J. Allergy Clin. Immunol.* **2009**, *123*, 1163–1169. [CrossRef]
78. Irei, A.V.; Sato, Y.; Lin, T.L.; Wang, M.F.; Chan, Y.C.; Hung, N.T.; Kunii, D.; Sakai, T.; Kaneda, M.; Yamamoto, S. Overweight is associated with allergy in school children of Taiwan and Vietnam but not Japan. *J. Med. Investig.* **2005**, *52*, 33–40. [CrossRef]
79. Hagerhed-Engman, L.; Bornehag, C.G.; Sundell, J.; Aberg, N. Day-care attendance and increased risk for respiratory and allergic symptoms in preschool age. *Allergy* **2006**, *61*, 447–453. [CrossRef]
80. Kansen, H.M.; Lebbink, M.A.; Mul, J.; van Erp, F.C.; van Engelen, M.; de Vries, E.; Prevaes, S.M.P.J.; Le, T.M.; van der Ent, C.K.; Verhagen, L.M. Risk factors for atopic diseases and recurrent respiratory tract infections in children. *Pediatr. Pulmonol.* **2020**, *55*, 3168–3179. [CrossRef] [PubMed]
81. Koplin, J.J.; Martin, P.E.; Tang, M.L.K.; Gurrin, L.C.; Lowe, A.J.; Osborne, N.J.; Robinson, M.N.; Ponsonby, A.; Dharmage, S.C.; Allen, K.J.; et al. Do factors known to alter infant microbial exposures alter the risk of food allergy and eczema in a population-based infant study? *J. Allergy Clin. Immunol.* **2012**, *129*, AB231. [CrossRef]
82. Custovic, A.; Rothers, J.; Stern, D.; Simpson, A.; Woodcock, A.; Wright, A.L.; Nicolaou, N.C.; Hankinson, J.; Halonen, M.; Martinez, F.D. Effect of day care attendance on sensitization and atopic wheezing differs by Toll-like receptor 2 genotype in 2 population-based birth cohort studies. *J. Allergy Clin. Immunol.* **2011**, *127*, 390–397. [CrossRef] [PubMed]

Review

Update on the Role of Allergy in Pediatric Functional Abdominal Pain Disorders: A Clinical Perspective

Craig Friesen *, Jennifer Colombo and Jennifer Schurman

Division of Gastroenterology, Hepatology and Nutrition, Children's Mercy Kansas City, School of Medicine, University of Missouri-Kansas City, Kansas City, MO 64108, USA; jmcolombo@cmh.edu (J.C.); jschurman@cmh.edu (J.S.)
* Correspondence: cfriesen@cmh.edu

Abstract: Both functional abdominal pain disorders (FAPDs) and food allergies are relatively common in children and adolescents, and most studies report an association between FAPDs and allergic conditions. FAPDs share pathophysiologic processes with allergies, including both immune and psychological processes interacting with the microbiome. No conclusive data are implicating IgE-mediated reactions to foods in FAPDs; however, there may be patients who have IgE reactions localized to the gastrointestinal mucosa without systemic symptoms that are not identified by common tests. In FAPDs, the data appears stronger for aeroallergens than for foods. It also remains possible that food antigens initiate an IgG reaction that promotes mast cell activation. If a food allergen is identified, the management involves eliminating the specific food from the diet. In the absence of systemic allergic symptoms or oral allergy syndrome, it appears unlikely that allergic triggers for FAPDs can be reliably identified by standard testing. Medications used to blunt allergic reactions or symptomatically treat allergic reactions may be useful in FAPDs. The purpose of the current manuscript is to review the current literature regarding the role of allergy in FAPDs from a clinical perspective, including how allergy may fit in the current model of FAPDs.

Keywords: functional abdominal pain disorders; functional dyspepsia; irritable bowel syndrome; food allergy

Citation: Friesen, C.; Colombo, J.; Schurman, J. Update on the Role of Allergy in Pediatric Functional Abdominal Pain Disorders: A Clinical Perspective. *Nutrients* **2021**, *13*, 2056. https://doi.org/10.3390/nu13062056

Academic Editor: Alexandra Papadopoulou

Received: 13 May 2021
Accepted: 12 June 2021
Published: 16 June 2021

1. Introduction

Chronic or recurrent abdominal pain is common in children and adolescents, with a worldwide prevalence estimated at 13.5% [1]. Most youth with chronic abdominal pain will fulfill the functional abdominal pain disorder (FAPD) criteria as defined by the Rome criteria [2]. The recognized FAPDs include irritable bowel syndrome (IBS), functional dyspepsia (FD), abdominal migraine, and functional abdominal pain, with IBS and FD being the two most common [2–4].

There has been increasing interest in the role of diet in FAPDs. Perceived food intolerances are common in pediatric FAPD patients, with over 90% identifying at least one food they associate with worsening symptoms [5]. These patients frequently avoid specific foods and self-implement dietary strategies [5]. There are a variety of mechanisms by which specific foods can increase symptoms, including food allergy (immunologic reactions), food intolerances (non-immunologic reactions, e.g., malabsorption), and reactions created by hypervigilance and anticipation of symptoms in patients with perceived intolerances that may increase anxiety with consumption of the suspected food [6,7]. Perceived intolerances may also be influenced by underlying psychological factors [8].

The separation between food allergy and intolerance has become increasingly blurred as some food intolerances can start a chain of events resulting in mucosal immune activation. For example, lactose malabsorption is a well-recognized food intolerance. In most studies, lactose restriction does not result in clinical improvement even in patients with demonstrated malabsorption [9–12]. It is now recognized that this malabsorption is associated with increased mucosal mast cells and increased colonic eosinophils and lymphocytes, which may persist after lactose elimination [13–16]. Non-absorbed sugars and fructooligosaccharides alter the intestinal microbiome and production of short-chain fatty acids, both of which affect the development of food allergies [6,13,14,17–20]. The altered microbiome can interrupt the intestinal epithelial barrier (another factor highly implicated in FAPDs) with a subsequent increase in the immune system's exposure to luminal food and microbial antigens [21]. Lastly, it is recognized that dietary compounds (or metabolic byproducts) can modulate mast cell function [22]. For example, the benefits of fiber supplements are in part due to slow fermentation, producing short-chain fatty acids which preserve the intestinal barrier and decrease inflammation, including inhibition of MC activation [23–25].

Food allergy refers to developing symptoms resulting from an immune reaction (generally involving mast cells and eosinophils) to an ingested antigen. Food allergies are divided into two categories: IgE- mediated and non-IgE-mediated. IgE-mediated reactions are associated with more rapid onset of symptoms, while non-IgE-mediated reactions typically result in delayed onset of symptoms [26,27]. Food allergy in children has an estimated worldwide prevalence of 6–8%, with estimates of 10% in high-income countries [27]. Approximately 50% of food allergy reactions will produce systemic symptoms (e.g., wheezing, hives, anaphylaxis), and 50% will produce only or primarily gastrointestinal symptoms [26]. Multiple physiologic factors prevent immune reactions to foods, termed tolerance, including microbiome features and the intestinal barrier [28]. Importantly, "outgrowing" a food allergy is associated with the development of food-specific IgG rather than IgE [29].

Both FAPDs and food allergies are relatively common in children and adolescents and may be linked in at least a subset of patients. The purpose of the current manuscript is to review the current literature regarding the role of allergy in FAPDs from a clinical perspective, including how allergy may fit in the current model of FAPDs. Although the focus is on pediatric FAPDs, we will also incorporate the more abundant adult literature relevant to adolescents.

2. Inflammation and the Biopsychosocial Model

The complex nature of chronic abdominal pain has long been viewed within the context of the biopsychosocial model, which recognizes that various interacting factors contribute to the initiation and maintenance of pain. These contributors include biologic factors (e.g., genetics, visceral hypersensitivity, inflammation, dysbiosis), psychologic factors (e.g., anxiety, depression), and social factors (e.g., poor relationships with parents, teachers, or peers). There appear to be four main host systems involved in symptom generation which interact readily with each other, including psychologic, neurologic, immunologic, and endocrinologic systems, all of which interact with the gastrointestinal microbiome. A central mechanism appears to be visceral hypersensitivity, an exaggerated response to stimuli such as gastrointestinal distension. Hypersensitivity to distension has been demonstrated in youth with chronic abdominal pain [30–32].

Given that allergy involves an immunologic reaction, the role of inflammation, particularly mast cell-related, within the biopsychosocial model appears to be most relevant to the current discussion (See Figure 1). Mast cells are generally positioned at interfaces between the host and environment, providing a connecting link between the neurologic and immunologic systems and, in part, a link between the enteric and central nervous systems [33]. As will be discussed, mast cells are also an important link to the psychologic system.

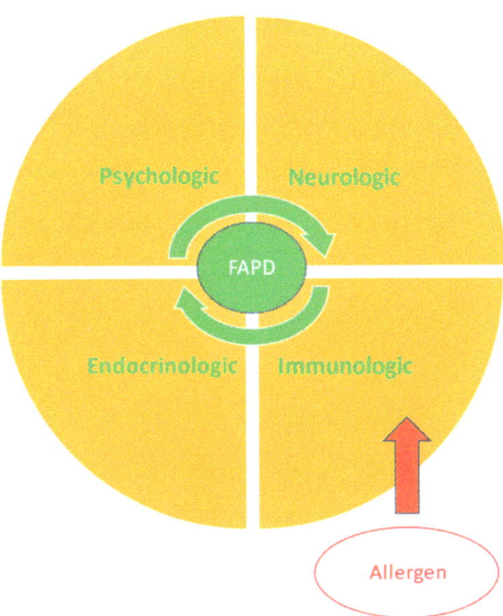

Figure 1. Four primary interacting systems generate the symptoms of functional abdominal pain disorders (FAPDs). This process can be initiated, maintained, or exacerbated by allergens through activation of the immune system.

There is considerable evidence implicating inflammation in FAPDs, particularly mast cells (and to a lesser degree, eosinophils) in IBS and both mast cells and eosinophils in FD [34–38]. With activation, both mast cells and eosinophils release mediators with biologic effects relevant to FAPDs. These mediators can stimulate afferent nerves sending a pain signal, sensitize afferent nerves inducing visceral hyperalgesia, and alter electromechanical function [36]. Mast cells are highly implicated in IBS, with increased density reported in the colon and ileum in over 80% of published studies investigating this relationship [34,39]. In addition, IBS is associated with an increased density of degranulating mast cells and mast cells in proximity to nerves which correlate with abdominal pain frequency and severity [40]. Both mast cell and eosinophil densities have been shown to be increased in pediatric IBS [41]. Increased densities of both mast cells and eosinophils have been demonstrated in FD, as has increased activation of mast cells and/or eosinophils in both adults and youth with FD [35,42–44]. Mast cell degranulation in the proximal stomach may be associated with visceral hyperalgesia in adults with FD [45]. To what degree mucosal inflammation may result from allergic reactions is unclear, but a history of allergy has been associated with increased duodenal eosinophils in adults with FD [46]. In children with cow's milk allergy, mucosal application of milk results in increased mast cell and eosinophil density and activation, as well as an increase in mast cells in proximity to nerves, findings similar to those reported in FAPDs in general [47].

Psychologic function is an important component of the biopsychosocial model and may directly relevant to a discussion of allergy. FAPDs are associated with psychologic dysfunction, including anxiety, depression, and maladaptive coping [48]. Psychologic disturbances are associated with greater abdominal pain severity and predict worse outcomes and persistence into adulthood [48–50]. In addition, asthma, allergic rhinitis, atopic dermatitis, and food allergy are associated with increased stress, changes in mood, and emotional dysfunction [51,52]. Psychologic functioning interacts with biologic functioning (in a bi-directional fashion) particularly with the immunologic system; anxiety and depression have been associated with increased mast cell and/or eosinophil density in youth and adults with FAPDs [41,53–55]. Psychologic functioning also interacts with the endocrinologic system as anxiety can trigger a stress response initiated by corticotropin-releasing hormone (CRH) release. Through activation of CRH receptors, stress results in mast cell degranulation, disrupting the epithelial barrier, increasing antigenic exposure [56]. As stress can exacerbate symptoms, its presence can also create difficulty determining a patient's response to food restrictions. Santos and colleagues studied a group of adults with documented food allergies [57]. Under conditions of cold stress, these patients exhibited increased luminal release of tryptase and histamine in the jejunum at a magnitude comparable to that induced by food allergen exposure [57]. Thus, a patient with food allergies may have symptoms triggered by other factors even after eliminating the food allergen. While allergies may have a role in FAPDs in at least a subset of patients, it is important to recognize the complex nature of chronic abdominal pain and the other factors that may be active in symptom generation.

3. Allergy and Functional Abdominal Pain Disorders

Most, but not all, studies have shown an association between FAPDs and allergic conditions [58]. FD and functional abdominal pain have been associated with asthma in adolescents [58]. Likewise, asthma and food allergy, allergic rhinitis, and eczema have been associated with FAPDs in adults [59–64]. In an extensive primary care study, both FD and IBS were associated with allergic conditions, and the relationship was partially explained by a common association with anxiety and depression [63]. Allergic conditions early in life also appear to predispose to later development of FAPDs. Pre-schoolers with allergic disease have an increased risk of IBS when they reach school-age, with earlier development of IBS in those with food allergies [65]. The highest risk was associated with allergic rhinitis [65]. In another study, the risk of childhood IBS was significantly increased in those with a history of atopic dermatitis [66]. Allergic proctocolitis early in life is also a risk factor for subsequent FAPD development [67]. The association of allergies early in life and subsequent FAPD development is not well understood. Still, there is some evidence that allergy may alter the microbiome, which could predispose to FAPDs, or allergy, and FAPD could be epiphenomena related to dysbiosis. Children with food sensitization have lower microbiota diversity overall with lower Bacteroides and higher Firmicutes colonization [18]. The microbial signature can distinguish between those with IgE-mediated and those with non-IgE-mediated food allergies in infants [19].

Interestingly, a placebo-controlled trial of a probiotic (Bifidobacteria) in children with allergic rhinitis in association with intermittent asthma demonstrated significant improvement in allergic rhinitis symptoms [68]. These findings suggest shared pathophysiology related to the gastrointestinal microbiome. Lastly, bacteria-derived (and host-derived) proteases have been implicated in disruption of the intestinal barrier, increasing antigen exposure, and may also directly stimulate mast cells and sensory neurons [69].

3.1. IgE-Mediated Allergies

No conclusive data is implicating IgE-mediated reactions to foods in FAPDs [70,71]. The gold standard for diagnosis of food allergy is a double-blind, placebo-controlled food challenge that can be time-consuming and are most likely to be helpful in FAPD patients who also experience systemic reactions with food ingestion [72]. The only proven clinically utilized diagnostic techniques for IgE-mediated reactions are skin prick tests (SPT), which have high sensitivity and low specificity, and measurement of serum food-specific IgE, which also have low specificity [72]. Both tests are indicative of sensitization but by themselves are not diagnostic of clinical allergy. The low specificity of food-specific IgE can be particularly problematic when ordering large panels. In a study of 220 adults with IBS and/or FD, food-specific IgE tests were positive in 38% [73]. On an elimination diet, a positive response was seen in 8 of 19 patients, all of whom relapsed on reintroduction, yielding an overall prevalence of 4% for IgE-medicated food allergy [73]. Not only is this frequency similar to that seen in the general population, but the study highlights the limited ability of a positive test to predict clinical symptoms. Another study of adults with FD and IBS found no differences in food-specific IgE compared to controls [74]. We previously found no increase in immunoreactivity (including IgE, SPT, IgG, IgG4, and atopy patches) to common food allergens in children with FD and duodenal eosinophilia [75].

There may be patients who have IgE reactions localized to the gastrointestinal mucosa without systemic symptoms who are not identified by SPT or serum food-specific IgE. Methods are available to evaluate localized reactions in the gastrointestinal tract, including the colonoscopic allergen provocation test (COLAP; mucosal testing akin to the SPT) and visualization utilizing confocal laser endomicroscopy (CLE). Utilizing COLAP, positive reactions to food antigens are recognized by the wheal and flare reactions occurring within 20 min of antigen application. Reactions are associated with mast cell degranulation eosinophil activation histologically [76]. In a study of 70 adults with gastrointestinal symptoms suspected to be related to food allergy, COLAP was positive in 97/210 (46%) of challenges in patients and not in any challenges in controls [76]. Reactions correlated with patient histories of food reactions but not SPT results or food-specific IgE [75]. In sum, these findings may indicate a higher rate of IgE-mediated food allergy in FAPD patients and cast some doubt on the sensitivity of standard allergy tests and their ability to rule out allergies to specific foods in the absence of systemic symptoms.

Recent studies in a mouse model demonstrate potentially important interactions with bacterial infection or colonization in predisposing to these localized intestinal allergic reactions [77]. In this model, the bacterial infection causes a loss of oral tolerance resulting in mucosal food-specific IgE and increased visceral pain via IgE- and mast cell-mediated mechanisms [77]. Studies in this model also demonstrate a possible role for superantigens, which are microbes known to cause non-specific activation of T lymphocytes and which have been implicated in non-gastrointestinal atopic conditions. The primary superantigens, *Staphylococcus aureus* and *Streptococcus pyogenes*, are more commonly present in the microbiome of IBS patients [77]. Skin colonization with *Staphylococcus aureus* in patients with atopic dermatitis has been associated with an increased risk of food allergy [78]. Likewise, a loss of balance and diversity in the intestinal microbiome increases food allergy risk [21].

In FAPDs, the data appears stronger for aeroallergens than for foods [79]. As discussed previously, FAPD is associated with allergic rhinitis, and allergic rhinitis early in life increases FAPD risk. In addition, adults with IBS have an increased risk of seasonal allergies (and consequently pollen-food syndrome), and seasonal allergic rhinitis is associated with greater IBS severity [61]. Aeroallergens enter the nasal and oral cavities with breathing and may be swallowed, or they may be ingested following food contamination [58]. In children with FAPDs, local pollen counts are associated with the onset of pain and are as strong a predictor as are affect or sleep disturbances [80]. Birch pollen, in particular, has been well studied. During birch pollen season, adults with birch pollen allergy demonstrate an increase in duodenal eosinophils and IgE-carrying mast cells along with oral allergy syndrome [81]. Oral allergy syndrome is a hypersensitivity to raw plant proteins, often proteins that cross-react with pollen proteins, resulting in oropharyngeal symptoms (e.g., itching, tingling, swelling) [26]. In a separate study, birch exposure was associated with increased intestinal eosinophil and mast cell densities [82]. Patients with gastrointestinal symptoms had increased IgE to birch (rBet v 1), hazelnuts, and apple [82]. In a study of patients with birch pollinosis, COLAP with rBet v 1, a positive reaction was seen in 81% where there was a history of gastrointestinal symptoms. There were 22% of those with pollinosis and no gastrointestinal symptoms and none in the healthy controls [83]. Aeroallergens might not only be a trigger for symptoms in FAPD patients but an indicator of potential food triggers.

3.2. Non-IgE-Mediated Allergies

Food antigens may precipitate gastrointestinal symptoms through cell-mediated processes (type IV hypersensitivity reactions [84]. The classic examples include food protein-induced allergic proctocolitis (FPIAP), a benign condition generally presenting in early infancy, and food protein-induced enterocolitis syndrome (FPIES), generally presenting with severe symptoms within the first 6 months of life [85]. While not completely characterized, FPIES results from food-induced immune activation, including activation of the innate immune system limited to the gastrointestinal tract [86,87]. FPIES is associated with disruption or a lack of development of tolerance [84]. FPIES most often presents with severe bouts of abdominal pain, nausea, vomiting, and diarrhea, symptoms also seen in FAPDs. While FPIES most often resolves by one year of age, it can persist into or develop during adolescence or adulthood [84,87–89]. Reactions in older children, adolescents, and adults are most frequently described in relation to seafood ingestion, but it is possible that milder cell-mediated reactions to other foods could contribute to FAPDs [87–89]. Increased density and activation of T lymphocytes and indirect evidence for TH17 activation have been demonstrated in adults with FAPDs [38]. While TH17 cells may have a pro-inflammatory role, they may also serve a protective role, depending on the inflammatory milieu, microbiome composition, and epigenetic modifications [90–94]. Under specific conditions, TH17 cells can induce eosinophil infiltration and activation and, potentially, mast cell accumulation [95–98]. Increased mucosal TH17 density has been demonstrated in pediatric IBS and pediatric FD associated with chronic gastritis [41,99]. In FD associated with chronic gastritis (but not in the absence of chronic gastritis), gastric and duodenal TH17 density was greater than controls and comparable to that seen in Helicobacter-pylori-associated gastritis and Crohn's-associated gastritis [99]. Lastly, in another study utilizing confocal laser endomicroscopy (CLE), 155 adults were challenged with four common food antigens, all of which were negative on SPT and without elevations of specific IgE. However, a localized mucosal IgE reaction was not assessed [100]. Of the 108 completers in the study, 70% had a positive test, and patients with positive tests were 4X more likely to have another atopic disorder than controls [100]. A positive response was associated with increased permeability, increasing mucosal lymphocyte density, and eosinophil (but not mast cell) activation [100]. Thus, there appears to be a role for lymphocytes in immunologic food reactions in the absence of IgE secretion.

There have been multiple studies, primarily in adults, evaluating food-specific IgG in patients with FAPDs. At present, IgG testing is discouraged by most major allergy organizations as it lacks proven clinical utility [101]. It is believed to be indicative of exposure, and as previously discussed, tolerance to a specific food is associated with the development of food-specific IgG [29,102]. However, the utility of food-specific IgG testing specifically in FAPDs awaits further evaluation, and, although not proven, there is data suggesting a potential for clinical utility. Multiple studies have reported increased food-specific IgG or IgG4 in adults with FAPD, particularly IBS [43,74,101,103,104]. Multiple studies have also reported clinical improvement on elimination diets guided by IgG or IgG4 testing [104–109]. However, only one of these studies restricted foods in a blinded fashion, and there were some other significant differences between the true and sham restricted diets in this study [107]. However, a greater benefit was seen with better compliance [107]. Another study found improvement in both compliant and non-compliant adults [106].

Additionally, while IgG titers are not related to symptom severity, they correlate with mast cell density and degranulation [74,108]. It remains possible that increase antigen exposure through an impaired epithelial barrier instigates a food-specific IgG reaction that promotes mast cell activation. However, further studies are needed before IgG testing should be considered.

Atopy patch testing is a diagnostic procedure for identifying delayed hypersensitivity reactions that have been primarily useful in identifying allergens in contact dermatitis [72]. In very limited, uncontrolled studies of adults with IBS, patch testing has been reported to identify foods that, when restricted, resulted in clinical improvement [110,111]. There is currently insufficient data to support patch testing in FAPDs.

4. Management

If a food allergen is identified, allergy management is relatively straightforward and involves eliminating the specific food from the diet. However, it appears quite unlikely that many patients with FAPDs will benefit from a standard approach that includes routine allergy testing to identify the cause of symptoms which, when eliminated from the diet, will result in the resolution of the FAPD. More likely, indiscriminate testing, particularly utilizing large panels identifying food-specific IgE (or IgG) elevations, will result in unnecessary diet restrictions without long-term benefit and could lead to nutritional deficiencies. In patients with only gastrointestinal symptoms, allergy testing is not likely to be helpful, either in identifying the culprit or in providing a list of safe foods. Patients who develop systemic allergic symptoms or those with oral allergy syndrome should be referred to an allergist for directed testing and management of identified allergens. This should not be a transfer of care but the formation of a collaboration to manage the patient. Another approach is an empiric restriction of the most common allergens used in the treatment of eosinophilic esophagitis. Unfortunately, there is very little data assessing this approach in patients with eosinophilic gastroenteritis, let alone FAPDs, and no randomized trials [112,113]. Institution of highly restrictive diets is not without risks, including decreased quality of life and potential nutritional deficiencies, and should only be instituted long-term in collaboration with a dietician [114].

While the patient would be expected to benefit from removing food allergens or managing aeroallergens if identified, it remains unlikely that gastrointestinal symptoms will completely resolve without addressing other biopsychosocial aspects of these disorders as allergens are only one trigger for activating gastrointestinal mast cells. See Figure 2 for an overview of therapeutic targets within FAPD pathophysiology. Anxiety/stress and depression may be important therapeutic targets as both FAPDs and allergies are associated with psychologic dysfunction, triggering symptoms. Treatment of stress has been shown to benefit allergic conditions and in children with FD in association with duodenal eosinophilia [52,115].

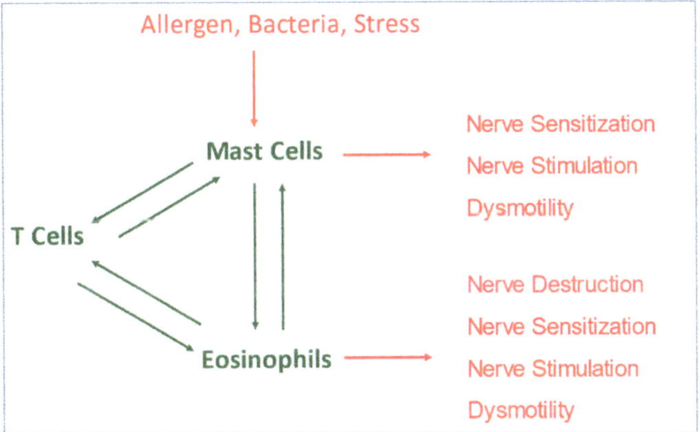

Figure 2. Within FAPD pathophysiology, particularly related to allergy, there are several therapeutic targets, including factors that stimulate or enhance an immunologic response, factors related to inflammatory cell infiltration or activation, or downstream effects following mediator release, either blocking receptors for released mediators or counteracting the physiologic effects of these mediators.

For patients with primarily gastrointestinal symptoms who do not have systemic symptoms typical of allergic reactions or oral allergy syndrome, it is unlikely that specific triggering allergens will be identified even if they exist as current tests likely lack adequate sensitivity or specificity. The challenge is that end-organ inflammatory processes do not necessarily identify patients who specifically have allergic triggers, as mast cells and eosinophils appear to be a common component of FAPDs. Another approach would be to direct treatment at the mast cells and/or eosinophils without regard to whether their activation is due to an allergen. As with non-gastrointestinal allergic diseases, treatment can be directed at mast cell activators (e.g., anti-IgE, anti-IL-5), mast cell mediator release (e.g., mast cell stabilizers, anti-siglec-8), or inhibition of mast cell mediators (e.g., histamine and cys-leukotrienes) at their effector sites [36]. Most have been studied in the context of FAPDs or mucosal eosinophilia with some data in patients with demonstrated food allergy.

There are no FDA-approved drugs for the treatment of gastrointestinal eosinophils [116]. Multiple biologics have been developed or are in development, which targets upstream mediators associated with mast cell and eosinophil infiltration and/or activation [See Pesek [116] and Wechsler [117] for a full review]. The most studied, in general, are omalizumab, a monoclonal antibody directed at IgE, and mepolizumab and reslizumab, monoclonal antibodies directed at interleukin 5. Omalizumab has demonstrated efficacy in providing a degree of protection from peanut allergy [118]. It has also been studied in 9 patients with FAPD symptoms and mucosal eosinophilia [119]. It was associated with symptom improvement, but decreases in antral and duodenal eosinophil densities did not reach statistical significance; the study was likely underpowered for assessment of changes in eosinophil density [119]. Anti-interleukin 5 antibodies have not been studied in eosinophilic gastroenteritis. Currently, stress appears to be the most viable upstream treatment target.

The largest body of data, albeit somewhat meager, exists for mast cell stabilizing medications, including oral cromolyn and ketotifen. Two placebo-controlled trials have assessed the response to ketotifen, a mast cell stabilizer and H1 antagonist, in adults with IBS, with both demonstrating improvement in abdominal pain and other IBS symptoms [120,121]. Visceral hypersensitivity improved in both while decreases in mast cell density and activation were seen in only one study [120,121]. Whether effects resulted from mast cell stabilization or an antihistamine effect, or both is not clear. Oral cromolyn has been studied in patients with IBS, frequently associated with positive allergy testing, and, in comparison to restricted diets [122–126]. There have been two pediatric studies [122,123]. In a study of children with abdominal pain and diarrhea, cromolyn was shown to be as effective as diet restriction guided by SPT [122]. A study of 10 children with egg allergy found no benefit from cromolyn in preventing egg reactions [123]. A study of 20 adults with IBS and documented food intolerances found that oral cromolyn allowed patients to tolerate their offending foods [124]. Another study in adults found oral cromolyn to be comparable to an elimination diet, with both treatments performing better in patients with positive food SPTs [125]. Lastly, another study in adults found oral cromolyn to improve mast cell activation, abdominal pain, and stool consistency [126]. A newer biologic targeting siglec-8 (which is present only on mast cells and eosinophils) has been shown to eliminate eosinophils and inactivate mast cells [127]. It has shown positive results in phase II trials, decreasing antral and duodenal eosinophils and improving symptoms [127]. Prevention of mast cell activation may be a viable strategy in FAPDs, including possibly, patients with food allergies who do not exhibit systemic allergic symptoms.

Medications used in other allergic conditions to inhibit the actions of mediators released from mast cells and eosinophils may also be beneficial in FAPDs, although the current evidence is limited. (See Table 1) Most studies have been undertaken in patients with mucosal eosinophilia or increased mucosal mast cells. Histamine has received significant attention as it has been shown to sensitize TRPV1 receptors, promoting visceral hypersensitivity [128]. H1 antagonists alone or in combination with H2 antagonists have been reported to be effective in children and adults with FD in uncontrolled studies [129–131]. In adults, the response was predicted by elevated duodenal eosinophils [131]. Ebastine, an H1 antagonist, has decreased visceral sensitivity in adults with IBS [128]. Another potential downstream target is cysteinyl- leukotriene (cysLT) receptors. Montelukast, a cysLT receptor antagonist, effectively treats pain in children with FD and duodenal eosinophilia. However, the effect is independent of changes in mast cell or eosinophil density or activation [132,133]. While not an anti-allergy drug, per se, proton pump inhibitors have proven efficacy in eosinophilic esophagitis and may be effective for treating duodenal eosinophilia [134–136]. In a prospective study of adults with FD and duodenal eosinophilia, treatment with pantoprazole was associated with improved symptoms and reduced mucosal eosinophil density [136].

Table 1. Trials in functional abdominal pain disorder patients utilizing medications with reported benefits in allergic conditions.

Medication	Mode of Action	Population	Study Type	Result
Diphenhydramine [129]	H1 antagonist	Adults with FD and mucosal mast cell density elevation	Open-label trial	Symptomatic improvement in 79%
Ebastine [128]	H1 antagonist	Adults with IBS	Randomized, double-blind placebo-controlled trial	Symptomatic improvement and reduced visceral sensitivity
Hydroxyzine/Ranitidine [130]	H1/H2 antagonists	Children with FD and mucosal eosinophilia	Retrospective case series	Symptomatic improvement in 50%
Loratidine/Ranitidine [131]	H1/H2 antagonists	Adults with FD	Retrospective case series	Symptomatic improvement in 71%
Montelukast [132]	Cys-Leukotriene antagonist	Children with FD and mucosal eosinophilia	Randomized, double-blind placebo-controlled cross-over trial	Superior to placebo in pain relief
Montelukast [133]	Cys-Leukotriene antagonist	Children with FD and mucosal eosinophilia	Open-label trial	Symptomatic improvement unrelated to changes in mucosal eosinophilia or mast cell density
Budesonide [137]	Steroid	Adults with FD and mucosal eosinophilia	Randomized, double-blind placebo-controlled trial	Symptomatic response not different from placebo
Unspecified PPI [135]	Proton pump inhibitor	Adults with FD and mucosal eosinophilia	Case-control study	Lower eosinophil density without symptomatic improvement
Pantoprazole [136]	Proton pump inhibitor	Adults with FD	Open-label trial	Symptomatic improvement and decreased mucosal eosinophil and mast cell densities

FD = functional dyspepsia; IBS = irritable bowel syndrome; PPI = proton pump inhibitor.

5. Conclusions

FAPDs share pathophysiologic processes with allergies, including both immune and psychological processes interacting with the microbiome. Although there is significant overlap in the medications that can be used for allergic disorders and FAPDs (particularly IBS and FD in association with mucosal eosinophilia), it is unclear to what degree allergens play a role FAPDs. In the absence of systemic allergic symptoms or oral allergy syndrome, it appears unlikely that allergic triggers for FAPDs can be reliably identified by standard testing. There is a need for high-quality studies assessing dietary strategies and anti-allergy medication to better understand the efficacy and the value of allergy tests to predict response.

Author Contributions: All authors participated in concept development, writing, and editing of the manuscript. All authors have read and agreed to the published version of the manuscript.

Funding: This research received no external funding.

Conflicts of Interest: Declare conflicts of interest or state the authors declare no conflict of interest, except CF has received a consulting fee from Allakos.

References

1. Korterink, J.J.; Diederen, K.; Benninga, M.A.; Tabbers, M.M. Epidemiology of pediatric functional abdominal pain disorders: A meta-analysis. *PLoS ONE* **2015**, *10*, e0126982. [CrossRef]
2. Hyams, J.S.; Di Lorenzo, C.; Saps, M.; Shulman, R.J.; Staiano, A.; van Tilburg, M. Childhood gastrointestinal disorders: Child/adolescent. *Gastroenterology* **2016**, *150*, 1456–1468. [CrossRef]
3. Walker, L.S.; Lipani, T.A.; Greene, J.W.; Caines, K.; Stutts, J.; Polk, D.B.; Caplan, A.; Rasquin-Weber, A. Recurrent abdominal pain: Symptom subtypes based on the Rome II criteria for pediatric functional gastrointestinal disorders. *J. Pediatr. Gastroenterol. Nutr.* **2004**, *38*, 187–191. [CrossRef]
4. Schurman, J.V.; Friesen, C.A.; Danda, C.E.; Andre, L.; Welchert, E.; Lavenbarg, T.; Cocjin, J.T.; Hyman, P.E. Diagnosing functional abdominal pain with the Rome II criteria: Parent, child, and clinician agreement. *J. Pediatr. Gastroenterol. Nutr.* **2005**, *41*, 291–295. [CrossRef] [PubMed]
5. Chumpitazi, B.P.; Weidler, E.M.; Lu, D.Y.; Tsai, C.M.; Shulman, R.J. Self-perceived food intolerances are common and associated with clinical severity in childhood irritable bowel syndrome. *J. Acad. Nutr. Diet.* **2016**, *116*, 1458–1464. [CrossRef] [PubMed]
6. Crowe, S.E. Food allergy vs food intolerance in patients with irritable bowel syndrome. *Gastroenterol. Hepatol.* **2019**, *15*, 38–40.
7. Pesce, M.; Cargiolli, M.; Cassarano, S.; Polese, B.; De Conno, B.; Aurino, L.; Mancino, N.; Sarnelli, G. Diet and functional dyspepsia: Clinical correlates and therapeutic perspectives. *World J. Gastroenterol.* **2020**, *26*, 456–465. [CrossRef]
8. De Petrillo, A.; Hughes, L.D.; McGuinness, S.; Roberts, D.; Godfrey, E. A systematic review of psychological, clinical and psychosocial correlates of perceived food intolerance. *J. Psychosom. Res.* **2021**, *141*, 110344. [CrossRef]
9. Chumpitazi, B.P.; Shulman, R.J. Dietary carbohydrates and childhood functional abdominal pain. *Ann. Nutr. Metab.* **2016**, *68*, 8–17. [CrossRef]
10. Krieger-Grübel, C.; Hutter, S.; Hiestand, M.; Brenner, I.; Güsewell, S.; Borovicka, J. Treatment efficacy of a low FODMAP diet compared to a low lactose diet in IBS: A randomized, cross-over designed study. *Clin. Nutr. ESPEN* **2020**, *40*, 83–89. [CrossRef]
11. Posovszky, C.; Roesler, V.; Becker, S.; Iven, E.; Hudert, C.; Ebinger, F.; Calvano, C.; Warschburger, P. Roles of lactose and fructose malabsorption and dietary outcomes in children presenting with chronic abdominal pain. *Nutrients* **2019**, *11*, 3063. [CrossRef]
12. Gijsbers, C.F.M.; Kneepkens, C.M.F.; Büller, H.A. Lactose and fructose malabsorption in children with recurrent abdominal pain: Results of double-blinded testing. *Acta Paediatr.* **2012**, *101*, e411–e415. [CrossRef]
13. Yang, J.; Fox, M.; Cong, Y.; Chu, H.; Zheng, X.; Long, Y.; Fried, M.; Dai, N. Lactose intolerance in irritable bowel syndrome patients with diarrhoea: The roles of anxiety, activation of the innate mucosal immune system and visceral sensitivity. *Aliment. Pharmacol. Ther.* **2014**, *39*, 302–311. [CrossRef]
14. Kamphuis, J.B.J.; Guiard, B.; Leveque, M.; Olier, M.; Jouanin, I.; Yvon, S.; Tondereau, V.; Rivière, P.; Guéraud, F.; Chevolleau, S.; et al. Lactose and fructo-oligosaccharides increase visceral sensitivity in mice via glycation processes, increasing mast cell density in colonic mucosa. *Gastroenterology* **2020**, *158*, 652–663. [CrossRef]
15. Singh, M.; Singh, V.; Friesen, C.A. Colonic mucosal inflammatory cells in children and adolescents with lactase deficiency. *Pathol. Res. Pract.* **2020**, *216*, 152971. [CrossRef] [PubMed]
16. Sun, J.; Lin, J.; Parashette, K.; Zhang, J.; Fan, R. Association of lymphocytic colitis and lactase deficiency in pediatric population. *Pathol. Res. Pract.* **2015**, *211*, 138–144. [CrossRef] [PubMed]
17. El-Salhy, M.; Hatlebakk, J.G.; Hausken, T. Diet in irritable bowel syndrome (IBS): Interaction with gut microbiota and gut hormones. *Nutrients* **2019**, *11*, 1824. [CrossRef] [PubMed]
18. Chen, B.-R.; Du, L.-J.; He, H.-Q.; Kim, J.J.; Zhao, Y.; Zhang, Y.-W.; Luo, L.; Dai, N. Fructo-oligosaccharide intensifies visceral hypersensitivity and intestinal inflammation in a stress-induced irritable bowel syndrome mouse model. *World J. Gastroenterol.* **2017**, *23*, 8321–8333. [CrossRef] [PubMed]
19. Ling, Z.; Li, Z.; Liu, X.; Cheng, Y.; Luo, Y.; Tong, X.; Yuan, L.; Wang, Y.; Sun, J.; Li, L.; et al. Altered fecal microbiota composition associated with food allergy in infants. *Appl. Environ. Microbiol.* **2014**, *80*, 2546–2554. [CrossRef] [PubMed]
20. Luu, M.; Monning, H.; Visekruna, A. Exploring the molecular mechanisms underlying protective effects of microbial SCFAs on intestinal tolerance and food allergy. *Front. Immunol.* **2020**, *11*, 1225. [CrossRef] [PubMed]
21. Ali, A.; Tan, H.Y.; Kaiko, G.E. Role of the intestinal epithelium and its interaction with the microbiota in food allergy. *Front. Immunol.* **2020**, *11*, 604054. [CrossRef]
22. Uranga, J.A.; Martinez, V.; Abalo, R. Mast cell regulation and irritable bowel syndrome: Effects of food components with potential nutraceutical use. *Molecules* **2020**, *25*, 4314. [CrossRef] [PubMed]
23. Leonel, A.J.; Alvarez-Leite, J.I. Butyrate: Implications for intestinal function. *Curr. Opin. Clin. Nutr. Metab. Care* **2012**, *15*, 474–479. [CrossRef]
24. Sivaprakasam, S.; Prasad, P.D.; Singh, N. Benefits of short-chain fatty acids and their receptors in inflammation and carcinogenesis. *Pharmacol. Ther.* **2016**, *164*, 144–151. [CrossRef] [PubMed]
25. Folkerts, J.; Stadhouders, R.; Redegeld, F.A.; Tam, S.Y.; Hendriks, R.W.; Galli, S.J.; Maurer, M. Effect of dietary fiber and metabolites on mast cell activation and mast cell-associated diseases. *Front. Immunol.* **2018**, *9*, 1067. [CrossRef] [PubMed]
26. Onyimba, F.; Crowe, S.E.; Johnson, S.; Leung, J. Food allergies and intolerances: A clinical approach to the diagnosis and management of adverse reactions to foods. *Clin. Gastroenterol. Hepatol.* **2021**, in press. [CrossRef] [PubMed]
27. Corica, D.; Aversa, T.; Caminiti, L.; Lombardo, F.; Wasniewska, M.; Battista Pajno, G. Nutrition and avoidance diets in children with food allergy. *Front. Pediatr.* **2020**, *8*, 518. [CrossRef]

28. Vickery, B.P.; Scurlock, A.M.; Jones, S.M.; Burks, A.W. Mechanisms of immune tolerance relevant to food allergy. *J. Allergy Clin. Immunol.* **2011**, *127*, 576–584. [CrossRef]
29. Schmiechen, Z.C.; Weissler, K.A.; Frischmeyer-Guerrerio, P.A. Recent developments in understanding the mechanisms of food allergy. *Curr. Opin. Pediatr.* **2019**, *31*, 807–814. [CrossRef]
30. Di Lorenzo, C.; Youssef, N.N.; Sigurdsson, L.; Scharff, L.; Griffiths, J.; Wald, A. Visceral hyperalgesia in children with functional abdominal pain. *J. Pediatr.* **2001**, *139*, 838–843. [CrossRef] [PubMed]
31. Van Ginkel, R.; Voskuijl, W.P.; Benninga, M.A.; Taminiau, J.A.; Boeckxstaens, G.E. Alterations in rectal sensitivity and motility in childhood irritable bowel syndrome. *Gastroenterology* **2001**, *120*, 31–38. [CrossRef]
32. Halac, U.; Noble, A.; Faure, C. Rectal sensory threshold for pain is a diagnostic marker of irritable bowel syndrome and functional abdominal pain in children. *J. Pediatr.* **2010**, *156*, 60–65.e1. [CrossRef]
33. Traina, G. The role of mast cells in the gut and brain. *J. Integr. Neurosci.* **2021**, *20*, 185–196. [CrossRef]
34. Krammer, L.; Sergeevna Sowa, A.; Lorentz, A. Mast cells in irritable bowel syndrome: A systematic review. *J. Gastrointest. Liver Dis.* **2019**, *28*, 463–472. [CrossRef]
35. Du, L.; Chen, B.; Kim, J.J.; Chen, X.; Dai, N. Micro-inflammation in functional dyspepsia: A systematic review and meta-analysis. *Neurogastroenterol. Motil.* **2018**, *30*, e13304. [CrossRef]
36. Friesen, C.A.; Schurman, J.V.; Colombo, J.M.; Abdel-Rahman, S.M. Eosinophils and mast cells as therapeutic targets in pediatric functional dyspepsia. *World J. Gastrointest. Pharmacol. Ther.* **2013**, *4*, 86–96. [CrossRef] [PubMed]
37. Bashashati, M.; Moossavi, S.; Cremon, C.; Barbaro, M.R.; Moraveji, S.; Talmon, G.; Rezaei, N.; Hughes, P.A.; Bian, Z.X.; Choi, C.H.; et al. Colonic immune cells in irritable bowel syndrome: A systematic review and meta-analysis. *Neurogastroenterol. Motil.* **2018**, *30*, e13192. [CrossRef] [PubMed]
38. Burns, G.; Carroll, G.; Mathe, A.; Horvat, J.; Foster, P.; Walker, M.M.; Talley, N.J.; Keely, S. Evidence for local and systemic immune activation in functional dyspepsia and irritable bowel syndrome: A systematic review. *Am. J. Gastroenterol.* **2019**, *114*, 429–436. [CrossRef]
39. Robles, A.; Perez Ingles, D.; Myneedu, K.; Deoker, A.; Sarosiek, I.; Zuckerman, M.J.; Schmulson, M.J.; Bashashati, M. Mast cells are increased in the small intestinal mucosa of patients with irritable bowel syndrome: A systematic review and meta-analysis. *Neurogastroenterol. Motil.* **2019**, *31*, e13718.
40. Barbara, G.; Stanghellini, V.; De Giorgio, R.; Cremon, C.; Cottrell, G.S.; Santini, D.; Pasquinelli, G.; Morselli-Labate, A.M.; Grady, E.F.; Bunnett, N.W.; et al. Activated mast cells in proximity to colonic nerves correlate with abdominal pain in irritable bowel syndrome. *Gastroenterology* **2004**, *126*, 693–702. [CrossRef]
41. Singh, M.; Singh, V.; Schurman, J.V.; Colombo, J.M.; Friesen, C.A. The relationship between mucosal inflammatory cells, specific symptoms, and psychological functioning in youth with irritable bowel syndrome. *Sci. Rep.* **2020**, *10*, 11988. [CrossRef]
42. Wauters, L.; Nightingale, S.; Talley, N.J.; Sulaiman, B.; Walker, M.M. Functional dyspepsia is associated with duodenal eosinophilia in an Australian paediatric cohort. *Aliment. Pharmacol. Ther.* **2017**, *45*, 1358–1364. [CrossRef] [PubMed]
43. Ren, X.; Li, Z.; Li, X.; Li, X.; Yuan, H.; Wang, X. Association between food allergy and duodenal mast cells in patients with functional dyspepsia. *Biomed. Res.* **2017**, *28*, 7274–7280.
44. Friesen, C.A.; Andre, L.; Garola, R.; Hodge, C.; Roberts, C. Activated duodenal mucosal eosinophils in children with dyspepsia: A pilot transmission electron microscopic study. *J. Pediatr. Gastroenterol. Nutr.* **2002**, *35*, 329–333. [CrossRef]
45. Hou, X.H.; Zhu, L.R.; Li, Q.X.; Chen, J.D.Z. Alterations in mast cells and 5-HT positive cells in gastric mucosa in functional dyspepsia patients with hypersensitivity. *Neurogastroenterol. Motil.* **2001**, *13*, 398–399.
46. Walker, M.M.; Salehian, S.S.; Murray, C.E.; Rajendran, A.; Hoare, J.M.; Negus, R.; Powell, N.; Talley, N.J. Implications of eosinophilia in the normal duodenal biopsy- an association with allergy and functional dyspepsia. *Aliment. Pharmacol. Ther.* **2010**, *31*, 1229–1236. [CrossRef]
47. Schäppi, M.G.; Borrelli, O.; Knafelz, D.; Williams, S.; Smith, V.V.; Milla, P.J.; Lindley, K.J. Mast cell-nerve interactions in children with functional dyspepsia. *J. Pediatr. Gastroenterol. Nutr.* **2008**, *47*, 472–480. [CrossRef]
48. Hollier, J.M.; van Tilburg, M.A.L.; Liu, Y.; Czyzewski, D.I.; Self, M.M.; Weidler, E.M.; Heitkemper, M.; Shulman, R.J. Multiple psychological factors predict abdominal pain severity in children with irritable bowel syndrome. *Neurogastroenterol. Motil.* **2019**, *31*, e13509. [CrossRef] [PubMed]
49. Newton, E.; Schosheim, A.; Patel, S.; Chitkara, D.K.; van Tilburg, M.A.L. The role of psycohological factors in pediatric functional abdominal pain disorders. *Neurogastroenterol. Motil.* **2019**, *31*, e13538. [CrossRef] [PubMed]
50. Deacy, A.D.; Friesen, C.A.; Staggs, V.S.; Schurman, J.V. Evaluation of clinical outcomes in an interdisciplinary abdominal pain clinic: A retrospective, exploratory review. *World J. Gastroenterol.* **2019**, *25*, 3079–3090. [CrossRef]
51. Liu, L.Y.; Coe, C.L.; Swenson, C.A.; Kelly, E.A.; Kita, H.; Busse, W.W. School examinations enhance airway inflammation to antigen challenge. *Am. J. Respir. Crit. Care Med.* **2002**, *165*, 1062–1067. [CrossRef]
52. Oland, A.A.; Booster, G.D.; Bender, B.G. Integrated behavioral health care for management of stress in allergic diseases. *Ann. Allergy Asthma Immunol.* **2018**, *121*, 31–36. [CrossRef]
53. Schurman, J.V.; Singh, M.; Singh, V.; Neilan, N.; Friesen, C.A. Symptoms and subtypes in pediatric functional dyspepsia: Relation to mucosal inflammation and psychological functioning. *J. Pediatr. Gastroenterol. Nutr.* **2010**, *51*, 298–303. [CrossRef]
54. Yuan, H.-P.; Li, Z.; Zhang, Y.; Li, X.-P.; Li, F.-K.; Li, Y.-Q. Anxiety and depression are associated with increased counts and degranulation of duodenal mast cells in functional dyspepsia. *Int. J. Clin. Exp. Med.* **2015**, *8*, 8010–8014.

55. Piche, T.; Saint-Paul, M.C.; Dainese, R.; Marine-Barjoan, E.; Iannelli, A.; Montoya, M.L.; Peyron, J.F.; Czerucka, D.; Cherikh, F.; Filippi, J.; et al. Mast cells and cellularity of the colonic mucosa correlated with fatigue and depression in irritable bowel syndrome. *Gut* **2008**, *57*, 468–473. [CrossRef]
56. He, S.-H. Key role of mast cells and their major secretory products in inflammatory bowel disease. *World J. Gastroenterol.* **2004**, *10*, 309–318. [CrossRef] [PubMed]
57. Santos, J.; Saperas, E.; Nogueiras, C.; Mourelle, M.; Antolín, M.; Cadahia, A.; Malagelada, J.R. release of mast cell mediators into the jejunum by cold pain stress in humans. *Gastroenterology* **1998**, *114*, 640–648. [CrossRef]
58. Loo, E.X.L.; Wang, D.Y.; Siah, K.T.H. Association between irritable bowel syndrome and allergic diseases: To make a case for aeroallergen. *Int. Arch. Allergy Immunol.* **2020**, *181*, 31–42. [CrossRef] [PubMed]
59. Kumari, M.V.; Devanarayana, N.M.; Amarasiri, L.; Rajindrajith, S. Association between functional abdominal pain disorders and asthma in adolescents: A cross-sectional study. *World J. Clin. Cases* **2018**, *6*, 944–951. [CrossRef]
60. Koloski, N.; Jones, M.; Walker, M.M.; Veysey, M.; Zala, A.; Keely, S.; Holtmann, G.; Talley, N.J. Population based study: Atopy and autoimmune diseases are associated with functional dyspepsia and irritable bowel syndrome, independent of psychological distress. *Aliment. Pharmacol. Ther.* **2019**, *49*, 546–555. [CrossRef] [PubMed]
61. Patel, K.; Vila-Nadal, G.; Shah, J.; Shamji, M.; Swan, L.; Durham, S.R.; Patel, K.; Skypala, I.J. Is pollen-food syndrome a frequent comorbidity in adults with irritable bowel syndrome? *Allergy* **2020**, *75*, 1780–1783. [CrossRef]
62. Tobin, M.C.; Moparty, B.; Farhadi, A.; DeMeo, M.T.; Bansal, P.J.; Keshavarzian, A. Atopic irritable bowel syndrome: A novel subgroup of irritable bowel syndrome with allergic manifestations. *Ann. Allergy Asthma Immunol.* **2008**, *100*, 49–53. [CrossRef]
63. Jones, M.P.; Walker, M.M.; Ford, A.C.; Talley, N.J. The overlap of atopy and functional gastrointestinal disorders among 23,471 patients in primary care. *Aliment. Pharmacol. Ther.* **2014**, *40*, 382–391. [CrossRef]
64. Deshmukh, F.; Vasudevan, A.; Mengalie, E. Association between irritable bowel syndrome and asthma: A meta-analysis and systematic review. *Ann. Gastroenterol.* **2019**, *32*, 570–577. [CrossRef]
65. Tan, T.-K.; Chen, A.-C.; Lin, C.-L.; Shen, T.-C.; Li, T.-C.; Wei, C.-C. Preschoolers with allergic diseases have an increased risk of irritable bowel syndrome when reaching school age. *J. Pediatr. Gastroenterol. Nutr.* **2017**, *64*, 26–30. [CrossRef] [PubMed]
66. Tsai, J.-D.; Wang, I.-C.; Shen, T.-C.; Lin, C.-L.; Wei, C.-C. A 8-year population-based cohort study of irritable bowel syndrome in childhood with a history of atopic dermatitis. *J. Investig. Med.* **2018**, *66*, 755–761. [CrossRef] [PubMed]
67. Di Nardo, G.; Cremon, C.; Frediani, S.; Lucarelli, S.; Pia Villa, M.; Stanghellini, V.; La Torre, G.; Martemucci, L.; Barbara, G. Allergic proctocolitis is a risk factor for functional gastrointestinal disorders in children. *J. Pediatr.* **2018**, *195*, 128–133. [CrossRef] [PubMed]
68. Del Giudice, M.M.; Indolfi, C.; Capasso, M.; Maiello, N.; Decimo, F.; Ciprandi, G. Bificobacterium mixture (B longum BB536, B infantis M-63, M breve M-16v) treatment in children with seasonal allergic rhinitis and intermittent asthma. *Ital. J. Pediatr.* **2017**, *43*, 25. [CrossRef]
69. Van Spaendonk, H.; Ceuleers, H.; Witters, L.; Patteet, E.; Joossens, J.; Augustyns, K.; Lambeir, A.-M.; De Meester, I.; De Man, J.G.; De Winter, B.Y. Regulation of intestinal permeability: The role of proteases. *World J. Gastroenterol.* **2017**, *23*, 2106–2123. [CrossRef]
70. Mansueto, P.; D'Alcamo, A.; Seidita, A.; Carroccio, A. Food allergy in irritable bowel syndrome: The case of non-celiac wheat sensitivity. *World J. Gastroenterol.* **2015**, *21*, 7089–7109. [CrossRef]
71. Choung, R.S.; Murray, J.A. The role for food allergies in the pathogenesis of irritable bowel syndrome: Understanding mechanisms of intestinal mucosal responses against food antigens. *Gastroenterology* **2019**, *157*, 15–17. [CrossRef]
72. Hammond, C.; Lieberman, J.A. Unproven diagnostic tests for food allergy. *Immunol. Allergy Clin. N. Am.* **2018**, *38*, 153–163. [CrossRef] [PubMed]
73. Ismail, F.W.; Abid, S.; Awan, S.; Lubna, F. Frequency of food hypersensitivity in patients with functional gastrointestinal disorders. *Acta Gastroenterol. Belg.* **2018**, *81*, 253–256. [CrossRef] [PubMed]
74. Zuo, X.L.; Li, Y.Q.; Li, W.J.; Guo, Y.T.; Lu, X.F.; Li, J.M.; Desmond, P.V. Alterations of food antigen-spedific serum immunoglobulins G and E antibodies in patients with irritable bowel syndrome and functional dyspepsia. *Clin. Exp. Allergy* **2007**, *37*, 823–830. [CrossRef] [PubMed]
75. Neilan, N.A.; Dowling, P.J.; Taylor, D.L.; Ryan, P.; Schurman, J.V.; Friesen, C.A. Useful biomarkers in pediatric eosinophilic duodenitis and their existence: A case-control, single-blind, observational pilot study. *J. Pediatr. Gastroenterol. Nutr.* **2010**, *50*, 377–384. [CrossRef]
76. Bischoff, S.C.; Mayer, J.; Meier, P.N.; Zeck-Kapp, G.; Manns, M.P. Clinical significance of the colonoscopic allergen provocation test. *Int. Arch. Allergy Immunol.* **1997**, *113*, 348–351. [CrossRef]
77. Aguilera-Lizarraga, J.; Florens, M.V.; Viola, M.F.; Jain, P.; Decraecker, L.; Appeltans, I.; Cuende-Estevez, M.; Fabre, N.; Van Beek, K.; Perna, E.; et al. Local immune response to food antigens drives meal-induced abdominal pain. *Nature* **2021**, *590*, 151–156. [CrossRef]
78. Jones, A.L.; Curran-Everett, D.; Leung, D.Y.M. Food allergy is associated with Staphylococcus aureus colonization in children with atopic dermatitis. *J. Allergy Clin. Immunol.* **2016**, *137*, 1247–1248. [CrossRef]
79. Vivinus-Nébot, M.; Dainese, R.; Anty, R.; Saint-Paul, M.C.; Nano, J.L.; Gonthier, N.; Marjoux, S.; Frin-Mathy, G.; Bernard, G.; Hébuterne, X.; et al. Combination of allergic factors can worsen diarrheic irritable bowel syndrome: Role of barrier defects and mast cells. *Am. J. Gastroenterol.* **2012**, *107*, 75–81. [CrossRef]

80. Schurman, J.V.; Friesen, C.A. Identifying potential pediatric chronic abdominal pain triggers using ecological momentary assessment. *Clin. Pract. Pediatr. Psychol.* **2015**, *3*, 131–141. [CrossRef]
81. Magnusson, J.; Lin, X.P.; Dahlman-Höglund, A.; Hanson, L.A.; Esbjörn, T.; Magnusson, O.; Bengtsson, U.; Ahlstedt, S. Seasonal intestinal inflammation in patients with birch pollen allergy. *J. Allergy Clin. Immunol.* **2003**, *112*, 45–50. [CrossRef] [PubMed]
82. Rentzos, G.; Lundberg, V.; Stotzer, P.O.; Pullerits, T.; Telemo, E. Intestinal allergic inflammation in birch pollen allergic patients n relation to pollen season, IgE sensitization profile and gastrointestinal symptoms. *Clin. Transl. Allergy* **2014**, *4*, 19. [CrossRef] [PubMed]
83. Pickert, C.N.; Lorentz, A.; Manns, M.P.; Bischoff, S.C. Colonoscopic allergen provocation test with rBet v 1 in patients with pollen-associated allergy. *Allergy* **2012**, *67*, 1308–1315. [CrossRef] [PubMed]
84. Berin, M.C. Advances in understanding immune mechanisms of food protein-induced enterocolitis syndrome. *Ann. Allergy Asthma Immunol.* **2021**, *126*, 478–481. [CrossRef]
85. Mehr, S.; Brown-Whitehorn, T. What do allergists in practice need to know about non-IgE-mediated food allergies. *Ann. Allergy Asthma Immunol.* **2019**, *122*, 589–597. [CrossRef]
86. Hoffman, N.V.; Ahmed, A.; Fortunato, J.E. Food protein-induced enterocolitis syndrome. Dynamic relationship among gastrointestinal symptoms, immune response, and the autonomic nervous system. *Ann. Allergy Asthma Immunol.* **2021**, *126*, 498–505.
87. Caubet, J.C.; Ford, L.S.; Sickles, L.; Järvinen, K.M.; Sicherer, S.H.; Sampson, H.A.; Nowak-Węgrzyn, A. Clinical features and resolution of food protein-induced enterocolitis syndrome: 10-year experience. *J. Allergy Clin. Immunol.* **2014**, *134*, 382–389. [CrossRef]
88. Nowak-Wegrzyn, A.; Berin, M.C.; Mehr, S. Food protein-induced enterocolitis syndrome. *J. Allergy Clin. Immunol. Pract.* **2020**, *8*, 24–35. [CrossRef]
89. Gonzalez-Delgado, P.; Caparrós, E.; Moreno, V.; Cueva, B.; Fernández, J. Food protein-induced enterocolitis-like syndrome in a population of adolescents and adults caused by seafood. *J. Allergy Clin. Immunol. Pract.* **2019**, *7*, 670–672. [CrossRef]
90. Bauché, D.; Marie, J.C. Transforming growth factor β: A master regulator of the gut microbiota and immune cell interactions. *Clin. Transl. Immunol.* **2017**, *6*, e136. [CrossRef]
91. Stockinger, B.; Omenetti, S. The dichotomous nature of T helper 17 cells. *Nat. Rev. Immunol.* **2017**, *17*, 535–544. [CrossRef]
92. Luo, A.; Leach, S.T.; Barres, R.; Hesson, L.B.; Grimm, M.C.; Simar, D. The microbiota and epigenetic regulation of T helper 17/regulatory T cells: In search of a balanced immune system. *Front. Immunol.* **2017**, *8*, 417. [CrossRef]
93. Sugawara, R.; Lee, E.-J.; Jang, M.S.; Jeun, E.-J.; Hong, C.-P.; Kim, J.-H.; Park, A.; Yun, C.H.; Hong, S.-W.; Kim, Y.-M.; et al. Small intestinal eosinophils regulate Th17 cells by producing IL-1 receptor antagonist. *J. Exp. Med.* **2016**, *213*, 555–567. [CrossRef]
94. Bin Dhuban, K.; d'Hennezel, E.; Ben-Shoshan, M.; McCusker, C.; Clarke, A.; Fiset, P.; Mazer, B.; Piccirillo, C.A. Altered T helper 17 responses in children with food allergy. *Int. Arch. Allergy Immunol.* **2013**, *162*, 318–322. [CrossRef]
95. Wakashin, H.; Hirose, K.; Maezawa, Y.; Kagami, S.; Suto, A.; Watanabe, N.; Saito, Y.; Hatano, M.; Tokuhisa, T.; Iwakura, Y.; et al. IL-23 and TH17 cells enhance Th2-cell-medicated eosinophilic airway inflammation in mice. *Am. J. Respir. Crit. Care Med.* **2008**, *178*, 1023–1032. [CrossRef]
96. Dias, P.M.; Banerjee, G. The role of Th17/IL-17 on eosinophilic inflammation. *J. Autoimmun.* **2013**, *40*, 9–20. [CrossRef]
97. Cheung, P.F.Y.; Wong, C.K.; Lam, C.W.K. Molecular mechanisms of cytokine and chemokine release from eosinophils activated by IL-17A, IL-17F, and IL-23: Implications for Th17 lymphocytes-mediated allergic inflammation. *J. Immunol.* **2008**, *180*, 5625–5635. [CrossRef]
98. Cho, K.-A.; Park, M.; Kim, Y.-H.; Woo, S.-Y. Th17 cell-mediated immune responses promote mast cell proliferation by triggering stem cell factor in keratinocytes. *Biochem. Biophys. Res. Commun.* **2017**, *487*, 856–861. [CrossRef]
99. Singh, M.; Singh, V.; Schurman, J.V.; Friesen, C.A. Mucosal Th17 cells are increased in pediatric functional dyspepsia associated with chronic gastritis. *Dig. Dis. Sci.* **2020**, *65*, 3184–3190. [CrossRef]
100. Fritscher-Ravens, A.; Pflaum, t.; Mösinger, M.; Ruchay, Z.; Röchen, C.; Milla, P.J.; Das, M.; Böttner, M.; Wedel, T.; Schuppan, d. Many patients with irritable bowel syndrome have atypical food allergies not associated with immunoglobulin E. *Gastroenterology* **2019**, *157*, 109–118. [CrossRef]
101. Lee, H.S.; Lee, K.J. Alterations of food-specific serum IgG4 titers to common food antigens in patients with irritable bowel syndrome. *J. Neurogastroenterol. Motil.* **2017**, *23*, 578–584. [CrossRef]
102. Kanagaratham, C.; El Ansari, Y.S.; Lewis, O.L.; Oettgen, H.C. IgE and IgG antibodies as regulators of mast cell and basophil functions in food allergy. *Front. Immunol.* **2020**, *11*, 603050. [CrossRef]
103. Karakula-Juchnowicz, H.; Gałęcka, M.; Rog, J.; Bartnicka, A.; Łukaszewicz, Z.; Krukow, P.; Morylowska-Topolska, J.; Skonieczna-Zydecka, K.; Krajka, T.; Jonak, K.; et al. The food-specific serum IgG reactivity in major depressive disorder patients, irritable bowel syndrome patients and healthy controls. *Nutrients* **2018**, *10*, 548. [CrossRef]
104. Xie, Y.; Zhou, G.; Xu, Y.; He, B.; Wang, Y.; Ma, R.; Chang, Y.; He, D.; Xu, C.; Xiao, Z. Effects of diet based on IgG elimination combined with probiotics on migraine plus irritable bowel syndrome. *Pain Res. Manag.* **2019**, *2019*, 7890461. [CrossRef]
105. Ostrowska, L.; Wasiluk, D.; Lieners, C.F.J.; Gałęcka, M.; Bartnicka, A.; Tveiten, D. IgG food antibody guided elimination-rotation diet was more effective than FODMAP diet and control diet in the treatment of women with mixed IBS-Results from an open label study. *Res. Sq.* **2020**, in press. [CrossRef]

106. Cappelletti, M.; Tognon, E.; Vona, L.; Basello, K.; Costanzi, A.; Speciani, M.C.; Speciani, A.F. Food-specific serum IgG and sumptom reduction with a personalized, unrestricted-calorie diet of six weeks in irritable bowel syndrome (IBS). *Nutr. Metab.* **2020**, *17*, 101. [CrossRef]
107. Atkinson, W.; Sheldon, T.A.; Shaath, N.; Whorwell, P.J. Food elimination based on IgG antibodies in irritable bowel syndrome: A randomized controlled trial. *Gut* **2004**, *53*, 1459–1464. [CrossRef]
108. Zar, S.; Mincher, L.; Benson, M.J.; Kumar, D. Food-specific IgG4 antibody-guided exclusion diet improves symptoms and rectal compliance in irritable bowel syndrome. *Scand. J. Gastroenterol.* **2005**, *40*, 800–807. [CrossRef]
109. Guo, H.; Jiang, T.; Wang, J.; Chang, Y.; Guo, H.; Zhang, W. The value of eliminating foods according to food-specific immunoglobin G antibodies in irritable bowel syndrome with diarrhoea. *J. Int. Med. Res.* **2012**, *40*, 204–210. [CrossRef]
110. Stierstorfer, M.B.; Sha, C.T.; Sasson, M. Food patch testing for irritable bowel syndrome. *J. Am. Acad. Dermatol.* **2013**, *68*, 377–384. [CrossRef]
111. Shin, G.H.; Smith, M.S.; Toro, B.; Ehrlich, A.C.; Luther, S.; Midani, D.; Hong, I.; Stierstorfer, M. Utility of food patch testing in the evaluation and management of irritable bowel syndrome. *Skin* **2018**, *2*, 96–110. [CrossRef]
112. Lucendo, A.J.; Serrano-Montalbán, B.; Arias, Á.; Redondo, O.; Tenias, J.M. Efficacy of dietary treatment for inducing disease remission in eosinophilic gastroenteritis. *J. Pediatr. Gastroenterol. Nutr.* **2015**, *61*, 56–64. [CrossRef]
113. Gutiérrez-Junquera, C.; Zevit, N. Dietary treatment of eosinophilic gastrointestinal disorders in children. *Curr. Opin. Clin. Nutr. Metab. Care* **2020**, *23*, 210–216. [CrossRef]
114. Kalmpourtzidou, A.; Xinias, I.; Agakidis, C.; Mavroudi, A.; Mouselimis, D.; Tsarouchas, A.; Agakidou, E.; Karagiozoglou-Lampoudi, T. Diet quality: A neglected parameter in children with food allergies; A cross-sectional study. *Front. Pediatr.* **2021**, *9*, 658778. [CrossRef]
115. Schurman, J.V.; Wu, Y.P.; Grayson, P.; Friesen, C.A. A pilot study to assess the efficacy of biofeedback-assisted relaxation training as an adjunct treatment for pediatric functional dyspepsia associated with duodenal eosinophilia. *J. Pediatr. Psychol.* **2010**, *35*, 837–847. [CrossRef]
116. Pesek, R.D.; Gupta, S.K. Future therapies for eosinophilic gastrointestinal disorders. *Ann. Allergy Asthma Immunol.* **2020**, *124*, 219–226. [CrossRef]
117. Wechsler, J.B.; Hirano, I. Biological therapies for eosinophilic gastrointestinal diseases. *J. Allergy Clin. Immunol.* **2018**, *142*, 24–31. [CrossRef]
118. Navinés-Ferrer, A.; Serrano-Candelas, E.; Molina-Molina, G.J.; Martin, M. EgE-related diseases and anti-IgE-based treatments. *J. Immunol. Res.* **2016**, *2016*, 8163803. [CrossRef]
119. Foroughi, S.; Foster, B.; Kim, N.; Bernadino, L.B.; Scott, L.M.; Hamilton, R.G.; Metcalfe, D.D.; Mannon, P.J.; Prussin, C. Anti-IgE treatment of eosinophil-associated gastrointestinal disorders. *J. Allergy Clin. Immunol.* **2007**, *120*, 594–601. [CrossRef]
120. Wang, J.; Wang, Y.; Zhou, H.; Gu, W.; Wang, X.; Yang, J. Clinical efficacy and safety of ketotifen in treating irritable bowel syndrome with diarrhea. *Eur. J. Gastroenterol. Hepatol.* **2020**, *32*, 706–712. [CrossRef]
121. Klooker, T.K.; Braak, B.; Koopman, K.E.; Welting, O.; Wouters, M.M.; van der Heide, S.; Schemann, M.; Bischoff, S.C.; van den Wijngaard, R.M.; Boeckxstaens, G.E. The mast cell stabiliser ketotifen decreases visceral hypersensitivity and improves intestinal symptoms in patients with irritable bowel syndrome. *Gut* **2010**, *59*, 1213–1221. [CrossRef]
122. Grazioli, I.; Melzi, G.; Balsamo, V.; Castellucci, G.; Castro, M.; Catassi, C.; Rätsch, J.M.; Scotta, S. Food intolerance and irritable bowel syndrome of childhood: Clinical efficacy of oral sodium cromoglycate and elimination diet. *Minerva Pediatr.* **1993**, *45*, 253–258.
123. Burks, A.W.; Sampson, H.A. Double-blind placebo-controlled trial of oral cromolyn in children with atopic dermatitis and documented food hypersensitivity. *J. Allergy Clin. Immunol.* **1988**, *81*, 417–423. [CrossRef]
124. Lunardi, C.; Bambara, L.M.; Biasi, D.; Cortina, P.; Peroli, P.; Nicolis, F.; Favari, F.; Pacor, M.L. Double-blind cross-over trial of oral sodium cromoglycate in patients with irritable bowel syndrome due to food intolerance. *Clin. Exp. Allergy* **1991**, *21*, 569–572. [CrossRef]
125. Stefanini, G.F.; Saggioro, A.; Alvisi, V.; Angelini, G.; Capurso, L.; di Lorenzo, G.; Dobrilla, G.; Dodero, M.; Galimberti, M.; Gasbarrini, M.; et al. Oral cromolyn sodium in comparison with elimination diet in the irritable bowel syndrome, diarrheic type. Multicenter study of 428 patients. *Scand. J. Gastroenterol.* **1995**, *30*, 535–541. [CrossRef]
126. Lobo, B.; Ramos, L.; Martínez, C.; Guilarte, M.; González-Castro, A.M.; Alonso-Cotoner, C.; Pigrau, M.; de Torres, I.; Rodiño-Janeiro, B.K.; Salvo-Romero, E.; et al. Downregulation of mucosal mast cell activation and immune response in diarrhoea-irritable bowel syndrome by oral disodium cromoglycate: A pilot study. *United Eur. Gastroenterol. J.* **2017**, *5*, 887–897. [CrossRef]
127. Dellon, E.S.; Peterson, K.A.; Murray, J.A.; Falk, G.W.; Gonsalves, N.; Chehade, M.; Genta, R.M.; Leung, J.; Khoury, P.; Klion, A.D.; et al. Anti-siglec-8 antibody for eosinophilic gastritis and duodenitis. *N. Engl. J. Med.* **2020**, *383*, 1624–1634. [CrossRef]
128. Wouters, M.M.; Balemans, D.; Van Wanrooy, S.; Dooley, J.; Cibert-Goton, V.; Alpizar, Y.A.; Valdez-Morales, E.E.; Nasser, Y.; Van Veldhoven, P.P.; Vanbrabant, W.; et al. Histamine receptor H1-mediated sensitization of TRPV1 mediates visceral hypersensitivity and symptoms in patients with irritable bowel syndrome. *Gastroenterology* **2016**, *150*, 875–887.e9. [CrossRef]
129. Matter, S.E.; Bhatia, P.S.; Miner, P.B., Jr. Evaluation of antral mast cells in nonulcer dyspepsia. *Dig. Dis. Sci.* **1990**, *35*, 1358–1363. [CrossRef]
130. Friesen, C.A.; Sandridge, L.; Andre, L.; Roberts, C.C.; Abdel-Rahman, S.M. Mucosal eosinophilia and response to H1/H2 antagonist and cromolyn therapy in pediatric dyspepsia. *Clin. Pediatr.* **2006**, *45*, 143–147. [CrossRef]

131. Potter, M.D.E.; Goodsall, T.M.; Walker, M.M.; Talley, N.J. Dual histamine blockade for the treatment of adult functional dyspepsia: A single centre experience. *Gut* **2020**, *69*, 966. [CrossRef]
132. Friesen, C.A.; Kearns, G.L.; Andre, L.; Neustrom, M.; Roberts, C.C.; Abdel-Rahman, S.M. Clinical efficacy and pharmacokinetics of montelukast in dyspeptic children with duodenal eosinophilia. *J. Pediatr. Gastroenterol. Nutr.* **2004**, *38*, 343–351. [CrossRef]
133. Friesen, C.A.; Neilan, N.A.; Schurman, J.V.; Taylor, D.L.; Kearns, G.L.; Abdel-Rahman, S.M. Montelukast in the treatment of duodenal eosinophilia in children with dyspepsia: Effect on eosinophil density and activation in relation to pharmacokinetics. *BMC Gastroenterol.* **2009**, *9*, 32. [CrossRef]
134. Gutiérrez-Junquera, C.; Fernández-Fernández, S.; Cilleruelo, M.L.; Rayo, A.; Echeverría, L.; Borrell, B.; Román, E. Long-term treatment with proton pump inhibitors is effective in children with eosinophilic esophagitis. *J. Pediatr. Gastroenterol. Nutr.* **2018**, *67*, 210–216. [CrossRef]
135. Potter, M.D.E.; Wodk, N.K.; Walker, M.M.; Jones, M.P.; Talley, N.J. Proton pump inhibitors and suppression of duodenal eosinophilia in functional dyspepsia. *Gut* **2019**, *68*, 1339–1340. [CrossRef]
136. Wauters, L.; Ceulemans, M.; Frings, D.; Lambaerts, M.; Accarie, A.; Toth, J.; Mols, R.; Augustijns, P.; De Hertogh, G.; Van Oudenhove, L.; et al. Proton pump inhibitors reduce duodenal eosinophilia, mast cells, and permeability in patients with functional dyspepsia. *Gastroenterology* **2021**, *160*, 1521–1531. [CrossRef]
137. Talley, N.J.; Walker, M.M.; Jones, M.; Keely, S.; Koloski, N.; Cameron, R.; Fairlie, T.; Burns, G.; Shah, A.; Hansen, T.; et al. Letter: Budesonide for functional dyspepsia with duodenal eosinophilia- randomised, double-blind, placebo-controlled parallel-group trial. *Aliment. Pharmacol. Ther.* **2021**, *53*, 1332–1333. [CrossRef]

Article

Investigation of Sensitization Potential of the Soybean Allergen Gly m 4 by Using Caco-2/Immune Cells Co-Culture Model

Ivan V. Bogdanov *, Ekaterina I. Finkina, Daria N. Melnikova, Rustam H. Ziganshin and Tatiana V. Ovchinnikova

M. M. Shemyakin and Yu. A. Ovchinnikov Institute of Bioorganic Chemistry, Russian Academy of Sciences, Miklukho-Maklaya Str., 16/10, 117997 Moscow, Russia; finkina@mail.ru (E.I.F.); d_n_m_@mail.ru (D.N.M.); ziganshin@mail.ru (R.H.Z.); ovch@ibch.ru (T.V.O.)

* Correspondence: contraton@mail.ru; Tel.: +7-495-335-42-00

Abstract: The soybean allergen Gly m 4 is known to cause severe allergic reactions including anaphylaxis, unlike other Bet v 1 homologues, which induce mainly local allergic reactions. In the present study, we aimed to investigate whether the food Bet v 1 homologue Gly m 4 can be a sensitizer of the immune system. Susceptibility to gastrointestinal digestion was assessed *in vitro*. Transport through intestinal epithelium was estimated using the Caco-2 monolayer. Cytokine response of different immunocompetent cells was evaluated by using Caco-2/Immune cells co-culture model. Absolute levels of 48 cytokines were measured by multiplex xMAP technology. It was shown that Gly m 4 can cross the epithelial barrier with a moderate rate and then induce production of IL-4 by mature dendritic cells *in vitro*. Although Gly m 4 was shown to be susceptible to gastrointestinal enzymes, some of its proteolytic fragments can selectively cross the epithelial barrier and induce production of Th2-polarizing IL-5, IL-10, and IL-13, which may point at the presence of the T-cell epitope among the crossed fragments. Our current data indicate that Gly m 4 can potentially be a sensitizer of the immune system, and intercommunication between immunocompetent and epithelial cells may play a key role in the sensitization process.

Keywords: allergen; sensitization; Gly m 4; Caco-2/Immune cells co-culture; cytokine

Citation: Bogdanov, I.V.; Finkina, E.I.; Melnikova, D.N.; Ziganshin, R.H.; Ovchinnikova, T.V. Investigation of Sensitization Potential of the Soybean Allergen Gly m 4 by Using Caco-2/Immune Cells Co-Culture Model. *Nutrients* **2021**, *13*, 2058. https://doi.org/10.3390/nu13062058

Academic Editors: Sara Manti, Gian Luigi Marseglia and Salvatore Leonardi

Received: 15 May 2021
Accepted: 12 June 2021
Published: 16 June 2021

1. Introduction

Soy-induced allergic symptoms can be systemic and even fatal in some cases [1]. Gly m 4, belonging to the family of Bet v 1 homologues, is one of the most clinically significant allergens isolated from soybeans *Glycine max*, together with other major allergens, such as Gly m 8 [2]. The birch pollen allergen Bet v 1 is a sensitizer responsible for the development of pollen and food allergic cross-reactions. It is known that many other food Bet v 1 homologues tend to cause mild local symptoms, like oral allergy syndrome, in Bet v 1-sensitized individuals [3]. However, Gly m 4 is able to induce severe reactions in allergic patients [4]. That is why Gly m 4 has been selected as a marker allergen for severe food-allergic reactions to soy [5].

Bet v 1 homologues share common structural features including a large internal hydrophobic cavity able to accommodate different ligands *in vitro* [4]. Recently, data supporting a key role of natural ligands binding to allergens in sensitization were reported [6]. Natural ligands of the birch Bet v 1 and hazelnut Cor a 1 allergens–quercetin-3-*O*-sophoroside and quercetin-3-*O*-(2″-*O*-β-D-glucopyranosyl)-β-D-galactopyranoside, respectively, have been identified [7], and an assumption that the natural Bet v 1 ligand can play an important role in the inflammation response has been proposed [8].

The present study aims to elucidate whether the soybean Gly m 4 allergen can be a sensitizer of the immune system. Here, we used quercetin-3,4′-diglucoside (Que-3,4′-di-Glc) as a ligand structurally close to natural ligands of Bet v 1 homologues to evaluate its possible role in a sensitization process. In this investigation, we focused on a possible impact of Que-3,4′-di-Glc on gastrointestinal digestion of Gly m 4 and looked at transport of its fragments through the Caco-2 epithelial barrier and cytokine/chemokine production by immunocompetent cells.

2. Materials and Methods

2.1. Heterologous Expression of Gly m 4 in E. coli

Recombinant plasmid pET-His8-TrxL-Gly m 4 (6231 bp) was constructed by ligating the 5253 bp BglII/XhoI fragment of pET-31b(+) vector (Novagen) with an insert containing T7 promoter, the ribosome binding site, lac-operator, and the sequence encoding the fusion recombinant protein. The last one included an octahistidine tag, TrxL carrier protein (*E. coli* thioredoxin A with Met37Leu mutation), and mature Gly m 4.0101 sequence [GenBank X60043, UniProt P26987]. The culture of BL21(DE3)/pET-His8-TrxL-Gly m 4 was grown in LB medium with 100 μg/mL ampicillin and 20 mM D(+)glucose at 37 °C. When culture reached OD_{600} of 0.7, expression was induced by the addition of 0.2 mM isopropyl β-D-1-thiogalactopyranoside (Sigma-Aldrich, St. Louis, MO, USA), and incubation was continued for 5 h at 30 °C. The cells, harvested by centrifugation at 6000 *g*, were sonicated on ice in the binding buffer (50 mM Tris-HCl, pH 7.8, 0.5 M NaCl, 20 mM imidazole and 1 mM phenylmethylsulfonyl fluoride (Calbiochem, Los Angeles, CA, USA)). After centrifugation for 20 min at 25,000 *g*, the supernatant containing the soluble fusion protein was collected and loaded onto a Ni^{2+}-sepharose (GE Healthcare, Chicago, IL, USA) column, which was prewashed with the binding buffer. The fusion protein was eluted with 0.5 M imidazole and dialyzed overnight against deionized water before lyophilization. Cyanogen bromide cleavage of the fusion protein was performed by using the standard cleavage protocol in 80% trifluoroacetic acid (TFA) (Sigma-Aldrich). In order to purify the target protein from the carrier and unreacted fusion proteins, a repeated IMAC in the same buffer system was performed. Then the target Gly m 4 allergen was purified by two steps of reversed phase high performance liquid chromatography (RP-HPLC). First step was carried out on Reprosil-Pur C18-AQ, d 5 μm, 120Å, 10 × 250 mm (Dr. Maisch GmbH, Ammerbuch, Germany) column by using a linear gradient from 5 to 80% acetonitrile for 60 min with 0.1% TFA at a flow rate of 2 mL/min. Second RP-HPLC step was performed on Luna C18, d 5 μm, 120Å, 4.6 × 250 mm (Phenomenex, Torrance, CA, USA) column by using a linear gradient: 0–40% solution B (0.1% (*v*/*v*) TFA, 80% (*v*/*v*) acetonitrile) for 5 min, 40–60% B for 25 min, 60–100% B for 5 min at a flow rate of 0.7 mL/min.

Endotoxin level was evaluated by the Limulus amebocyte lysate (LAL) test using E-TOXATE Kit (Sigma-Aldrich). The endotoxin level in cell cultures with a final protein concentration was of <0.02 EU/mL.

2.2. Ligand-Binding Fluorescence Assay

Gly m 4 was tested for ligand binding by displacement of fluorescent 2-p-toluidinonap hthalene-6-sulphonate (TNS) (Sigma-Aldrich) as previously described [9]. Fluorescence experiments were performed on F-2710 spectrophotometer (Hitachi, Tokyo, Japan). Concentrations of the Gly m 4 and TNS stock solutions were determined spectrophotometrically. A base-line fluorescence of the initial sample of TNS diluted to the concentration of 4 μM with 10 mM phosphate buffer, pH 7.4, was measured by excitation at 320 nm and the emission spectrum was recorded from 330 to 550 nm. Contributions of the buffer, Gly m 4, and the ligand to the measured fluorescence were subtracted. After equilibrating TNS (4 μM) in 10 mM phosphate buffer, pH 7.4, for 2 min with gentle mixing, 2 mM Que-3,4′-di-Glc was titrated into 2 mL of 4 μM Gly m 4 solution in 1 μL aliquots. A simple binding model was employed to express the affinity of the ligand:

$$F_{obs} = \Delta F \times (1 - (IC_{50}/(IC_{50} + [L])) + F_{basiline}, \quad (1)$$

where F_{obs} is the observed fluorescence, ΔF is the fluorescence change, $F_{baseline}$ is the fluorescence at saturation, and L denotes ligand [10]. IC_{50}, ΔF, and $F_{baseline}$ are fitted as free parameters by non-linear least squares regression analysis.

2.3. Bioinformatic Approach to Study Interaction of Que-3,4′-di-Glc with Gly m 4

NMR solution structure of Gly m 4 [PDB ID: 2K7H] was used for study *in silico* of the interaction between Gly m 4 and quercetin-3,4′-diglucoside. 3D conformer of Que-3,4′-di-Glc was obtained from the PubChem database [PubChem CID: 5320835]. Preparation of Gly m 4 and Que-3,4′-di-Glc structures for molecular docking was carried out using the DockPrep tool of the UCSF Chimera v.1.4 software package (San Francisco, CA, USA) [11]. The docking box was chosen so that the whole protein molecule in the ribbon representation was entirely inside this box. Blind docking of Que-3,4′-di-Glc based on the Lamarckian genetic algorithm (LGA) into Gly m 4 molecule was carried out using the AutoDock Vina tool of the UCSF Chimera v.1.4 software [12]. The structure of the complex Gly m 4-Que-3,4′-di-Glc was visualized with the Discovery Studio Visualizer v20.1.0.19295 software [13].

2.4. Simulation of Gastrointestinal Digestion In Vitro

Gastrointestinal digestion of the recombinant Gly m 4 *in vitro* was performed as previously reported [14]. Briefly, gastric digestion was performed for 2 h using 50 ng (0.1 U) of pepsin (Sigma-Aldrich) per 1 μg of Gly m 4 in 0.05 M HCl, pH 2.0 (final protein concentration 0.05 mM). For duodenal digestion, pH of the mixture resulting from gastric digestion was adjusted to 8.0 by addition of ammonium bicarbonate. The obtained mixture was incubated for 2 h at 37 °C with 2.5 ng (0.03 U) of trypsin (Promega, Madison, WI, USA) and 10 ng (0.4×10^{-3} U) of α-chymotrypsin (Sigma-Aldrich) per 1 μg of the substrate. In order to investigate the effect of Que-3,4′-di-Glc on proteolytic cleavage of Gly m 4, the allergen was preincubated with the ligand at protein-to-ligand molar ratio of 1:4 for 10 min. The extent of proteolysis was monitored by sodium dodecyl sulfate polyacrylamide gel electrophoresis [15]. For experiments with cytokines production by cell cultures, 0.5 mg of Gly m 4 was subjected to proteolysis in a similar manner, except duodenal digestion was conducted for 1 h. The obtained digest was frozen at −70 °C. Afterwards, the frozen digest was thawed, diluted with the complete culture medium and used in the experiment.

2.5. Human Cell Lines and Cultures

Colorectal adenocarcinoma Caco-2 cell line (ATCC HTB-37) was cultured in complete DMEM/F12 (1:1) medium containing 10% fetal bovine serum (FBS, Invitrogen, Waltham, MA, USA) and 1X antibiotic-antimycotic solution (Invitrogen, Waltham, MA, USA) in a humidified CO_2-incubator (5% CO_2, 37 °C). The acute monocytic leukemia THP-1 line (ATCC TIB-202) was cultured in complete RPMI 1640 medium, containing 10% FBS, 1X antibiotic-antimycotic solution, and 0.05 mM β-mercaptoethanol, in the CO_2-incubator (5% CO_2, 37 °C).

THP-1 cells were differentiated into pro-inflammatory macrophages (MΦ1) and mature dendritic cells (mDCs) according to previously reported protocols [16,17]. Primary peripheral blood mononuclear cells (PBMC) collected from healthy donor were purchased from American Type Culture Collection (ATCC PCS-800-011), thawed, and seeded into wells of 24- and 96-well plates 2 days prior to the experiment. Two different cell subpopulations (Monocytes and T-/B-/NK-lymphocytes) were isolated from PBMC based on their adherence ability.

For growing cells, mimicking epithelial barriers *in vitro*, Caco-2 cells were seeded onto 24-well polycarbonate Millicell cell culture inserts (0.4 μm, 0.6 cm^2 surface area) (Millipore, Burlington, MA, USA), precoated with 0.2% bovine gelatin (Sigma-Aldrich), at a density of 7.5×10^4 cells/cm^2. The cells were grown for 21–29 days in complete DMEM/F12 medium with re-feeding every 2–3 days with a fresh complete medium. The integrity of the Caco-2 cell monolayer was checked by measuring the transepithelial electrical resistance (TEER) using a Millicell-ERS Voltohmmeter (Millipore, Burlington, MA, USA). Only cell monolayers with TEER > 400 Ω cm^2 (ohm per cm^2, after subtracting TEER in blank inserts without Caco-2 cells) were used in transport and cytokine production experiments between the 21st and 29th days.

2.6. Labeling of Gly m 4 with FITC

The recombinant Gly m 4 was labelled with fluorescein isothiocyanate isomer I (FITC) (Sigma-Aldrich). For this, 1.5 mg of Gly m 4 was reconstituted in 50 μL of DMSO, then added to 300 μL of the buffer for coupling (0.1 M sodium carbonate, 0.1 M sodium bicarbonate, pH 9.6) and 2.9 mg of FITC in 100 μL of DMSO. The coupling reaction was conducted for 2 h at 20 °C in the dark. In order to purify FITC-Gly m 4, the reaction mixture was loaded onto PD10 gel-filtration column (GE Healthcare, Chicago, IL, USA) previously equilibrated with distilled water.

2.7. Transport of FITC-Gly m 4 across the Caco-2 Epithelial Barrier

Transport of FITC-Gly m 4 with or without Que-3,4′-di-Glc across the Caco-2 epithelial barrier *in vitro* was performed in the transport buffer (Hank's balanced salt solution, containing 1 mM $MgCl_2$, 1 mM $CaCl_2$, and 10 mM D(+)glucose, pH 7.4). Bidirectional "apical-to-basolateral" (A→B) and "basolateral-to-apical" (B→A) transport of FITC-Gly m 4 with or without Que-3,4′-di-Glc in the transport buffer across epithelial barrier was investigated. The "apical-to-basolateral" assay was initiated by adding 0.4 mL of 2 μM Gly m 4 with or without 5 μM Que-3,4′-di-Glc to the apical (luminal) side of the monolayer and 0.7 mL of the transport buffer (pH 7.4) to the basolateral side of the monolayer. The "basolateral-to-apical" assay was performed in a similar manner, except that 0.4 mL of the transport buffer (pH 7.4) was added to the apical side and 0.7 mL of 2 μM Gly m 4 with or without 5 μM Que-3,4′-di-Glc to the basolateral (serosal) side. All solutions were pre-warmed to 37 °C before taking into the transport experiment. Transport *in vitro* across Caco-2 barriers was conducted for 90 min in 4 independent inserts for each studied transport variant (16 inserts in total).

An apparent permeability coefficient (P_{app}) was calculated for each insert according to the following equation:

$$P_{app} = (V/(A \times C_i)) \times \Delta C/\Delta t, \quad (2)$$

where V is a volume of the acceptor chamber, A is the area of the membrane insert, C_i is the initial concentration of Gly m 4, $\Delta C/\Delta t$ is the solute flux across the barrier. Uptake ratios:

$$UR = P_{app}(A \rightarrow B)/P_{app}(B \rightarrow A), \quad (3)$$

and efflux ratios:

$$ER = P_{app}(B \rightarrow A)/P_{app}(A \rightarrow B), \quad (4)$$

for Gly m 4 with or without Que-3,4′-di-Glc were calculated from averaged apparent permeability coefficients measured in 4 independent inserts. Monolayer integrity was checked by measuring TEER before and after the end of the experiment.

2.8. LC-MS/MS

Thawed gastroduodenal digest was loaded on a home-made trap column 20 × 0.1 mm packed with Inertsil ODS3 3 μm (GL Sciences, Torrance, CA, USA) in the loading buffer (2% acetonitrile, 97.9% H_2O, 0.1% trifluoroacetic acid (TFA)) at 10 μL/min flow rate and separated at 20 °C in a home-packed [18] fused-silica column 300 × 0.1 mm, packed with Reprosil Pur C18 AQ 1.9 (Dr. Maisch, Ammerbuch, Germany) and pulled into an emitter using a P2000 Laser Puller (Sutter, Atlanta, GA, USA).

Preparation of each sample from several independent basolateral chambers was performed in the presence of sodium deoxycholate as follows. The sample solution (500 μL) was added to 50 μL of the buffer solution containing 100 mM Tris, pH 8.5, 1% sodium deoxycholate (SDC). The solution was heated at 95 °C for 20 min, cooled to 20 °C, and centrifuged at 16,000 *g* for 15 min. The supernatant was transferred into a preconditioned VIVASPIN spin filter (Sartorius, Göttingen, Germany) with a 10 kDa MWCO PES membrane (cat. no. VS0102). The sample was centrifuged at 15,000 *g* until the volume reached ~50 μL.

The filtrate was collected to a clean tube and washed with 200 µL of 0.5 M NaCl. The filter was preconditioned by washing (5 min, 15,000 *g*) with 400 µL of 100 mM Tris, pH 8.5, and then with 400 µL of 100 mM Tris, pH 8.5, containing 1% SDC. The ultrafiltrate was acidified with TFA to the final concentration of 1%. The deoxycholic acid precipitate was extracted with ethyl acetate (3 × 500 µL) under active stirring. Ethyl acetate and the aqueous phase were separated by centrifugation (15,000 *g*, 4 min), upon which ethyl acetate was removed. The peptides contained in the aqueous phase were desalted on Empore SDB-RPS StageTips microcolumns (3M, St. Paul, MN, USA) as described earlier [19], with minor modifications. The samples were applied to a microcolumn (200 *g*, 10 min), and washed with a mixture of 50 µL of 1% TFA and 50 µL of ethyl acetate, then 100 µL of 0.1% TFA. The peptides were eluted with 60 µL of solution containing 5% ammonium hydroxide and 80% acetonitrile. The eluates were spin-dried and stored until the LC-MS analysis at −85 °C.

Reverse-phase chromatography was performed with an Ultimate 3000 Nano LC System (Thermo Fisher Scientific, Waltham, MA, USA), which was coupled to the Q Exactive Plus Orbitrap mass spectrometer (Thermo Fisher Scientific) via a nanoelectrospray source (Thermo Fisher Scientific). The peptides were loaded in a loading solution A (0.1% (*v*/*v*) formic acid, 2% (*v*/*v*) acetonitrile) and eluted with a linear gradient: 3–35% solution B (0.1% (*v*/*v*) formic acid, 80% (*v*/*v*) acetonitrile) for 105 min; 35–55% B for 18 min, 55–99% B for 0.1 min, 99% B during 10 min, 99–2% B for 0.1 min at a flow rate of 500 nl/min. After each gradient cycle, the column was reequilibrated with solution A (0.1% (*v*/*v*) formic acid, 2% (*v*/*v*) acetonitrile) for 10 min. MS1 parameters were as follows: 60 K resolution, 350–2000 scan range, max injection time—30 ms, AGC target—3×10^6. Ions were isolated with 1.4 *m*/*z* window, preferred peptide match and isotope exclusion. Dynamic exclusion was set to 30 s. MS2 fragmentation was carried out in the HCD mode at 17.5 K resolution with the HCD collision energy value of 29%, max injection time–80 ms, AGC target–1×10^5. Other settings: charge exclusion—unassigned, 1, >7.

2.9. *Cytokines/Chemokines/Growth Factors Production by Cell Cultures*

PBMC, T/B/NK, Monocytes, MΦ1 and mDCs were seeded into the wells of 24- and 96-well plates in the complete RPMI 1640 medium 48 h prior to the experiment. Caco-2 cells were seeded into wells of a 96-well plate 3 weeks before the experiment. Then, 24 h after the seeding of all cell lines and cultures, other than Caco-2, into 24- and 96-well plates, Millicell inserts with Caco-2 monolayers with TEER > 400 Ω cm^2 were placed into the wells of the 24-well plate, containing PBMC, T/B/NK, Monocytes, MΦ1 and mDCs cultures in their basolateral chambers. Then, media in all basolateral chambers were replaced by fresh medium, and each well of the 96-well plate or apical chamber of Caco-2-containing inserts was replaced by fresh complete RPMI 1640 medium with or without compounds under the investigation: fresh medium alone for the control wells, or fresh medium with 5 µM Gly m 4 for 24- and 96-well plates, or fresh medium with 2.5 µM Que-3,4′-di-Glc for the 96-well plate or 5 µM for apical chambers of 24-well plate inserts, or fresh medium with 5 µM Gly m 4 + 2.5 µM Que-3,4′-di-Glc for the 96-well plate or 5 µM Gly m 4 + 5 µM Que-3,4′-di-Glc for apical chambers of 24-well plate inserts, or fresh medium with Gly m 4 digest corresponding to 5 µM of the intact Gly m 4 allergen (Table 1). Cell cultures were kept in CO_2-incubator (5% CO_2, 37 °C) for 24 h. Culture supernatants from the 96-well plate and basolateral chambers of 24-well plate were collected 24 h later and stored at −70 °C degrees less than one week prior to analytes assessment. Monolayer integrity was checked by measuring TEER before and after the end of an incubation period.

Table 1. Two stimulation ways, which were applied to each cell culture, except Caco-2 line (only direct stimulation because Caco-2 cells were on the inserts).

Stimulation Way	Control	Gly m 4 Alone	Que-3,4′-di-Glc Alone	Gly m 4 + Que-3,4′-di-Glc	Gly m 4 Digest
Direct stimulation (into 96-well plate)	–	5 μM	2.5 μM	5 μM + 2.5 μM	5 μM
Transepithelial stimulation (into 24-well plate with Caco-2 inserts)	–	5 μM	5 μM	5 μM + 5 μM	5 μM

2.10. Assessment of Absolute Levels of Cytokines, Chemokines, and Growth Factors in Cell Cultures

Absolute levels of the following 48 analytes were measured by multiplex xMAP technology using the MILLIPLEX MAP Cytokine/Chemokine/Growth Factor Panel A kit (HCYTA-60K-PXBK48, Merck, Darmstadt, Germany): sCD40L, EGF, Eotaxin-1/CCL11, FGF-2/FGF-basic, Flt-3 ligand, Fractalkine/CX3CL1, G-CSF, GM-CSF, GROα, IFNα2, IFNγ, IL-1α, IL-1β, IL-1RA, IL-2, IL-3, IL-4, IL-5, IL-6, IL-7, IL-8/CXCL8, IL-9, IL-10, IL-12(p40), IL-12(p70), IL-13, IL-15, IL-17A/CTLA8, IL-17E/IL-25, IL-17F, IL-18, IL-22, IL-27, IP-10/CXCL10, MCP-1/CCL2, MCP-3/CCL7, M-CSF, MDC/CCL22, MIG/CXCL9, MIP-1α/CCL3, MIP-1β/CCL4, PDGF-AA, PDGF-AB/BB, RANTES/CCL5, TGFα, TNFα, TNFβ, and VEGF-A. Multiplex-based assay was carried out using MAGPIX system (Merck, Darmstadt, Germany) with the xPONENT 4.2 software (Merck) in accordance with the manufacturer's instruction. Final analysis was performed with the MILLIPLEX Analyst v5.1 software (Merck). Measurements were performed twice for each sample.

2.11. Statistics

Absolute values of the analytes in cell culture supernatants were normalized using a logarithmic transformation by LN function [20] in Microsoft Excel. LN-transformed values were used for comparing the analyte levels in control and experimental samples by unpaired two-sample *t*-test using Statistica v.10.0.1011.0 analytic package (StatSoft, Tulsa, OK, USA). The normality of P_{app} coefficients distribution was assessed using Shapiro-Wilk (*W*-test) and Lilliefors-corrected Kolmogorov-Smirnov tests. P_{app} coefficients for Gly m 4 alone and Gly m 4 with Que-3,4′-di-Glc in both A→B and B→A directions were compared by one-way ANOVA using Statistica v.10.0.1011.0.

3. Results

3.1. Gly m 4 Is Able to Bind Quercetin-3,4′-Diglucoside

Previously, it has been shown that Bet v 1 homologues can bind different ligands [21]. To substantiate this finding, we tested Gly m 4 binding with Que-3,4′-di-Glc. At the first stage, the binding of Gly m 4 with Que-3,4′-di-Glc was investigated by means of blind molecular docking. The AutoDock Vina software calculated 10 conformations of the ligand with affinity energy ranges between −8.1 and−6.8 kcal mol^{-1}. These two best conformations differed from the others and had lower affinity energies −8.1 and −7.9 kcal mol^{-1}, while the rest 8 conformations had affinity energies in the range between −7.3 and −6.8 kcal mol^{-1}. In the case of these two most energetically favorable conformations, Que-3,4′-di-Glc is located completely inside the hydrophobic cavity of Gly m 4 (purple) or partially immersed in the cavity near its entrance (green) (Figure 1A). To confirm the ability of Gly m 4 to bind Que-3,4′-di-Glc, we used an extrinsic fluorescent probe, TNS (Figure 1B). TNS is highly fluorescent when bound to the hydrophobic cavity of the protein and competed with lipid molecules for binding with the allergen.

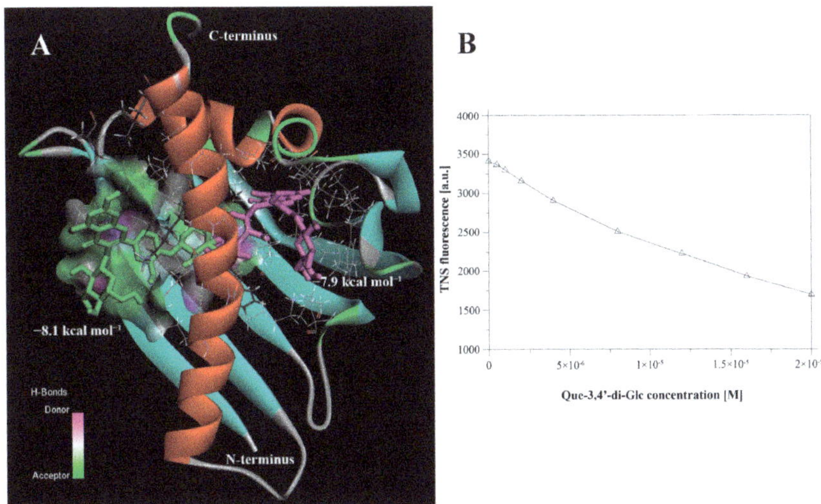

Figure 1. (**A**) Gly m 4 complexed with two best Que-3,4′-di-Glc conformations calculated by means of molecular docking. Green conformation has affinity energy -8.1 kcal mol^{-1}, magenta conformation -7.9 kcal mol^{-1}. (**B**) Que-3,4′-di-Glc binding to Gly m 4. Titration of 4 μM Gly m 4 and 4 μM TNS with Que-3,4′-di-Glc in 10 mM phosphate buffer, pH 7.4, 25 °C. Fitting the data to Equation (1), IC_{50} yields Kd of 30.2 ± 0.2 μM for Que-3,4′-di-Glc. TNS was excited at 320 nm; the emission at 423 nm for Que-3,4′-di-Glc is displayed.

3.2. *Gly m 4 Can Effectively cross the Caco-2 Epithelial Barrier*

It is known that polarized Caco-2 monolayers represent a reliable model for studies of absorption of drugs and other compounds after oral intake in humans [22]. Proteins labelled with fluorescent probes are widely used for an assessment of permeability of Caco-2 monolayers mimicking the gastrointestinal epithelial barrier [23,24]. Here, we used the FITC-labelled recombinant allergen Gly m 4 for an assessment of "apical-to-basolateral" (A→B, absorptive) and "basolateral-to-apical" (B→A, secretory) bidirectional transport of the allergen across the Caco-2 epithelial barrier. After 90 min around 0.3 μg of Gly m 4 was transported from apical to the basolateral side of the monolayer. Apparent permeability A→B coefficients (P_{app}) for Gly m 4 alone measured in 4 independent inserts were within the range of 2–4.5 × 10^{-6} cm/s (Figure 2), which predicts a moderate transepithelial absorption of the Gly m 4 allergen in human gut. The established relationship between the *in vivo* absorption of drugs in humans and P_{app} values allows to correlate P_{app} values ~1–10 × 10^{-6} cm/s with a 20–70% absorption in gut which could be expected in humans, however, in the case of protein allergens it is still to be validated [25].

The uptake ratios were of 1.88±0.022 for Gly m 4 and Gly m 4 with Que-3,4′-di-Glc which suggests active transport, e.g., endocytosis, of the allergen across the Caco-2 epithelial barrier [26]. At the same time, in both cases much lower P_{app} in the B→A direction was observed. The efflux ratios (ER) of 0.532 ± 0.006 for Gly m 4 and Gly m 4 with Que-3,4′-di-Glc argued for not involving active efflux pumps shown to be present in Caco-2 cells, such as P-glycoprotein (ABCB1), ABCG2 or ABCC2, in the Gly m 4 transport across Caco-2 epithelial barrier. The presence of 5 μM Que-3,4′-di-Glc had no significant effect (p = 0.13) on the Gly m 4 permeability across the Caco-2 epithelial barrier in both directions (Figure 2). Neither Gly m 4 nor Que-3,4′-di-Glc affected the monolayer integrity which was checked by measuring of TEER following the end of the experiment.

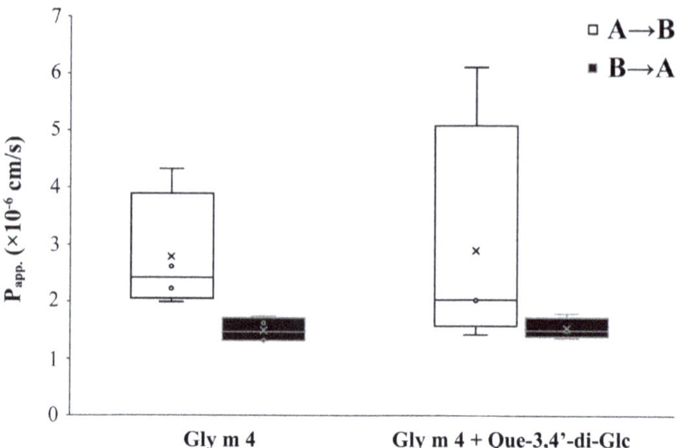

Figure 2. Bidirectional "apical-to-basolateral" (A→B) and "basolateral-to-apical" (B→A) transport of Gly m 4 across the Caco-2 epithelial barrier.

3.3. *Gly m 4 Is Susceptible to Proteolytic Cleavage Mimicking Gastrointestinal Digestion In Vitro*

It is known that Bet v 1 homologues, such as apple Mal d 1, hazelnut Cor a 1, and celery Api g 1 allergens, are rapidly degraded by pepsin during gastric digestion and have moderate susceptibility to trypsin [27]. However, experimental data on the susceptibility of Gly m 4 to gastrointestinal enzymes were not available untill now. Here, Gly m 4 also showed a high susceptibility to cleavage with pepsin mimicking the gastric digestion which resulted in a ~9 kDa fragment that was completely digested by subsequent cleavage with duodenal enzymes *in vitro* (Figure 3). Preincubation of Gly m 4 with Que-3,4′-di-Glc did not affect the rate of gastrointestinal digestion.

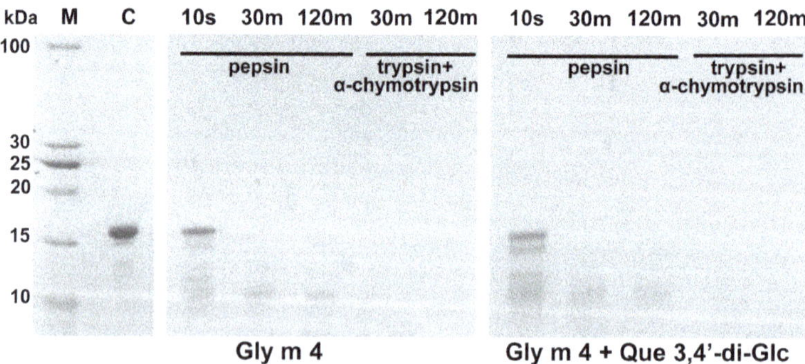

Figure 3. SDS-PAGE analysis of proteolytic cleavage mimicking gastrointestinal digestion *in vitro* of Gly m 4 with or without Que-3,4′-di-Glc. M—molecular mass standards; C—an intact Gly m 4 (control); 10 s, 30 m, 120 m—the allergen fragmentation after incubation with pepsin mimicking gastric digestion during 10 s, 30 min, 120 min, respectively, and digests after the subsequent allergen incubation with the mixture of trypsin and α-chymotrypsin during 30 min and 120 min, respectively.

We also studied whether resulting proteolytic fragments of the allergen can cross the gastrointestinal epithelial barrier. Gly m 4 proteolytic fragments were analyzed by LC-MS/MS in samples taken from an apical side before and from a basolateral side 24 h after loading the resulted digest onto the insert with the Caco-2 monolayer. Eight clusters of the fragments, covering almost all the amino acid sequence of Gly m 4, have been found after simulated gastroduodenal digestion *in vitro*, which revealed the key sites of the gastrointestinal proteolysis (Figure 4, white background). However, only proteolytic fragments including amino acid residues 4–18, 37–54, 59–77, 91–99, and 104–136 were identified in basolateral chambers after passing of the digest across the Caco-2 monolayer (Figure 4, gray background). T-cell epitopes of birch allergen Bet v 1 have been previously reported by proliferation of short-term allergen-specific T-cell lines (TCLs) derived from a large number of patients (n = 57) with associated food allergy [28]. 7 distinct T cell-activating regions within Bet v 1 were recognized by at least 18% of the studied TCLs [28]. Regions, homologous to two out of these 7 T-cell epitopes, were found among the crossed Gly m 4 proteolytic fragments (Figure 4, in black frames). At the same time, the region, homologous to the immunodominant T-cell epitope Bet v $1_{142–156}$, which was recognized by 61% of the TCLs, has not been identified in basolateral chambers among the crossed fragments of Gly m 4 (Figure 4, in red frame). Interestingly, the entire region 142–156 homologous to the immunodominant T-cell epitope Bet v 1 was not found after simulated gastroduodenal digestion of Gly m 4 *in vitro*. Among all the identified fragments V_{66}LHKIESIDE$_{75}$ had the highest absorptive capacity (Table 2, Figure 5). As Gly m 4 proved to be susceptible to proteolytic enzymes, its digest after cleavage mimicking gastrointestinal digestion *in vitro* was used in the cytokines/chemokines production experiment.

Table 2. Gly m 4 proteolytic fragments identified by LC-MS/MS in basolateral chambers after passing them across the Caco-2 monolayer. The table is ranked based on highest-to-lowest peak area from the top to the bottom.

Peptide	−10lgP	Molecular Mass	ppm	*m/z*	RT	Peak Area
V_{66}LHKIESIDE$_{75}$	33.48	1181.6292	0.9	591.8224	38.81	1.50×10^8
H_{68}KIESIDE$_{75}$	22.38	969.4767	0.6	485.7459	23.92	2.07×10^7
L_{67}HKIESIDE$_{75}$	27.23	1082.5608	1.6	542.2885	33.2	1.76×10^7
E_{59}DGETKF$_{65}$	23.74	824.3552	2.1	413.1857	28.72	1.51×10^7
N_{42}VEGN(+0.98)GGPGTIKK$_{54}$	33.21	1270.6517	1.9	636.3344	21.63	1.05×10^7
Y_{119}ETKGDAEPNQDELKTGK$_{136}$	48.5	2021.9541	3.5	1011.988	27.7	9.75×10^6
K_{38}SVENVEGN(+0.98)GGPGTIKK$_{54}$	45.73	1713.8896	2.6	857.9543	25.6	9.26×10^6
S_{39}VENVEGN(+0.98)GGPGTIKK$_{54}$	34.66	1585.7947	2.2	529.6067	30.94	8.78×10^6
N_{42}VEGN(+0.98)GGPGTIK$_{53}$	27.83	1142.5568	2.5	572.2871	28.81	3.99×10^6
T_{116}VKYETKGDAEPNQDELKTGK$_{136}$	35.94	2350.1653	4	784.3989	29.17	2.32×10^6
S_{39}VENVEGNGGPGTIKK$_{54}$	31.09	1584.8107	3.1	793.415	29.33	2.30×10^6
F_{37}KSVENVEGN(+0.98)GGPGTIKK$_{54}$	39.96	1860.9581	−0.7	466.2465	38.79	2.07×10^6
N_{108}(+0.98)GGSAGKL$_{115}$	22.71	703.35	1.2	352.6827	23.07	1.86×10^6
N_{42}VEGNGGPGTIKK$_{54}$	26.72	1269.6677	1	424.2303	18.9	1.47×10^6
K_{38}SVENVEGN(+0.98)GGPGTIK$_{53}$	41.56	1585.7947	3.3	793.9072	32.29	1.31×10^6
N_{42}VEGNGGPGTIK$_{53}$	24.76	1141.5728	1.8	571.7947	26.04	1.01×10^6
V_{40}ENVEGN(+0.98)GGPGTIK$_{53}$	22.11	1370.6677	3.3	686.3434	35.55	9.07×10^5
E_{120}TKGDAEPNQDELKTGK$_{136}$	46.16	1858.8907	1.4	930.4539	22.67	8.76×10^5
E_{41}NVEGN(+0.98)GGPGTIK$_{53}$	30.24	1271.5994	1.8	636.8081	30.92	8.76×10^5
V_{66}LHKIESID$_{74}$	20.04	1052.5865	1.6	527.3014	36.84	8.58×10^5
N_{128}QDELKTGK$_{136}$	22.69	1031.5247	3.4	516.7714	15.08	7.29×10^5
V_{40}ENVEGN(+0.98)GGPGTIKK$_{54}$	24.92	1498.7627	3.2	750.391	27.38	7.26×10^5
T_4FEDEINSPVAPATL$_{18}$	36.32	1602.7777	3.6	802.399	97.49	5.50×10^5
G_{123}DAEPNQDELKTGK$_{136}$	35.57	1500.7056	1.7	751.3613	25.69	5.18×10^5
S_{39}VENVEGNGGPGTIK$_{53}$	23.14	1456.7157	5.1	729.3688	38.54	5.00×10^5
P_{91}DTAEKITF$_{99}$	30.33	1020.5128	2.5	511.265	64.51	3.78×10^5
S_{39}VENVEGN(+0.98)GGPGTIK$_{53}$	27.36	1457.6997	1.6	729.8583	40.55	3.75×10^5
V_{66}LHKIESIDEANL$_{78}$	33.81	1479.7932	0.9	740.9045	64.96	2.07×10^5
E_{120}TKGDAEPNQDELK$_{133}$	34.3	1572.7267	1.7	787.3719	24.41	1.43×10^5

Peptide—amino acid sequences of the peptides determined by the PEAKS search workflow. A modified residue is followed by a pair of parentheses enclosing the modification mass. −10lgP—the peptide −10lgP score. Ppm—the precursor mass error, calculated as $10^6 \times$ (precursor mass − peptide mass)/peptide mass. RT—retention time (elution time).

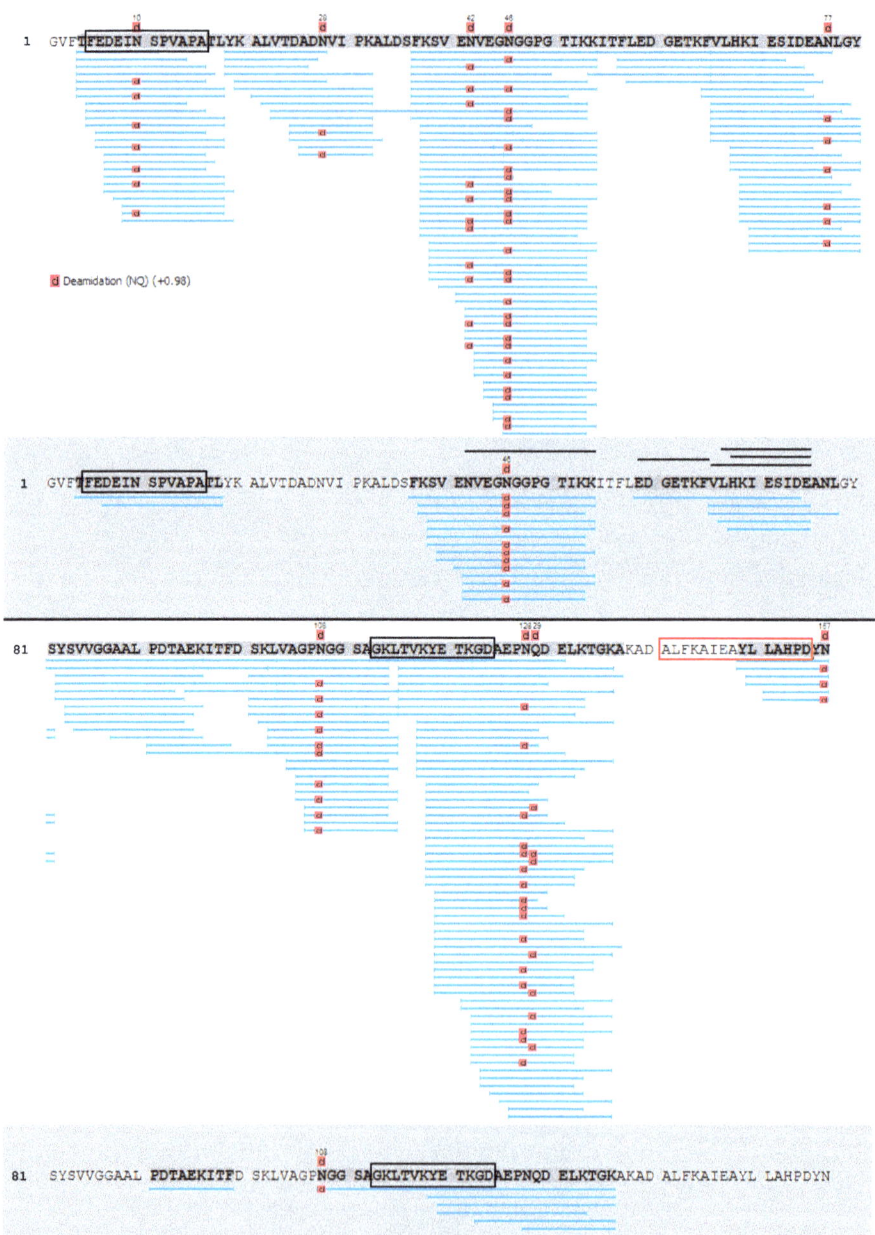

Figure 4. The fragments of Gly m 4 resulted from proteolysis mimicking gastrointestinal digestion *in vitro* (white background), and those ones crossed the Caco-2 epithelial barrier (grey background). Amino acid regions highlighted in bold are denoted sequence, covered by identified LC-MS/MS fragments. Regions, homologous to T-cell epitopes of Bet v 1 able to induce proliferation of Bet v 1-specific T-cell lines from more than 10 patients out of 57 ones, are framed in black, and the immunodominant T-cell epitope of Bet v 1 is framed in red [28]. The most abundant proteolytic fragments of Gly m 4 identified in basolateral chambers are marked with black lines above its sequences.

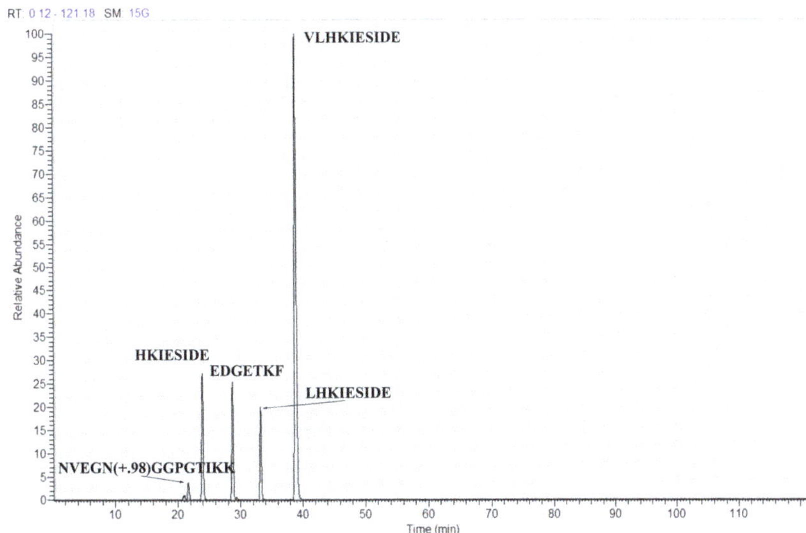

Figure 5. LC-MS/MS chromatogram of the 5 most abundant proteolytic fragments of Gly m 4 identified into basolateral chambers after crossing of the digest through Caco-2 barrier.

3.4. Intercommunication between Epithelial and Immune Cells Changes Cytokine Production in Response to the Intact Gly m 4 and Its Proteolytic Fragments

A pro-monocytic THP-1 line has proved to be a reliable model for obtaining and studying macrophages [16] and mature dendritic cells [17]. Here, we used THP-1 line to differentiate into pro-inflammatory macrophages (MΦ1) and mature dendritic cells (mDCs). Differentiated cells were observed by light microscopy with a CKX41 microscope (Olympus, Tokyo, Japan) equipped with a C310 digital camera and shown to have proper morphological properties (Figure 6). Macrophage MΦ1 polarization was assessed by expression of several classical pro-inflammatory MΦ1 markers, such as cytokines TNFα, IL-1β, IL-6, and chemokine CXCL10 (IP-10) [16]. The analyte levels in MΦ1-containing control well were of 224.9 pg/mL, 119.7 pg/mL, 367.4 pg/mL, and 111.2 pg/mL for TNF-α, IL-1β, IL-6, and CXCL10, respectively, while for THP-1-derived mDCs they were of 37 pg/mL, 14.4 pg/mL, 1.89 pg/mL, and 8.13 pg/mL, respectively (Table S1).

The experiment on production of the analytes by different cells included two parts. The first one was focused on the study of cytokines, chemokines, and growth factors production by Caco-2 cells (Figure 7A) and various immunocompetent cells (Figure 7B) in response to direct stimulation with Gly m 4, Que-3,4′-di-Glc, the Gly m 4 and Que-3,4′-di-Glc combination, or the Gly m 4 digest resulted from proteolytic cleavage mimicking gastroduodenal digestion *in vitro*. This part of the experiment was carried out into the wells of 96-well plate.

The second part of the experiment consisted of an evaluation of cytokines/chemokines/growth factors production by various immunocompetent cells at the basolateral side of the Caco-2 epithelial barrier after the same studied compounds crossed the barrier from an apical side in a 24-well plate (Figure 7C). Both parts of the experiment were performed at the same time in parallel plates.

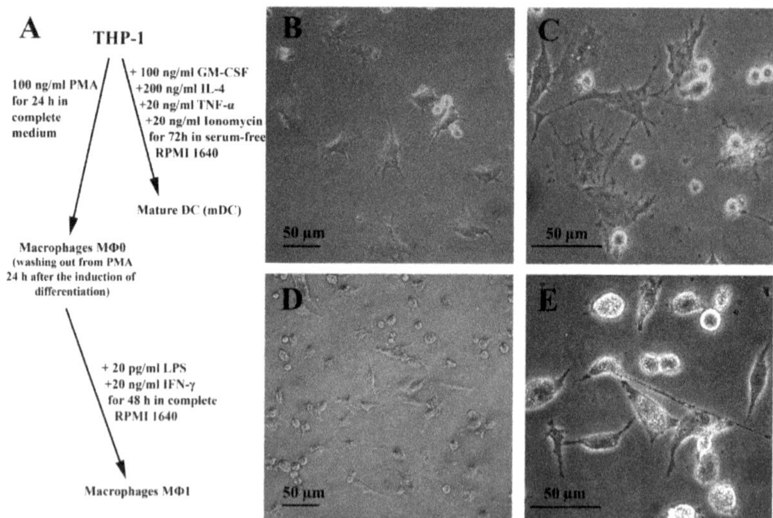

Figure 6. (**A**) Scheme of the differentiation protocol to obtain mature dendritic cells (mDCs) and pro-inflammatory macrophages (MΦ1) from THP-1 cells; Light microscopy of mDCs (**B**,**C**) and MΦ1 (**D**,**E**) cells under magnification of 200× and 400×, respectively.

It was shown that both Gly m 4 and its gastroduodenal digest induced production of pro-inflammatory chemokine CXCL10/IP-10 by Caco-2 cells from 16.87 pg/mL in control wells to 44.53 and 43.76 pg/mL in sample wells, respectively (Figure 8A). In Caco-2/immune cells co-culture system Gly m 4 increased production of several pro-inflammatory cytokines and chemokines: RANTES/CCL5 by Monocytes (from 161.32 to 541.41 pg/mL, $p < 0.005$), IL-1α by T/B/NK (from 8.4 to 47.16 pg/mL, $p < 0.005$), IL-6 by PBMC (from 3.76 to 15.02 pg/mL, $p < 0.01$) and T/B/NK (from 130.98 to 769.54 pg/mL, $p < 0.005$), MIP-1β/CCL4 (from 67.89 to 123.8 pg/mL, $p < 0.005$), MIG/CXCL9 (from 80.51 to 114.68 pg/mL, $p < 0.005$), GM-CSF (from 101.52 to 266.73 pg/mL, $p < 0.01$) and TNFα (from 37 to 66.12 pg/mL, $p < 0.005$) by mDCs, as well as anti-inflammatory cytokines: IL-4 by mDCs (from 137.49 to 349.49 pg/mL, $p<0.001$), IL-10 by T/B/NK (from 242.35 to 452.2 pg/mL, $p < 0.01$), and IL-13 by PBMC (from 13.14 to 36.50 pg/mL, $p < 0.005$). Production of the above mentioned pro-inflammatory cytokines and chemokines was not a result of nonspecific activation by residual LPS, which was checked by comparing IL-1β levels in control (12 pg/mL) and Gly m 4-containing (16.61 pg/mL) wells with monocytes in case of direct stimulation, as human monocytes represent a highly pyrogen-sensitive culture. At the same time, in the co-culture system Gly m 4 digest induced increased production of mainly anti-inflammatory cytokines: IL-1 receptor antagonist by mDCs (from 635.14 to 870.41 pg/mL, $p < 0.01$), IL-5 (from 0.48 to 0.76 pg/mL, $p < 0.05$) and IL-10 (from 242.35 to 426.28 pg/mL, $p < 0.05$) by T/B/NK, as well as IL-13 by PBMC (from 13.14 to 27.38 pg/mL, $p < 0.05$) and MΦ1 (from 38.97 to 50.77 pg/mL, $p < 0.001$).

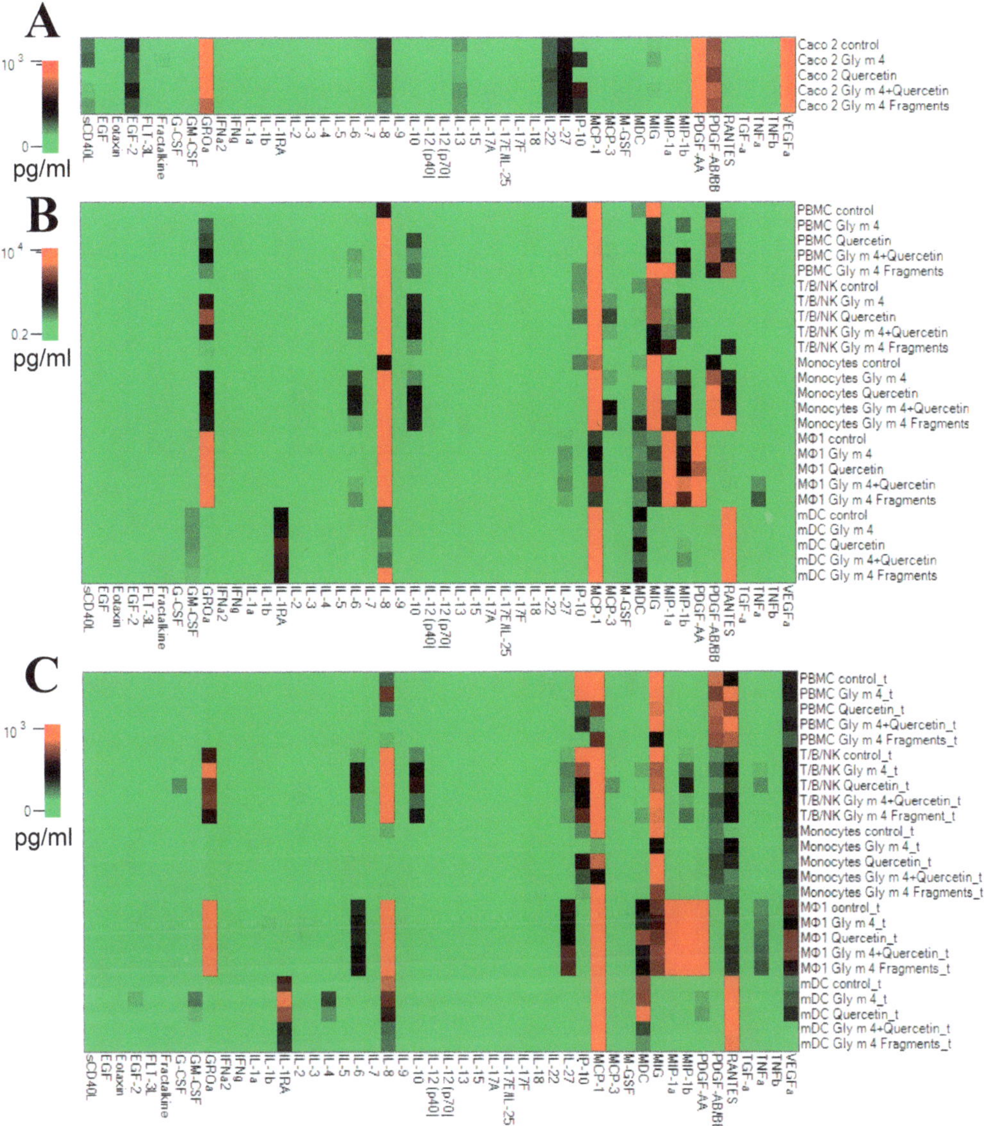

Figure 7. Heat map represented profiles of cytokines/chemokines/growth factors production by (**A**) Caco-2 cell line and (**B**) PBMC, T/B/NK, Monocytes, THP-1-derived MΦ1, and mDCs in response to the direct stimulation, or (**C**) to the transepithelial stimulation by incubation of the same cultures in basolateral chambers of 24-well Millicel inserts after the compounds passed across the Caco-2 epithelial barriers from an apical side.

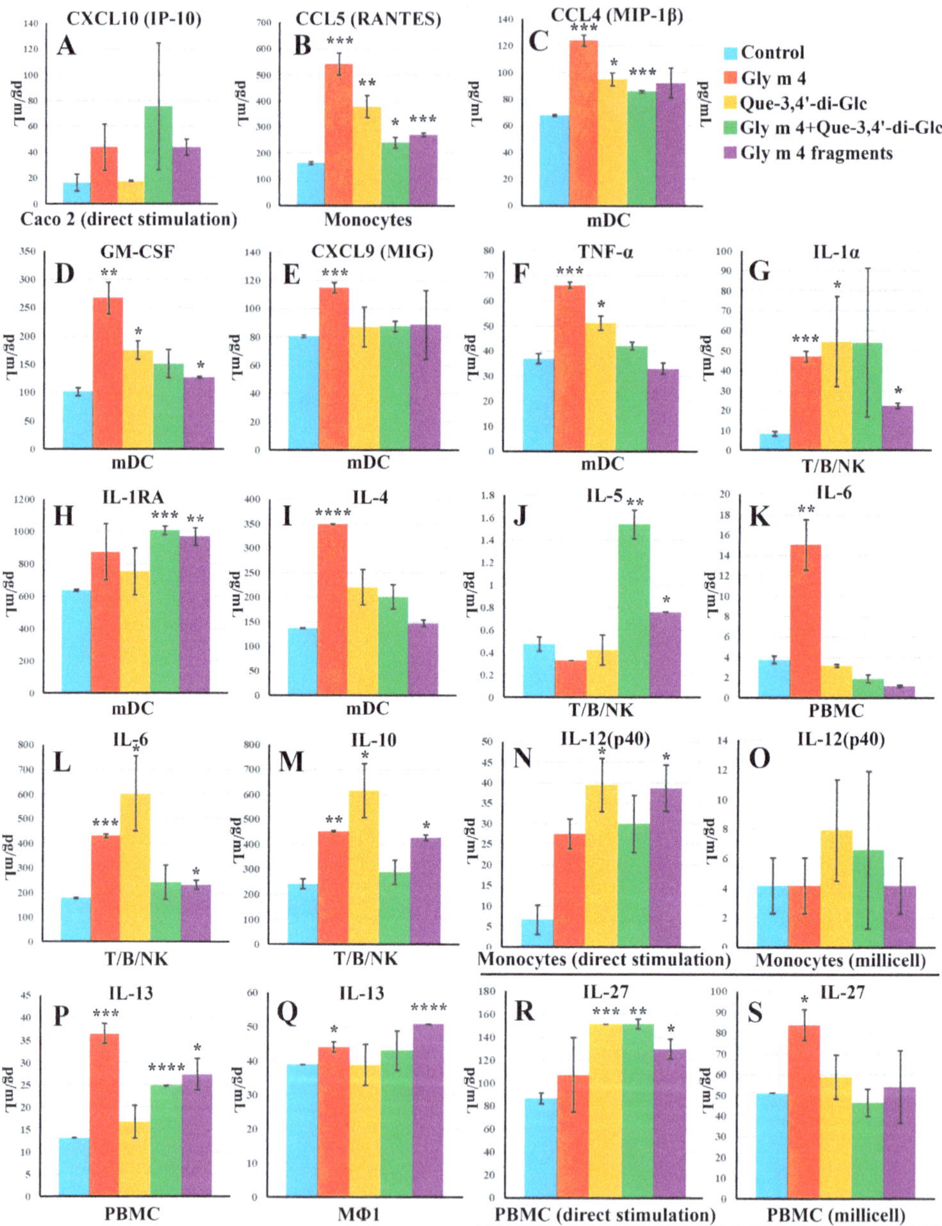

Figure 8. Cytokines and chemokines production by the Caco-2 cell line, Monocytes and PBMC (the direct stimulation) or by PBMC, T/B/NK, Monocytes, MΦ1 and mDCs cultures from basolateral chambers (the transepithelial stimulation). (**A**,**N**,**R**) represent cytokines and chemokines, which were assessed after direct stimulation; (**B**–**M**,**O**–**Q**,**S**) represent cytokines and chemokines, which were assessed after transepithelial stimulation. Error bars represent standard deviation between two technical replications (or biological replications for the direct stimulation). Significance levels are: * $p < 0.05$, ** $p < 0.01$, *** $p < 0.005$, **** $p < 0.001$.

Four patterns of production of cytokines/chemokines/growth factors were observed when comparing the 2 stimulation ways. The first one took place when both direct and transepithelial stimulations did not result in a significant effect on production of analytes. The second one occurred when a level of the same analyte was found to be increased by both stimulation ways compared to control wells. For example, after incubation of PBMC with Gly m 4, the concentration of IL-8 was elevated from 1613 to 7757 pg/mL in case of the direct stimulation and from 290 to 677 pg/mL in case of the transepithelial stimulation. In case of the stimulation with Gly m 4, the same production pattern was observed for IL-10 and IL-1α production by T/B/NK; for IL-6 production by PBMC and T/B/NK; for IL-1α production by Monocytes. The third production pattern was observed when a level of the same analyte was increased by the direct stimulation but remained unchanged when the transepithelial stimulation was carried out. In case of the stimulation with Gly m 4, this pattern was observed for G-CSF production by PBMC; for MCP-3 and IL-1β production by T/B/NK; for IL-6, IL-12(p40) and TNF-α production by Monocytes; for MIP-1α production by PBMC and Monocytes (Figure 8).

The Gly m 4 digest induced the strongest production of TNF-α by PBMC, Monocytes and MΦ1 cultures among all studied compounds by the direct stimulation; however, this effect was not observed in case of the transepithelial stimulation (Table S1). The last production pattern was observed when production of the same analyte remained unchanged after the direct stimulation but was increased in response to the transepithelial stimulation. For instance, in case of the transepithelial stimulation by Gly m 4, this pattern was observed for sCD40L, EGF-2, IL-1α, and IL-1β production by MΦ1; for IL-13 production by PBMC and MΦ1; and for IL-4, G-CSF, and GM-CSF production, by mDCs (Figure 7B,C). These changes in cytokines/chemokines production can be explained by communication between epithelial and immunocompetent cells in the Caco-2/immune cells system by soluble factors. Thus, using the Caco-2/immune cells co-culture model in study of food allergens makes the obtained results more reliable in context of the situation *in vivo*.

4. Discussion

The soybean allergen Gly m 4 is known to cause severe allergic reactions including anaphylaxis, unlike other Bet v 1 homologues, which mainly induce local allergic reactions [4]. This work aimed to elucidate mechanisms underlying the unique properties of this allergen. Complexity of the mucosal immune system causes difficulties in mimicking its properties *in vitro*, but a co-culture system makes it possible to elaborate mechanisms involved in communication between epithelial and immune system cells. The co-culture of Caco-2/immune cells was used in current study as a model system [29].

The Gly m 4 allergen can effectively pass across the Caco-2 polarized monolayer which was used in current study as a simplified model of the intestinal epithelium, and then can activate immunocompetent cells. Sensitization effects of Gly m 4 were interpreted according to data obtained by using the Caco-2/Immune cells co-culture as follows. First, passing of the allergen across the Caco-2 barrier activates epithelial cells that resulted in production of pro-inflammatory chemokine CXCL10/IP-10 (Figure 8A), which could activate and recruit leukocytes such as T-cells, eosinophils, monocytes, and NK-cells [29]. CXCL10 was previously proposed to play a role in chronic allergic inflammation [30]. Then, the invaded Gly m 4 might force dendritic cells (DCs), localized underneath the epithelium, to produce CCL4/MIP-1β (Figure 8C), CXCL9/MIG (Figure 8E), which predominantly mediated lymphocytic infiltration to the focal sites, as well as to promote TNF-α production (Figure 8F). These cytokines, apparently, may cause an allergic inflammation in the human gut after Gly m 4 invasion. The increase of CCL4/MIP-1β and CCL5/RANTES obtained in the current research was comparable with their observed increase in biological fluids during allergic inflammation *in vivo* [31,32]. The Gly m 4-induced inflammation might be sustained via IL-1α (Figure 8G), IL-6 (Figure 8K,L) and CCL5/RANTES (Figure 8B), produced by lymphocytes recruited through CXCL10 and CXCL9, and via GM-CSF produced by DCs

(Figure 8D). Later on, recruited lymphocytes might inhibit pro-inflammatory stimuli by IL-13 (Figure 8P,Q) [33]. At the same time, Gly m 4-stimulated mDCs produce a key Th2-associated cytokine—IL-4 at a high level (Figure 8I). Being activated by the Gly m 4 allergen, IL-4-producing mDCs apparently move to lymph nodes for the antigen presentation to naïve T-cells with subsequent differentiation of the latter into allergen-specific Th2-lymphocytes. The suggested mechanism coincides with the assumption that the Gly m 4 allergen is potentially able to induce sensitization in a lymph node after absorption in the human gut.

However, in this regard, a key question arises: whether Gly m 4 can reach the intestinal epithelium in its intact immunogenic form? It is known that binding of allergens with ligands may affect their properties and allergenicity. In our study, Que-3,4′-di-Glc had no significant effect on gastrointestinal digestion of Gly m 4, its transport across epithelium and production of cytokines, except IL-5 produced by T/B/NK cells. However, this cytokine by itself can apparently induce only eosinophilic inflammation [34]. Nevertheless, the sensitizing capacity of food allergens may depend, on the one hand, on their susceptibility towards proteolysis in the digestive tract and, on the other hand, on the abundance of T-cell epitopes with immunostimulating capacity [35]. Gly m 4 proved to be susceptible to gastrointestinal enzymes, which provided an evidence that it hardly could reach the intestinal epithelium *in vivo* in an intact form. However, the question is still open. Although some proteolytic fragments resulting from the gastrointestinal digestion of Gly m 4 are capable to pass through the epithelial barrier, they failed to induce IL-4 by mDCs and most of the abovementioned pro-inflammatory stimuli, except CXCL10/IP-10 produced by Caco-2 cells. Instead, proteolytic fragments of Gly m 4 able to cross the Caco-2 monolayer were found to be responsible for a strong anti-inflammatory response by induction of IL-1 receptor antagonist (IL-1RA), IL-5, IL-10, and IL-13 by MΦ1 or lymphocytes (Figure 8H,J,M,P,Q) that can be recruited by CXCL10. These anti-inflammatory stimuli could be responsible for the differentiation of naïve Th0 cells into Th2 after presentation of the crossed fragments by macrophages or dendritic cells in the human gut. Interestingly, Gly m 4 digest induced production of Th2-suppressing cytokines IL-12(p40) and IL-27 by direct stimulation, while transepithelial stimulation did not result in production of these cytokines (Figure 8N,O,R,S) [36,37]. It still remains unclear whether the observed suppression is induced by those Gly m 4 fragments which cannot cross the Caco-2 monolayer or intermediated by epithelial-immune cells communication. Strong anti-inflammatory response of immunocompetent cells toward those Gly m 4 fragments which could pass across the Caco-2 monolayer might speak for the presence of the T-cell epitope among the crossed fragments. Amino acid residues of several crossed fragments correspond to previously mapped T-cell epitopes of the birch Bet v 1 but not its immunodominant epitope Bet v $1_{142-156}$ [28]. However, Gly m 4 may contain its own T-cell epitopes. Our current data argue for an assumption that the Gly m 4 allergen can potentially act as a sensitizer of the immune system (Figure 9); thus, study of a cohort of Gly m 4-sensitized patients without sensitization to Bet v 1 is of special interest.

To verify our finding, mice models of sensitization by an intact Gly m 4 and its proteolytic fragments through oral administration have to be used in further investigation.

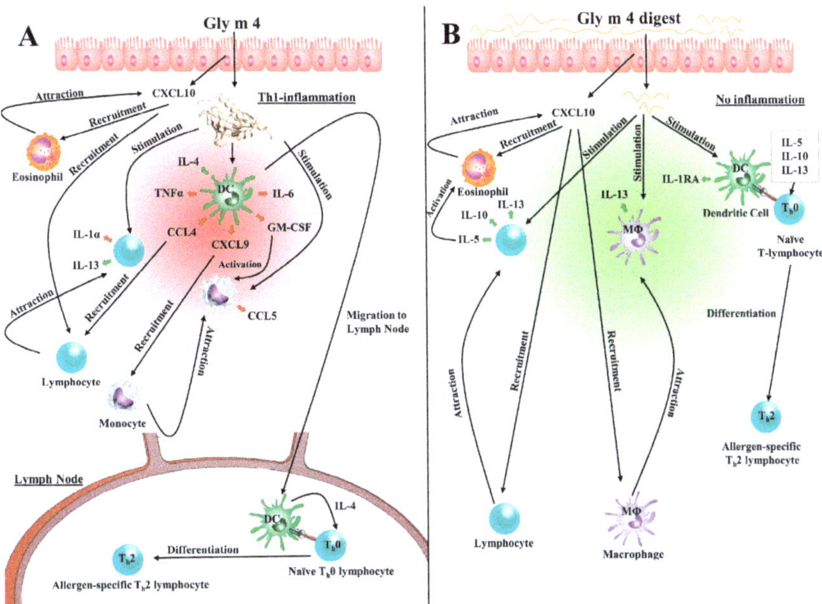

Figure 9. Two possible ways of sensitization with the soybean allergen Gly m 4 based on experimental data obtained in the current study. (**A**) If Gly m 4 can reach the human intestine in intact form, it is able to cross the intestinal epithelium and induce the production of several pro-inflammatory stimuli by different cells, as well as high levels of IL-4 by DCs. Gly m 4-stimulated DCs migrate to the lymph node from intestinal Th1-inflammatory site and induce there IL-4-dependent differentiation of naïve Th0 lymphocytes into allergen-specific Th2 lymphocytes. (**B**) The second proposed mechanism of sensitization is mediated by proteolytic fragments of Gly m 4 that resulted after gastrointestinal digestion. Some of the proteolytic fragments are able to cross the intestinal epithelium and induce the production of several anti-inflammatory stimuli, namely, (IL-1RA), IL-5, IL-10, and IL-13, which leads to the differentiation of naïve Th0 lymphocytes into allergen-specific Th2 lymphocytes in the intestinal lamina propria.

Supplementary Materials: The following are available online at https://www.mdpi.com/article/10.3390/nu13062058/s1, Table S1: Absolute levels of 48 cytokines, chemokines and growth factors assessed by multiplex xMAP technology.

Author Contributions: The study was designed by I.V.B.; plasmid construction, heterologous expression in *E. coli*, molecular docking, transport across Caco-2 monolayer, experiments with cell cultures and assessment of cytokine production were done by I.V.B.; ligand binding study was performed by D.N.M.; gastrointestinal digestion *in vitro* was performed by E.I.F.; LC-MS/MS was carried out by R.H.Z.; all authors analyzed obtained experimental data; I.V.B. drafted the manuscript; T.V.O. revised the manuscript critically and approved and prepared it for publication; T.V.O. obtained the research funding supporting this study and supervised the whole project. All authors have read and agreed to the published version of the manuscript.

Funding: This work was supported by the Russian Science Foundation (project No. 20-45-05002).

Institutional Review Board Statement: Not applicable.

Informed Consent Statement: Not applicable.

Data Availability Statement: All data generated and analyzed during this study are included in this published article and its supplementary information file "Table S1".

Conflicts of Interest: The authors declare no conflict of interest.

References

1. Kosma, P.; Sjölander, S.; Landgren, E.; Borres, M.P.; Hedlin, G. Severe reactions after the intake of soy drink in birch pollen-allergic children sensitized to Gly m 4. *Acta Paediatr.* **2011**, *100*, 305–306. [CrossRef] [PubMed]
2. Ebisawa, M.; Brostedt, P.; Sjölander, S.; Sato, S.; Borres, M.P.; Ito, K. Gly m 2S albumin is a major allergen with a high diagnostic value in soybean-allergic children. *J. Allergy Clin. Immunol.* **2013**, *132*, 976–978.e5. [CrossRef] [PubMed]
3. Treudler, R.; Simon, J.-C. Pollen-related food allergy: An update. *Allergo J. Int.* **2017**, *26*, 273–282. [CrossRef]
4. Finkina, E.I.; Melnikova, D.N.; Bogdanov, I.V.; Ovchinnikova, T.V. Plant pathogenesis-related proteins PR-10 and PR-14 as components of innate immunity system and ubiquitous allergens. *Curr. Med. Chem.* **2017**, *24*, 1772–1787. [CrossRef]
5. Berneder, M.; Bublin, M.; Hoffmann-Sommergruber, K.; Hawranek, T.; Lang, R. Allergen chip diagnosis for soy-allergic patients: Gly m 4 as a marker for severe food-allergic reactions to soy. *Int. Arch. Allergy Immunol.* **2013**, *161*, 229–233. [CrossRef] [PubMed]
6. Tordesillas, L.; Cubells-Baeza, N.; Gómez-Casado, C.; Berin, C.; Esteban, V.; Barcik, W.; O'Mahony, L.; Ramirez, C.; Pacios, L.F.; Garrido-Arandia, M.; et al. Mechanisms underlying induction of allergic sensitization by Pru p 3. *Clin. Exp. Allergy* **2017**, *47*, 1398–1408. [CrossRef]
7. Jacob, T.; von Loetzen, C.S.; Reuter, A.; Lacher, U.; Schiller, D.; Schobert, R.; Mahler, V.; Vieths, S.; Rösch, P.; Schweimer, K.; et al. Identification of a natural ligand of the hazel allergen Cor a 1. *Sci. Rep.* **2019**, *9*, 8714. [CrossRef]
8. Von Loetzen, C.S.; Hoffmann, T.; Hartl, M.J.; Schweimer, K.; Schwab, W.; Rösch, P.; Hartl-Spiegelhauer, O. Secret of the major birch pollen allergen Bet v 1: Identification of the physiological ligand. *Biochem. J.* **2014**, *457*, 379–390. [CrossRef]
9. Mogensen, J.E.; Wimmer, R.; Larsen, J.N.; Spangfort, M.D.; Otzen, D.E. The major birch allergen, Bet v 1, shows affinity for a broad spectrum of physiological ligands. *J. Biol. Chem.* **2002**, *277*, 23684–23692. [CrossRef]
10. Cheng, Y.; Prusoff, W.H. Relationship between the inhibition constant (K1) and the concentration of inhibitor which causes 50 per cent inhibition (I50) of an enzymatic reaction. *Biochem. Pharmacol.* **1973**, *22*, 3099–3108.
11. Pettersen, E.F.; Goddard, T.D.; Huang, C.C.; Couch, G.S.; Greenblatt, D.M.; Meng, E.C.; Ferrin, T.E. UCSF Chimera—A visualization system for exploratory research and analysis. *J. Comput. Chem.* **2004**, *25*, 1605–1612. [CrossRef]
12. Trott, O.; Olson, A.J. AutoDock Vina: Improving the speed and accuracy of docking with a new scoring function, efficient optimization, and multithreading. *J. Comput. Chem.* **2009**, *31*, 455–461. [CrossRef]
13. Dassault Systèmes BIOVIA. *Discovery Studio Visualizer*; v20.1.0.19295; Dassault Systèmes: San Diego, CA, USA, 2020.
14. Bogdanov, I.V.; Shenkarev, Z.O.; Finkina, E.I.; Melnikova, D.N.; Rumynskiy, E.I.; Arseniev, A.S.; Ovchinnikova, T.V. A novel lipid transfer protein from the pea Pisum sativum: Isolation, recombinant expression, solution structure, antifungal activity, lipid binding, and allergenic roperties. *BMC Plant Biol.* **2016**, *16*, 1–17. [CrossRef] [PubMed]
15. Backman, L.; Persson, K. The No-Nonsens SDS-PAGE. *Methods Mol. Biol.* **2018**, *1721*, 89–94. [CrossRef] [PubMed]
16. Genin, M.; Clement, F.; Fattaccioli, A.; Raes, M.; Michiels, C. M1 and M2 macrophages derived from THP-1 cells differentially modulate the response of cancer cells to etoposide. *BMC Cancer* **2015**, *15*, 577. [CrossRef] [PubMed]
17. Berges, C.; Naujokat, C.; Tinapp, S.; Wieczorek, H.; Höh, A.; Sadeghi, M.; Opelz, G.; Daniel, V. A cell line model for the differentiation of human dendritic cells. *Biochem. Biophys. Res. Commun.* **2005**, *333*, 896–907. [CrossRef]
18. Kovalchuk, S.I.; Jensen, O.N.; Rogowska-Wrzesinska, A. FlashPack: Fast and simple preparation of ultrahigh-performance capillary columns for LC-MS. *Mol. Cell Proteom.* **2019**, *18*, 383–390. [CrossRef] [PubMed]
19. Rappsilber, J.; Mann, M.; Ishihama, Y. Protocol for micro-purification, enrichment, pre-fractionation and storage of peptides for proteomics using StageTips. *Nat. Protoc.* **2007**, *2*, 1896–1906. [CrossRef]
20. Schmidt, F.M.; Weschenfelder, J.; Sander, C.; Minkwitz, J.; Thormann, J.; Chittka, T.; Mergl, R.; Kirkby, K.C.; Faßhauer, M.; Stumvoll, M.; et al. Inflammatory cytokines in general and central obesity and modulating effects of physical Activity. *PLoS ONE* **2015**, *10*, e0121971. [CrossRef] [PubMed]
21. McBride, J.K.; Cheng, H.; Maleki, S.J.; Hurlburt, B.K. Purification and Characterization of Pathogenesis Related Class 10 Panallergens. *Foods* **2019**, *8*, 609. [CrossRef] [PubMed]
22. Lea, T. Caco-2 cell line. In *The Impact of Food Bioactives on Health: In Vitro and Ex Vivo Models*; Springer International Publishing: Berlin/Heidelberg, Germany, 2015; pp. 103–111.
23. Liu, X.; Zheng, S.; Qin, Y.; Ding, W.; Tu, Y.; Chen, X.; Wu, Y.; Yanhua, L.; Cai, X. Experimental Evaluation of the Transport Mechanisms of PoIFN-α in Caco-2 Cells. *Front. Pharmacol.* **2017**, *8*, 781. [CrossRef] [PubMed]
24. Tordesillas, L.; Gómez-Casado, C.; Garrido-Arandia, M.; Murua-García, A.; Palacín, A.; Varela, J.; Konieczna, P.; Cuesta-Herranz, J.; Akdis, C.A.; O'Mahony, L.; et al. Transport of Pru p 3 across gastrointestinal epithelium—An essential step towards the induction of food allergy? *Clin. Exp. Allergy* **2013**, *43*, 1374–1383. [CrossRef] [PubMed]
25. Press, B. Optimization of the Caco-2 permeability assay to screen drug compounds for intestinal absorption and efflux. *Methods Mol. Biol.* **2011**, *763*, 139–154. [PubMed]
26. Hubatsch, I.; Ragnarsson, E.G.E.; Artursson, P. Determination of drug permeability and prediction of drug absorption in Caco-2 monolayers. *Nat. Protoc.* **2007**, *2*, 2111–2119. [CrossRef]
27. Schimek, E.M.; Zwölfer, B.; Briza, P.; Jahn-Schmid, B.; Vogel, L.; Vieths, S.; Ebner, C.; Bohle, B. Gastrointestinal digestion of Bet v 1-homologous food allergens destroys their mediator-releasing, but not T cell-activating, capacity. *J. Allergy Clin. Immunol.* **2005**, *116*, 1327–1333. [CrossRef]

28. Jahn-Schmid, B.; Radakovics, A.; Lüttkopf, D.; Scheurer, S.; Vieths, S.; Ebner, C. Bohle, B. Bet v $1_{142\text{-}156}$ is the dominant T-cell epitope of the major birch pollen allergen and important for cross-reactivity with Bet v 1-related food allergens. *J. Allergy Clin. Immunol.* **2005**, *116*, 213–219. [CrossRef]
29. Kleiveland, C.R. Co-culture Caco-2/Immune Cells. In *The Impact of Food Bioactives on Health: In Vitro and Ex Vivo Models*; Springer International Publishing: Berlin/Heidelberg, Germany, 2015; pp. 197–205.
30. Tworek, D.; Kuna, P.; Młynarski, W.; Górski, P.; Pietras, T.; Antczak, A. MIG (CXCL9), IP-10 (CXCL10) and I-TAC (CXCL11) concentrations after nasal allergen challenge in patients with allergic rhinitis. *Arch. Med. Sci.* **2013**, *9*, 849–853. [CrossRef]
31. König, K.; Klemens, C.; Eder, K.; Nicoló, M.S.; Becker, S.; Kramer, M.F.; Gröger, M. Cytokine profiles in nasal fluid of patients with seasonal or persistent allergic rhinitis. *Allergy Asthma Clin. Immunol.* **2015**, *11*, 26. [CrossRef]
32. Holgate, S.T.; Bodey, K.S.; Janezic, A.; Frew, A.J.; Kaplan, A.P.; Teran, L.M. Release of RANTES, MIP-1 alpha, and MCP-1 into asthmatic airways following endobronchial allergen challenge. *Am. J. Respir. Crit. Care Med.* **1997**, *156*, 1377–1383. [CrossRef]
33. Bieber, T. Interleukin-13: Targeting an underestimated cytokine in atopic dermatitis. *Allergy Eur. J. Allergy Clin. Immunol.* **2020**, *75*, 54–62. [CrossRef]
34. Nagase, H.; Ueki, S.; Fujieda, S. The roles of IL-5 and anti-IL-5 treatment in eosinophilic diseases: Asthma, eosinophilic granulomatosis with polyangiitis, and eosinophilic chronic rhinosinusitis. *Allergol. Int.* **2020**, *69*, 178–186. [CrossRef] [PubMed]
35. Pekar, J.; Ret, D.; Untersmayr, E. Stability of allergens. *Mol. Immunol.* **2018**, *100*, 14–20. [CrossRef]
36. Wills-Karp, M. IL-12/IL-13 axis in allergic asthma. *J. Allergy Clin. Immunol.* **2001**, *107*, 9–18. [CrossRef] [PubMed]
37. Chen, X.; Deng, R.; Chi, W.; Hua, X.; Lu, F.; Bian, F.; Gao, N.; Li, Z.; Pflugfelder, S.C.; de Paiva, C.S.; et al. IL-27 signaling deficiency develops Th17-enhanced Th2-dominant inflammation in murine allergic conjunctivitis model. *Allergy Eur. J. Allergy Clin. Immunol.* **2019**, *74*, 910–921. [CrossRef] [PubMed]

Article

LTP Allergy Follow-Up Study: Development of Allergy to New Plant Foods 10 Years Later

Diana Betancor [1], Alicia Gomez-Lopez [1], Carlos Villalobos-Vilda [1], Emilio Nuñez-Borque [2], Sergio Fernández-Bravo [2], Manuel De las Heras Gozalo [1,3], Carlos Pastor-Vargas [3,4,*], Vanesa Esteban [2,3,5,†] and Javier Cuesta-Herranz [1,3,†]

1 Department of Allergy, Instituto de Investigación Sanitaria Hospital Universitario Fundación Jiménez Díaz (IIS-FJD, UAM), 28040 Madrid, Spain; diana13_b@hotmail.com (D.B.); alicia.gomezl@quironsalud.es (A.G.-L.); c.villalobosvilda@gmail.com (C.V.-V.); mheras@fjd.es (M.D.l.H.G.); j.cuestaherranz@gmail.com (J.C.-H.)
2 Department of Immunology, Instituto de Investigación Sanitaria Hospital Universitario Fundación Jiménez Díaz (IIS-FJD, UAM), 28040 Madrid, Spain; enbli7@hotmail.com (E.N.-B.); sergio.fernandezb@quironsalud.es (S.F.-B.); vesteban@fjd.es (V.E.)
3 RETIC ARADyAL, Instituto de Salud Carlos III, 28029 Madrid, Spain
4 Department of Biochemistry and Molecular Biology, Universidad Complutense de Madrid, 28040 Madrid, Spain
5 Faculty of Medicine and Biomedicine, Alfonso X El Sabio University, Villanueva de la Cañada, 28691 Madrid, Spain
* Correspondence: cpasto01@ucm.es
† These authors have contributed equally to this work.

Citation: Betancor, D.; Gomez-Lopez, A.; Villalobos-Vilda, C.; Nuñez-Borque, E.; Fernández-Bravo, S.; De las Heras Gozalo, M.; Pastor-Vargas, C.; Esteban, V.; Cuesta-Herranz, J. LTP Allergy Follow-Up Study: Development of Allergy to New Plant Foods 10 Years Later. *Nutrients* **2021**, *13*, 2165. https://doi.org/10.3390/nu13072165

Academic Editors: Sara Manti, Gian Luigi Marseglia and Salvatore Leonardi

Received: 15 May 2021
Accepted: 22 June 2021
Published: 24 June 2021

Publisher's Note: MDPI stays neutral with regard to jurisdictional claims in published maps and institutional affiliations.

Abstract: Introduction: Allergy to nonspecific lipid transfer protein (nsLTP) is the main cause of plant-food allergy in Spain. nsLTPs are widely distributed in the plant kingdom and have high cross-reactivity but extremely variable clinical expression. Little is known about the natural evolution of this allergy, which complicates management. The objective of this study was to assess the development of allergy to new plant foods in nsLTP-sensitized patients 10 years after diagnosis. Methods: One hundred fifty-one patients showing specific IgE to nsLTP determined by ISAC (Thermofisher) were included. After clinical workup (i.e., anamnesis, skin test, and challenge when needed), these patients were divided into two groups: 113 patients allergic to one or more plant food (74.5%) and 38 patients not allergic to any plant food (25.1%). Ten years later, a telephone interview was conducted to check whether patients had developed additional allergic reactions to plant foods. Results: Ten years after diagnosis, 35 of the 113 (31%) plant-food-allergic patients sensitized to nsLTP reported reactions to new, previously tolerated plant foods, mainly *Rosaceae/Prunoideae* fruits and nuts followed by vegetables, *Rosacea/Pomoideae* fruits, legumes, and cereals. Five out of 38 (13.2%) patients previously sensitized to nsLTP but without allergy to any plant food had experienced allergic reactions to some plant food: two to *Rosaceae/Prunoideae* fruits, two to *Rosaceae/Prunoideae* fruit and nuts, and one to legumes. Conclusion: Patients sensitized to nsLTP developed allergic reactions to other plant foods, mainly *Rosaceae-Prunoideae* fruits and nuts. This was more frequent among plant-food-allergic patients than among those who had never had plant-food allergy.

Keywords: nsLTP; plant-food allergy; Pru p 3; peach; nut; *Rosaceae* fruit; ISAC

1. Introduction

Food allergy affects around 0.3% to 5.6% of the population, showing substantial geographical variation in prevalence and in terms of the culprit food [1]. Allergy to plant foods is the most common food allergy among older children and adults [1].

Nonspecific lipid transfer proteins (nsLTPs) are small, highly stable and conserved molecules involved in the plant defense against fungi and bacteria [2,3]. nsLTPs are found in high concentrations in the epidermal tissues of fruits and are the main allergens of fruits of the Rosaceae family. In addition, allergenic nsLTPs have been found in nuts, seeds, vegetables, pollen, and latex from *Hevea brasiliensis* [4]. Allergy to nsLTP involves several taxonomically unrelated plant-derived foods and heterogeneous sensitization profiles and can trigger severe systemic reactions. It has been reported to be responsible for a large number of plant-food-induced anaphylactic reactions in southern Europe [5–8].

Fruits of the Rosaceae family are the most frequently involved foods in allergic reactions among nsLTP-allergic patients [9]. Allergy to nsLTP occurs predominantly in the Mediterranean area (Spain, Italy, etc.) [5,6], although it has also been reported in other areas such as Australia [10] and China [11]; in contrast, nsLTP allergy is a rare finding in northern and central Europe [7,12] and the USA [5].

Patients with allergy to nsLTP exhibit considerable clinical heterogeneity, as some react to only one food (often peach), while others may experience symptoms to multiple nsLTPs from allergenic sources that are not taxonomically related and do not follow a defined pattern [13]. The extreme variability of nsLTP allergy in terms of the culprit plant food and the clinical expression of the allergy is still unexplained. Strict plant-food avoidance diets are sometimes recommended due to the unknown clinical course, though these measures have a significant negative impact on patients' quality of life and nutrition. Little is known about the natural evolution of this syndrome.

The management of patients allergic to nsLTP is complex and poses a major challenge for both allergists and patients. The problem lies in the fact that LTP is a panallergen, meaning that it is a ubiquitous protein that is widely distributed in plant foods and has wide cross-reactivity and a highly variable clinical expression, sometimes eliciting life-threatening reactions. Further complicating this situation is the possibility that patients sensitized to homologous nsLTPs of other plant foods can progress over time from mere sensitization (without clinical expression) to severe or even fatal allergic reactions, which has clear implications for the dietary recommendations given to nsLTP-allergic patients.

On the other hand, the LEAP study revealed that early food introduction can prevent the onset of allergy [14], the STOP study showed that induction of tolerance can halt allergy [15], and Pru p 3 SLIT induces an improvement not only in peach allergy but also acts upon other relevant food allergens causing severe reactions, such as peanut or tree nuts [16–18]. These facts could also have important implications for dietary recommendations for LTP-allergic patients. In this respect, intake of plant foods containing cross-reactive proteins that the patient tolerates and to which he/she is sensitized might improve LTP allergy in the future.

The management of such patient heterogeneity continues to challenge the expertise of allergists despite the study by Asero et al. [19] and the recommendations given by the EAACI Task Force on nsLTP Allergy Across Europe [4].

The aim of this study was to assess the development of allergy to new plant foods in nsLTP-sensitized patients over 10 years. The results reinforce key points that inform decision-making related to the management of this heterogenous and complex type of allergy.

2. Materials and Methods

2.1. Study Design

One hundred fifty-one out of 164 patients sensitized to nsLTP as determined by ImmunoCAP™ ISAC (Thermo Fisher Scientific, Uppsala, Sweden) performed during 2009–2011 in the allergy department of Fundación Jiménez Díaz (Madrid, Spain) were included in the study. Thirteen patients (7.9%) were excluded because they did not respond to the follow-up phone call, refused to answer, or did not give consent to participate in the study. After a clinical study (2009–2011) in real-life conditions (i.e., anamnesis, skin test or specific IgE and challenge test when needed), patients were divided into 2 groups: 113 patients allergic to plant food (74.8%) and 38 non-food-allergic patients (25.1%). Once a patient was diagnosed with an allergy to a plant food, they were advised to avoid the food in question and continue eating those they tolerated. Ten years later, in 2020–2021, a telephone interview was conducted to determine whether the patients had developed new allergic reactions to previously tolerated plant foods (Figure 1).

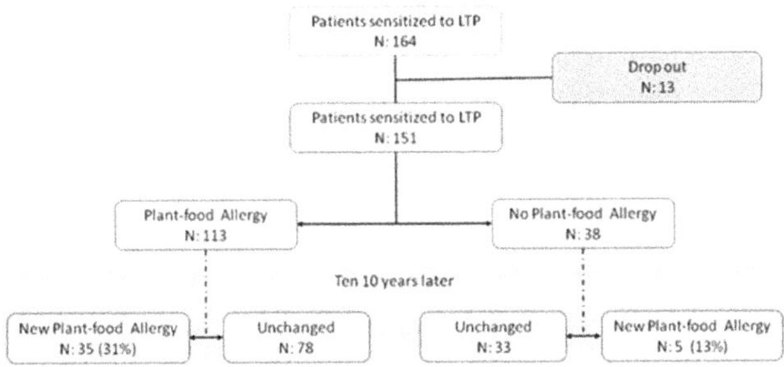

Figure 1. Study flow chart.

2.2. Specific IgE to LTP

All patients showed specific IgE to at least one nsLTP (Pru p 3, Cor a 8, Art v 3 before 2011 and Pru p 3, Cor a 8, Art v 3, Ara h 9, Jug r 3, Ole e 7, Pla a 3 after 2011) measured by ImmunoCAP™ ISAC following manufacturer recommendations. Results were expressed in ISU (ISAC standardized units).

2.3. Study Variables

On the one hand, data were collected at the time of diagnosis, including demographic and clinical characteristics of the patients; sensitization to common allergens (defined as at least 1 positive skin prick test or serum-specific IgE to common allergens); associated rhinitis or asthma; specific IgE to different nsLTPs, profilins, and PR-10 proteins as determined by ImmunoCAP™ ISAC microarray; and data related to plant-food allergy such as the plant food eliciting allergy and symptoms of the reactions, which were categorized into local symptoms, systemic symptoms, and anaphylaxis (two or more organs involved).

On the other hand, after the telephone interview, data collected at the time of diagnosis were re-evaluated to distinguish those patients who had developed allergy to nsLTP-related foods during follow-up so as to search for characteristics that could predict progression to allergy in nsLTP syndrome.

3. Statistical Analysis

Statistical analysis was performed with SPSS (SPSS Inc., Chicago, IL, USA). Qualitative variables were expressed as percentages and confidence intervals were calculated at 95%. For quantitative variables, means and standard deviation (SD) were calculated, and for specific IgE results, medians and 25th (Q1) and 75th (Q3) percentiles were given. A χ^2 test was used to compare frequencies. Values were considered significant at a p-value of less than 0.05.

4. Results

4.1. Patient Characteristics

One hundred fifty-one patients sensitized to nsLTP were selected and analyzed for this study. Thirty-eight patients were asymptomatic upon nsLTP-related food exposure and 113 patients were allergic to nsLTP-related plant foods. Characteristics of the patients are shown in Table 1.

Table 1. Demographic and clinical characteristics of patients sensitized to nsLTP (baseline data).

	Food Allergy Group (*n*:113)	Non-Food-Allergy Group (*n*:38)
Sex, male	59 (52.21%)	23 (60.52%)
Age (years) (mean, SD)	31.67 (14.36)	30.87 (12.5)
Previous atopy1 history	94 (83.18%)	28 (73.68%)
Allergic rhinitis	92 (81.41%)	29 (76.32%)
Asthma	58 (51.32%)	18 (47.37%)
Sensitization to common allergens	102 (90.26%)	35 (92.10%)
Pollen sensitization	101 (89.38%)	34 (89.47%)
Grass	92 (81.41%)	29 (76.31%)
Olive	64 (56.64%)	20 (52.63%)
Cypress	58 (51.33%)	18 (47.36%)
Platanus tree	65 (57.52%)	19 (50.00%)
Mugwort	72 (63.72%)	15 (39.47%)
Animal sensitization	47 (41.59%)	22 (57.89%)
Dust mite sensitization	28 (24.77%)	11 (28.95%)
Mold sensitization	21 (18.58%)	8 (21.05%)
Grass pollen immunotherapy	58 (51.33%)	24 (63.16%)
Panallergen sensitization	71 (62.83%)	22 (57.89%)
Profilin	38 (33.63%)	9 (23.68%)
Bet v 1	19 (16.81%)	10 (26.32%)

4.1.1. Plant-Food-Allergy Group (Baseline Data)

One hundred thirteen patients sensitized to nsLTP were allergic to plant food before the start of the follow-up period. Characteristics of the patients are shown in Table 1.

The frequency of sensitization to different nsLTPs (ISAC) at the beginning of the study was as follows: 87.6% to Pru p 3 (n = 99 out 113 patients tested), with a median positive test value of 3.3 ISU (1.15–5.5 Q1–Q3); 80.6% to Pla a 3 (n = 29/36), median 0.8 ISU (0.6–2.2); 75.9% to Jug r 3 (n = 23/29), median 1 ISU (0.55–1.75); 64.3% to Art v 3 (n = 72/112), median 1.6 ISU (0.6–3.2); 53.1%to Cor a 8 in (n = 59/111), median 1.3 ISU (0.7–3.13); 59.3% to Ara h 9 (n = 16/27), 0.9 ISU (0.5–1.6); and 30.8% to Ole e 7 (n = 8/26), median 1.3 ISU (0.4–2.3). These results are shown in Figure 2.

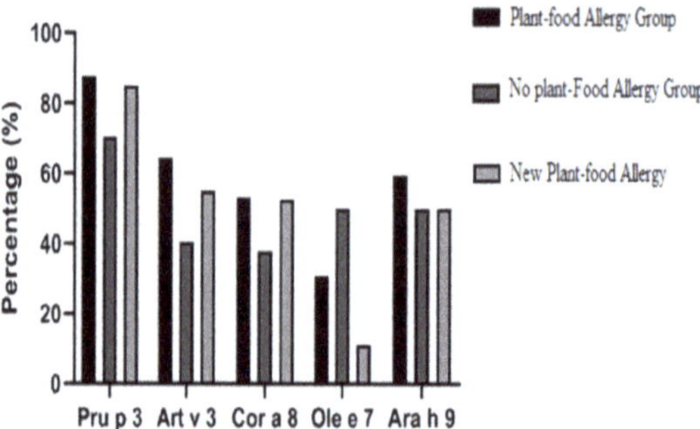

Figure 2. Percentage of sensitization to several nsLTPs in different groups of patients.

Foods eliciting allergy in this patient group are listed in Table 2. Peach and nut were the most frequently involved plant foods (70 and 54 patients, respectively) followed by apple (37 patients) and hazelnut and peanut (36 patients in both). Cofactors were associated in 11 patients (9.7%), 4 of whom had anaphylaxis.

Table 2. Plant foods involved in the allergic reactions of nsLTP allergy group (n = 113) at baseline. Results are shown in number of patients.

Plant Food	Food Allergy	Oral Tolerance	Not Known
Nuts	**77**	**30**	**6**
Walnut	54	21	38
Hazelnut	36	41	36
Peanut	36	41	36
Almond	29	48	36
Sunflower seed	15	55	43
Fruits	**95**	**18**	**0**
***Rosaceae* fruits**			
Peach	70	37	6
Peach (peel only)	35	37	41
Apricot	22	42	49
Cherry	18	45	50
Strawberry	8	58	47
Plum	20	46	47
***Pomoideae* fruits**			
Apple	37	50	26
Apple (peel only)	54	50	9
Pear	15	90	8
Other fruits			
Kiwi	18	67	28
Banana	11	67	35
Legumes	**12**	**78**	**23**
Lentil	7	95	11
Bean	4	80	29
Soybean	2	109	2
Chickpea	1	100	12
Vegetables	**23**	**89**	**1**
Tomato	12	99	2
Lettuce and derivates	10	84	19
Corn	3	91	19
Eggplant	2	78	35
Cauliflower	2	65	46
Seed	**9**	**64**	**40**
Mustard	8	54	51
Sesame	1	69	43
Cereal (Wheat)	**2**	**111**	**0**

Eighty-five patients (75.2%) developed systemic symptoms, 23 of whom (20%) experienced an anaphylactic reaction. The plant foods responsible for the anaphylactic reactions were as follows: nuts (39.1%), *Rosacea/Prunoideae* fruits (21.7%), *Rosacea/Pomoideae* fruits (17.4%), lettuce (13.0%), and legumes (8.7%). Sensitization to profilin in this anaphylaxis subgroup was 26% (6 patients) and 30.4% were sensitized to PR-10 (7 patients). The rate of anaphylaxis in profilin-sensitized patients was 7.8%, and 22% of profilin-negative patients presented anaphylaxis.

4.1.2. Non-Food-Allergy Group (Baseline Data)

At the start of the study, 38 out of 151 patients sensitized to any nsLTP had not experienced any plant-food allergy. These patients made up the group of non-plant-food-allergic patients. Characteristics of the patients are shown in Table 1.

The most common allergens identified through specific IgE (ISAC) were the following: Pru p 3 in 70.3% (26/37) of patients, median 1.85 ISU (0.8–3.4); Art v 3 in 40.5% (15/37), median value 0.8 ISU (0.6–1.45); Cor a 8 in 37.84% (14/37), 1.05 ISU (0.6–1.4). These results are shown in Figure 2.

There was no statistically significant difference in specific IgE to different nsLTPS between the 2 groups (nsLTPS-allergy group and the non-food-allergy group). However, there was a statistically significant difference between the percentage of positive patients between the 2 groups to Pru p 3 ($p = 0.012$) and Art v 3 ($p = 0.013$), but not to Ara h 9, Cor a 8, Jug r 3, Ole e 7, and Pla a 3. Comparisons are shown in Figure 2.

4.2. Characteristics of Patients Not Sensitized to Pru p 3

Twenty-five of 156 patients had negative specific IgE to Pru p 3: 14/113 patients (12.4%) from the plant-food-allergy group and 11/38 patients (28.9%) from the group without food allergy.

Focusing on the plant-food-allergy group, 5 out of 14 (20%) patients had a systemic reaction, one of which (4%) was an anaphylactic reaction. Despite the negative value for Pru p 3, 9 patients had allergy-related symptoms to peach. Sensitization to nsLTP among Pru p 3-negative patients was as follows: 7 patients (50%) monosensitized to Art v 3, 2 patients to Pla a 3, 1 patient to Cor a 8, and 1 patient to Ara h 9. The other 3 patients were polysensitized with Art v 3 involved in all cases: 1 patient to Cor a 8 and Art v 3, 1 patient to Jug r 3 and Art v 3, and 1 patient to Ara h 9, Jug r 3, Art v 3, and Pla a 3. These results are shown in Table 3.

Table 3. Characteristics of patients non-sensitized to Pru p 3 ($n = 25$).

	Plant-Food-Allergy Group ($n = 14$)	Non-Plant-Food-Allergy Group ($n = 11$)
Sensitization to nsLTP		
Art v 3	7 (50%)	6 (54.54%)
Ara h 9	1 (7.14%)	0 (0%)
Cor a 8	1 (7.14%)	3 (27.27%)
Pla a 3	2 (14.28%)	0 (0%)
Jug r 3	0 (0%)	1 (9.09%)
Cor a 8 + Art v 3	1 (7.14%)	1 (9.09%)
Jug r 3 + Art v 3	1 (7.14%)	0 (0%)
Ara h 9 + Jug r 3 + Art v 3 + Pla a 3	1 (7.14%)	0 (0%)
Panallergen sensitization		
Profilin	6 (42.9%)	1 (9.1%)
PR10	3 (21.4%)	2 (18.2%)
Allergy to new plant food (clinical progression)	5 (35.7%)	1 (9.1%)

Nine patients non-sensitized to Pru p 3 were positive for other panallergens: 6 to profilin and 3 to the PR-10 protein family.

In the group without food allergy, 5 patients (55.4%) were monosensitized to Art v 3, 3 patients to Cor a 8, and 1 patient to Jug r 3. The other patient was sensitized to both Cor a 8 and Art v 3.

Three patients were sensitized to other panallergens: 2 patients to allergens belonging to the PR-10 protein family and 1 patient to profilin.

4.3. Follow-Up Study: Allergy to New Plant Foods over the Years

Forty out of 151 patients sensitized to nsLTP (26.5%; 95% CI 20–34%) developed symptoms of allergy to new (previously tolerated) plant foods during the follow-up period. Patients in this group had a mean age of 31.4 years (range 2 to 62 years) with a higher prevalence of female patients (60%). In addition, 95% of patients had a history of atopy and 90.2% had current atopy.

The frequency of sensitization to different nsLTPs (ISAC) at the beginning of the study was as follows: 85% of 40 patients were sensitized to Pru p 3, with a median value of 2.4 ISU (1.04–4.7 Q1–Q3); 50% of 10 patients to Ara h 9, median 0.9 ISU (0.4–1); 52.5% of 40 patients to Cor a 8, median 1.2 ISU (0.6–0.8); 72.7% of 11 patients to Jug r 3, median 1 ISU (0.9–2.8); 55% of 40 patients to Art v 3, median 1.2 ISU (0.6–2.1); 77% of 13 patients to Pla a 3, mv 1.56 ISU (0.6–2.3); and 11.1% to ole e 7 (one patient). There was no statistical difference in the specific IgE rate to different LTPs between the group of patients that developed allergy to new plant foods or not (Figure 2). In addition to nsLTP sensitization, 8 patients (20%) that developed allergy to new plant foods were also sensitized to PR-10 and 9 patients (22.5%) to profilin.

4.3.1. Plant-Food-Allergy Group: Allergy to New Foods

Thirty-five (31%; 95% CI 23–40%) of the 113 patients from the plant-food-allergy group developed allergy to new plant foods: 16 patients to *Rosaceae* fruits (13 to *Rosaceae*/*Prunoideae* fruits and 3 to *Rosaceae*/*Pomoideae* fruits), 16 to nuts (5 patients shared *Rosaceae* fruits and nuts), 4 patients to vegetables, 2 to cereals, 1 to legumes, and 1 to seeds. The allergy symptoms in these patients were local reactions in 37.1% and systemic reactions in 62.9%; 8.6% (of the total) were anaphylactic reactions. All new plant foods that elicited allergic reactions during the follow-up period are shown in Table 4.

Table 4. Allergy to new plant foods on follow-up study in patients sensitized to nsLTP.

New Plant Food Eliciting Allergy	Plant-Food-Allergy Group (n = 35)	Non-Plant-Food-Allergy Group (n = 5)
Rosacea/*Prunoideae* fruit	7	2
Rosacea/*Pomoideae* fruit	3	0
Nuts	7	0
Vegetables	4	0
Cereals	2	0
Legumes	1	1
Seed	1	0
Rosaceae/*Prunoideae* fruit & nuts	5	2
Nuts & vegetables	3	0
Nuts & legumes	1	0
Rosaceae/*Prunoideae* fruit & legumes	1	0

Patients from this group had a mean age of 26.9 years (range 2 to 61 years) and were predominantly female (60%). Sensitization to common allergens was present in 91.43% of the patients; 85.7% of the patients had associated rhinitis while 62.8% presented asthma. Sensitization to profilin was 31.4% and 17.1% were sensitized to PR-10. Nineteen patients received grass pollen immunotherapy and none of them to birch pollen.

4.3.2. Non-Food-Allergy Group: Allergy to New Foods

Five out of 38 patients (13.2%; 95% CI 6–27%) from the non-food-allergy group, which comprised patients who had never experienced allergic reactions to any plant foods, developed allergy to new plant foods. The plant foods eliciting allergy in this subgroup were as follows: *Rosacea* fruits in 2 patients, nuts in 2 patients, and legumes in 1 patient. Two patients from this group developed allergy to both *Rosacea* fruits and nuts. The allergy symptoms in these patients were local in 60% and systemic in 40%. None experienced anaphylactic reactions.

Patients from this group ranged in age from 18 to 50 years (mean age 31.4 years), with a higher prevalence of females (60%). Sensitization to common allergens was present in 100% of the patients. Rhinitis and asthma were also prevalent comorbidities (80% and 60% of the patients, respectively). Profilin sensitization was not found in any patients and 2 patients were sensitized to PR-10 family protein (40%). Three patients received grass pollen immunotherapy and none of them to birch pollen.

5. Discussion

The present study focuses on the development of allergy to new plant foods among nsLTP-sensitized patients. We consider this unresolved issue to be a key point in the management of nsLTP-allergic patients. Our results reveal that 31% of nsLTP-allergic patients became allergic to new plant foods that had been tolerated at the time of diagnosis. We also found that, after 10 years, 13% of patients simply sensitized to nsLTP developed plant-food allergy.

This is a real-life study based on clinical allergy practice. Ten years following diagnosis, a telephone interview was conducted to determine whether patients developed allergies to new plant foods. Real-life studies and the results of the telephone interviews have both advantages and disadvantages which should be considered when interpreting these results. However, we found the results to be valuable as they provide interesting information on the development of allergy to new plant foods, both among patients with nsLTP allergy and among nsLTP-sensitized subjects who have never been allergic to plant foods (latent atopy).

To our knowledge, the report by Asero et al. [19] is the only study designed to evaluate the development of new food allergies in the follow-up of patients allergic to nsLTP. The results of our study, in which 31% of patients developed new plant-food allergies, are in agreement with those of Asero et al. (27%; 18/67 patients), which reinforces the results of both.

A literature search revealed no previous studies addressing the development of plant-food allergy among nsLTP-sensitized patients without previous plant-food allergy. We found that allergy to new plant foods among patients without previous plant-food allergy was not only less frequent, but also less severe, as no patients in the sensitized group had anaphylactic reactions. These data support current recommendations indicating that patients who experienced systemic reactions should always carry auto-injectable adrenaline with them.

Another finding of our study, and one that is found throughout the literature on nsLTP allergy, is that rosaceous fruits and nuts are the foods most frequently responsible for nsLTP allergic reactions [14,19,20], even when discussing new plant-food allergies in the evolution of these patients. We consider this issue relevant, as clinicians should not restrict all nsLTP-allergenic foods in the same way, but rather prioritize the most frequently involved foods when an avoidance diet is necessary.

We also found that sensitization to profilin and PR-10 allergens appears to decrease the risk of severe reactions [21,22], and that nsLTP-specific IgE levels do not predict the occurrence of new plant-food allergy [23], which is consistent with data published in multiple studies.

Asero et al. [19] provided useful recommendations for the management of patients with nsLTP-related food allergy, which we support fully. In addition to these recommendations, we believe nsLTP-allergic patients should undergo risk stratification, as this would allow for tailored management of the heterogeneous and highly variable population of patients with this type of allergy. Specifically, our findings lead us to recommend the following:

- Patients should avoid plant foods that provoke allergic reactions after an allergy study based on anamnesis, skin testing, and/or determination of specific IgE and challenge tests when necessary;
- Patients with systemic reactions should always carry self-injectable adrenaline on their person;
- Additional dietary restrictions should be based on patient risk stratification, as it is impossible to predict severity and/or allergy to new plant foods. In our opinion, key points to stratify the risk of the nsLTP-allergic patients are those appearing in Table 5. Thus, for patients who have developed a systemic reaction to peach peel but who tolerate other foods (even peach pulp), it would be sufficient to avoid peach peel and take self-injectable adrenaline. However, when traveling to the mountains, the countryside, or other remote locales, they should strictly avoid foods related to the nsLTP allergy and be vigilant with NSAIDs and other cofactors, since accessibility to emergency services may be limited and their quality of life would not be significantly altered by such a one-off situation. This is an example of how allergy-management recommendations should be adapted depending on risk stratification.

Table 5. Risk stratification of nsLTP-allergic patients.

Key Points.
- Severity of previous reactions
- New foods most frequently implicated
- Accessibility of emergency services
- Sensitivity to PR-10 and profilin
- Cofactors
- Quality of life

6. Conclusions

In summary, one-third of nsLTP-allergic patients developed allergy to novel plant foods, while one-tenth of nsLTP-sensitized patients without food allergy eventually developed reactions to novel plant foods, which were milder. Finally, risk stratification should be a cornerstone of individualized management for highly varied patients with nsLTP allergy.

Author Contributions: Conceptualization, D.B., E.N.-B., C.P.-V., V.E. and J.C.-H.; methodology, D.B., A.G.-L., C.V.-V., E.N.-B., S.F.-B., M.D.l.H.G.; software D.B. and J.C.-H.; validation, D.B., A.G.-L., C.V.-V., E.N.-B., S.F.-B., M.D.l.H.G., C.P.-V., V.E. and J.C.-H.; formal analysis, D.B., E.N.-B., C.P.-V., V.E. and J.C.-H.; in V.E. stigation, D.B., E.N.-B., C.P.-V., V.E. and J.C.-H.; writing—original draft preparation, D.B., E.N.-B. and J.C.-H.; writing—review and editing, D.B., A.G.-L., C.V.-V., E.N.-B., S.F.-B., M.D.l.H.G., C.P.-V., V.E. and J.C.-H.; supervision, C.P.-V., V.E. and J.C.-H.; funding acquisition, V.E. and J.C.-H. All authors have read and agreed to the published version of the manuscript.

Funding: This work was supported by grants from Instituto de Salud Carlos III co-founded by FEDER funds for the Thematic Networks and Co-operative Research Centers, ARADyAL RD16/0006/0013 and PI18/00348. Also supported by the SEAIC (19_A08) and by Comunidad de Madrid_P2018/BAAA-4574. Betancor D is supported by Rio Hortega Research Contract.

Institutional Review Board Statement: The study was conducted according to the guidelines of the Declaration of Helsinki, and approved by Fundación Jiménez Díaz Ethics Committee.

Informed Consent Statement: Informed consent was obtained from all subjects involved in the study.

Acknowledgments: Oliver Shaw (IIS-FJD) for editorial support.

Conflicts of Interest: The authors declare no conflict of interest.

References

1. Lyons, S.A.; Burney, P.G.J.; Ballmer-Weber, B.K.; Fernandez-Rivas, M.; Barreales, L.; Clausen, M.; Dubakiene, R.; Fernandez-Perez, C.; Fritsche, P.; Jedrzejczak-Czechowicz, M.; et al. Food Allergy in Adults: Substantial Variation in Prevalence and Causative Foods Across Europe. *J. Allergy Clin. Immunol. Pract.* **2019**, *7*, 1920–1928. [CrossRef]
2. Bogas, G.; Muñoz-Cano, R.; Mayorga, C.; Casas, R.; Bartra, J.; Pérez, N.; Pascal, M.; Palomares, F.; Torres, M.J.; Gómez, F. Phenotyping peach-allergic patients sensitized to lipid transfer protein and analysing severity biomarkers. *Allergy* **2020**, *75*, 3228–3236. [CrossRef]
3. Martín-Pedraza, L.; Mayorga, C.; Gómez, F.; Bueno-Díaz, C.; Blanca-López, N.; González, M.; Martinez-Blanco, M.; Cuesta-Herranz, J.; Molina, E.; Villalba, M.; et al. IgE-reactivity pattern of tomato seed and peel nonspecific Lipid-Transfer Proteins after in vitro gastrointestinal digestion. *Agric. Food Chem.* **2021**, *69*, 3511–3518. [CrossRef]
4. Skypala, I.J.; Bartra, J.; Ebo, D.G.; Faber, M.A.; Fernández-Rivas, M.; Gomez, F.; Luengo, O.; Till, S.J.; Asero, R.; Barber, D.; et al. The diagnosis and management of allergic reactions in patients sensitized to non-specific lipid transfer proteins. *Allergy* **2021**. [CrossRef]
5. Vereda, A.; van Hage, M.; Ahlstedt, S.; Ibañez, M.D.; Cuesta-Herranz, J.; van Odijk, J.; Wickman, M.; Sampson, H.A. Peanut allergy: Clinical and immunologic differences among patients from 3 different geographic regions. *J. Allergy Clin. Immunol.* **2011**, *127*, 603–607. [CrossRef]
6. Haroun-Díaz, E.; Azofra, J.; González-Mancebo, E.; Heras, M.D.L.; Pastor-Vargas, C.; Esteban, V.; Villalba, M.; Díaz-Perales, A.; Cuesta-Herranz, J. Nut Allergy in Two Different Areas of Spain: Differences in Clinical and Molecular Pattern. *Nutrients* **2017**, *9*, 909. [CrossRef]
7. Mothes-Luksch, N.; Raith, M.; Stingl, G.; Focke-Tejkl, M.; Razzazi-Fazeli, E.; Zieglmayer, R.; Wöhrl, S.; Swoboda, I. Pru p 3, a marker allergen for lipid transfer protein sensitization also in Central Europe. *Allergy* **2017**, *72*, 1415–1418. [CrossRef]
8. Asero, R.; Antonicelli, L.; Arena, A.; Bommarito, L.; Caruso, B.; Colombo, G.; Crivellaro, M.; de Carli, M.; Della Torre, E.; Della Torre, F.; et al. Causes of food-induced anaphylaxis in Italian adults: A mul-ti-centre study. *Int. Arch. Allergy Immunol.* **2009**, *150*, 271–277. [CrossRef]
9. Cuesta-Herranz, J.; Lázaro, M.; Figueredo, E.; Igea, J.M.; Umpiérrez, A.; De-Las-Heras, M. Allergy to plant-derived fresh foods in a birch- and ragweed-free area. *Clin. Exp. Allergy* **2000**, *30*, 1411–1416. [CrossRef]
10. Murad, A.; Katelaris, C.H.; Baumgart, K. A case study of apple seed and grape allergy with sensitisation to nonspecific lipid transfer protein. *Asia Pac. Allergy* **2016**, *6*, 129–132. [CrossRef]
11. Gao, Z.-S.; Yang, Z.; Wu, S.-D.; Wang, H.-Y.; Liu, M.-L.; Mao, W.-L.; Wang, J.; Gadermaier, G.; Ferreira, F.; Zheng, M.; et al. Peach allergy in China: A dominant role for mugwort pollen lipid transfer protein as a primary sensitizer. *J. Allergy Clin. Immunol.* **2013**, *131*, 224–226.e3. [CrossRef]
12. Faber, M.A.; van Gasse, A.L.; Decuyper, I.I.; Uyttebroek, A.; Sabato, V.; Hagendorens, M.M.; Bridts, C.H.; de Clerck, L.S.; Fernandez-Rivas, M.; Pascal, M.; et al. IgE-reactivity profiles to nonspecific lipid transfer proteins in a northwestern Eu-ropean country. *J. Allergy Clin. Immunol.* **2017**, *139*, 679–682. [CrossRef] [PubMed]
13. Pascal, M.; Cano, R.M.; Reina, Z.; Palacín, A.; Vilella, R.; Picado, C.; Juan, M.; Sánchez-López, J.; Rueda, M.; Salcedo, G.; et al. Lipid transfer protein syndrome: Clinical pattern, cofactor effect and profile of molecular sensitization to plant-foods and pollens. *Clin. Exp. Allergy* **2012**, *42*, 1529–1539. [CrossRef]
14. Du Toit, G.; Roberts, G.; Sayre, P.H.; Bahnson, H.T.; Radulovic, S.; Santos, A.F.; Brough, H.A.; Phippard, D.; Basting, M.; Feeney, M.; et al. LEAP Study Team. Randomized trial of peanut consumption in infants at risk for peanut allergy. *N. Engl. J. Med.* **2015**, *372*, 803–813. [CrossRef]
15. Anagnostou, K.; Islam, S.; King, Y.; Foley, L.; Pasea, L.; Bond, S.; Palmer, C.; Deighton, J.; Ewan, P.; Clark, A. Assessing the efficacy of oral immunotherapy for the desensitisation of peanut allergy in children (STOP II): A phase 2 randomised controlled trial. *Lancet* **2014**, *383*, 1297–1304. [CrossRef]
16. Gomez, F.; Bogas, G.; Gonzalez, M.; Campo, P.; Salas, M.; Diaz-Perales, A.; Rodriguez, M.J.; Prieto, A.; Barber, D.; Blanca, M.; et al. The clinical and immunological efects of Pru p 3 sublingual immunotherapy on peach and peanut allergy in patients with systemic reactions. *Clin. Exp. Allergy* **2017**, *47*, 339–350. [CrossRef]
17. González Pérez, A.; Carbonell Martínez, A.; Escudero Pastor, A.I.; Navarro Garrido, C.; Miralles López, J.C. Pru p 3 oral immuno-therapy eficacy, induced immunological changes and quality of life improvement in patients with LTP syndrome. *Clin. Transl. Allergy* **2020**, *10*, 20. [CrossRef]
18. Beitia, J.M.; Castro, A.V.; Cárdenas, R.; Peña-Arellano, M.I. Pru p 3 Sublingual Immunotherapy in Patients with Lipid Transfer Protein Syndrome: Is It Worth? *Int. Arch. Allergy Immunol.* **2021**, *182*, 447–454. [CrossRef] [PubMed]
19. Asero, R.; Piantanida, M.; Pravettoni, V. Allergy to LTP: To eat or not to eat sensitizing foods? A follow-up study. *Eur. Ann. Allergy Clin. Immunol.* **2018**, *50*, 156. [CrossRef] [PubMed]
20. Scala, E.; Till, S.J.; Asero, R.; Abeni, D.; Guerra, E.C.; Pirrotta, L.; Paganelli, R.; Pomponi, D.; Giani, M.; de Pità, O.; et al. Lipid transfer protein sensitization: Reactivity profiles and clinical risk assessment in an Italian cohort. *Allergy* **2015**, *70*, 933–943. [CrossRef] [PubMed]

21. Pastorello, E.A.C.; Farioli, L.; Pravettoni, V.; Scibilia, J.; Mascheri, A.; Borgonovo, L.; Piantanida, M.; Primavesi, L.; Stafylaraki, C.; Pasqualetti, S.; et al. Pru p 3-sensitised Italian peach-allergic patients are less likely to develop severe symptoms when also presenting IgE antibodies to Pru p 1 and Pru p 4. *Int. Arch. Allergy Immunol.* **2011**, *156*, 362–372. [CrossRef]
22. Scala, E.D.; Abeni, D.; Guerra, E.C.; Locanto, M.; Pirrotta, L.; Meneguzzi, G.; Giani, M.; Asero, R. Cosensitization to profilin is associat-ed with less severe reactions to foods in nsLTPs and storage proteins reactors and with less severe respiratory allergy. *Allergy* **2018**, *73*, 1921–1923. [CrossRef]
23. Pastorello, E.A.; Farioli, L.; Pravettoni, V.; Robino, A.M.; Scibilia, J.; Fortunato, D.; Conti, A.; Borgonovo, L.; Bengtsson, A.; Ortolani, C. Lipid transfer protein and vicilin are important walnut allergens in patients not allergic to pollen. *J. Allergy Clin. Immunol.* **2004**, *114*, 908–914. [CrossRef]

MDPI

Review

Monoclonal Antibodies in Treating Food Allergy: A New Therapeutic Horizon

Sara Manti [1], Giulia Pecora [1,†], Francesca Patanè [1,†], Alessandro Giallongo [1,*], Giuseppe Fabio Parisi [1], Maria Papale [1], Amelia Licari [2], Gian Luigi Marseglia [2] and Salvatore Leonardi [1]

1 Pediatric Respiratory Unit, Department of Clinical and Experimental Medicine, San Marco Hospital, University of Catania, Via Santa Sofia 78, 95123 Catania, Italy; saramanti@hotmail.it (S.M.); giupec87@hotmail.it (G.P.); francescapatane27@gmail.com (F.P.); giuseppeparisi88@hotmail.it (G.F.P.); mariellapap@yahoo.it (M.P.); leonardi@unict.it (S.L.)

2 Pediatric Clinic, Department of Pediatrics, Fondazione IRCCS Policlinico San Matteo, University of Pavia, 27100 Pavia, Italy; amelialicari@gmail.com (A.L.); glmarseglia@smatteo.pv.it (G.L.M.)

* Correspondence: alegiallongo@hotmail.it; Tel.: +39-095-4794-181

† These authors contributed equally to this work.

Abstract: Food allergy (FA) is a pathological immune response, potentially deadly, induced by exposure to an innocuous and specific food allergen. To date, there is no specific treatment for FAs; thus, dietary avoidance and symptomatic medications represent the standard treatment for managing them. Recently, several therapeutic strategies for FAs, such as sublingual and epicutaneous immunotherapy and monoclonal antibodies, have shown long-term safety and benefits in clinical practice. This review summarizes the current evidence on changes in treating FA, focusing on monoclonal antibodies, which have recently provided encouraging data as therapeutic weapons modifying the disease course.

Keywords: monoclonal antibodies; food allergy; biologics; children; adults

Citation: Manti, S.; Pecora, G.; Patanè, F.; Giallongo, A.; Parisi, G.F.; Papale, M.; Licari, A.; Marseglia, G.L.; Leonardi, S. Monoclonal Antibodies in Treating Food Allergy: A New Therapeutic Horizon. *Nutrients* **2021**, *13*, 2314. https://doi.org/10.3390/nu13072314

Academic Editor: Isabel Comino

Received: 30 May 2021
Accepted: 1 July 2021
Published: 5 July 2021

1. Introduction

Food Allergy (FA) is a pathological and potentially deadly immune response caused by exposure to an innocuous and specific food allergen [1]. Epidemiological global data suggest that FA prevalence ranges from 0.45% to 10% among children younger than five years old. It has been estimated that approximately 40% of patients with FA have experienced a life-threatening allergic reaction, and that 30% of children with FA show multiple FAs [2,3].

Based on the underlying immune mechanism, FA is broadly classified into immunoglobulin (Ig)E-mediated (characterized by immediate reactions), non-IgE mediated (characterized by delayed reactions), or mixed (characterized by both IgE-dependent and IgE-independent mechanisms). The main characteristics of IgE-mediated, non-IgE mediated, and mixed FAs are summarized in Table 1.

Affecting up to 10% of the pediatric population [4], IgE-mediated FA is the most common and costly FA subtype. Although the allergens triggering the FA vary with country and dietary habits, milk, egg, peanut, wheat, soy, and shellfish are currently the most common foods to induce IgE-mediated FA [5]. After exposure to the offending allergen, food allergen-derived epitopes bind to the IgE and, by binding with the FcεRI receptor expressed on the surface of mast cells and basophils, induce the IgE-mediated degranulation of the immune effector cells. The latter releases preformed histamine, leukotrienes (LTs), platelet-activating factor (PAF), and cytokines such as interleukin-4 (IL-4), IL-5, and IL-13, which are able to maintain the allergic immune response [1]. Clinically, early and rapid symptom onset can occur and may involve one or more systems among the cutaneous system (with flushing, urticaria, angioedema, pruritus), respiratory system (with bronchial hyperresponsiveness and/or wheezing), gastrointestinal system (with nausea, abdominal pain, diarrhea, and vomiting), and cardiovascular system (with hypotension). Whenever a severe allergic reaction involves multiple organ systems, the patient experiences anaphylaxis, which can potentially become a life-threatening event [1–6].

Table 1. Main clinical findings of Food Allergies in pediatric population.

			Ig-E Mediated Food Allergies		
Disorder	**World Prevalence**	**Common Allergens**	**Description**	**Diagnosis**	**Treatment**
Urticaria/angioedema Contact urticaria	Up to 14.5% for males and 16.2% for females 13.3–24.5%	Milk, egg, peanut, nuts, fish, shellfish Fresh fruit, fish, milk, egg	Immediate reaction to foods with erythema and wheals Urticaria resulting from direct contact with skin	SPT*, serum IgE * levels, and OFC *	Elimination diet and emergency medication Research: OIT *, SLIT *, EPIT *, and biologic drugs Contact avoidance and emergency medication
Oral allergy syndrome	5–8%	Fresh fruits and vegetables	Itching and mild edema of oral cavity	SPT or PBP *, serum IgE levels, and OFC	Elimination diet and emergency medication Research: OIT, SLIT, EPIT, and biologic drugs
Anaphylaxis	0.3%	Milk, egg, peanuts, nuts, fish, shellfish	Rapid reaction with involvement of skin, respiratory tract, and cardiocirculatory apparatus	SPT, serum IgE levels, and OFC	Elimination diet and emergency medication Research: OIT, SLIT, EPIT, and biologic drugs
Exercise-induced anaphylaxis	5–15%	Wheat, shellfish, celery	Food induces anaphylaxis only if ingestion is temporally followed by physical exercise	Anamnesis	Elimination diet, time interval between food consumption and exercise, and emergency medication Research: OIT, SLIT, EPIT, and biologic drugs
			Non IgE-Mediated Food Allergies		
Disorder	**Prevalence**	**Common Allergens**	**Description**	**Diagnosis**	**Treatment**
FPIES Food protein-induced proctocolitis	Few data Few data	Milk, egg, soy, oat, rice Milk protein through breast feeding or egg, soy, wheat	Immediate reaction to foods with vomiting, diarrhea, pallor, sweating, hypotension Mucus in stools	Clinical history and OFC Elimination diet and OFC	Elimination diet and drugs Elimination diet
Food protein enteropathy	Few data	Milk, egg, soy, and wheat	Malabsorption syndrome	Elimination diet or OFC with jejunal biopsy	Elimination diet
			Mixed Food Allergies		
Disorder	**Prevalence**	**Common Allergens**	**Description**	**Diagnosis**	**Treatment**
Atopic dermatitis Eosinophilic esophagitis	27–37% of patients with AD * Up to 50/100,000 patients	Mostly milk and egg Egg, milk, beef, chicken, soy, and wheat	Immediate reaction to foods with erythema and wheals Reflux symptoms including vomiting, dysphagia, cough, and food impaction	SPT, serum IgE levels, and OFC Eosinophil infiltrates on esophageal biopsies	Elimination diet Research: OIT, SLIT, EPIT, and biologic drugs Elimination diet or topical steroids
Eosinophilic gastroenteritis	Rare	Multiple allergens or may not have food allergy etiology	Nonspecific gastrointestinal disorders associated with eosinophilic infiltrate of gastrointestinal tract region and layer	Eosinophil infiltrates on gastrointestinal biopsies, eosinophils in ascites	Elimination diet or topical steroids

* SPT: skin prick test; IgE: immunoglobulin E; PBP: prick by prick; OFC: oral food challenge; OIT: oral immunotherapy; SLIT: Sublingual-swallow immunotherapy; EPIT: Epicutaneous Immunotherapy; AD: atopic dermatitis.

In addition to the classic IgE-mediated FA, two variants are worthy of consideration: oral allergy syndrome (OAS) and FA to the carbohydrate galactose-alpha-1,3-galactose (alpha-gal). OAS is characterized by the immediate onset of oral pruritus, mucosal angioedema, and/or abdominal pain in patients with allergic rhinitis who produce specific IgE for aero-allergens cross-reactive with fruit- or vegetable-protein epitopes. As plant-derived proteins are also sensitive to heat exposure, the same foods are typically tolerated after cooking. This aspect can help in diagnosing OAS [7,8].

The FA to alpha-gal occurs in patients producing specific IgE for the red meat carbohydrate alpha-gal after exposure to tick vectors Dermacentor variabilis (brown dog tick) and Amblyomma americanum (lone star tick). Although it is an IgE-mediated FA, unlike the classic IgE-mediated allergic reactions, the FA to the carbohydrate alpha-gal features a delayed reaction and lacks any relationship with other atopic diseases. Other mechanisms behind the type-2 immune response have been suggested to be involved in the pathogenesis of FA to alpha-gal [8].

Food protein-induced enterocolitis syndrome (FPIES), food protein-induced allergic proctocolitis (FPIAP), and food protein enteropathy (FPE) are the most widely known non-IgE-mediated FAs [9]. Generally, the median age at FPIES onset is 5.5 months. In accordance with the symptom onset, clinical features, duration and severity of symptoms, and offending foods, FPIES is classified into early- (primarily within three months of age) and late-onset (mostly four to seven months of age); typical or atypical type (in older patients, positive skin prick test results, and serum-specific IgE levels); acute and chronic symptoms; and milk or soy FPIES, solid food FPIES, and multiple food FPIES. Clinically, FPIES is primarily characterized by gastrointestinal symptoms, such as profuse vomiting, sometimes accompanied by diarrhea; however, a variable and atypical clinical presentation can also occur. FPIES can occur after the first or second ingestion of the offensive food, as a result of an inappropriate T-cell activation and proliferation, leading to the release of pro-inflammatory cytokines, such as tumor necrosis factor-a (TNF-a), interferon-y (IFN-y), IL-8, and IL-9, which, in turn, impair the permeability barrier, inducing local intestinal inflammation [10].

FPIAP is characterized by inflammatory injury in the distal colon in response to one or more offending food proteins, such as cow's milk or soy. Studies provide evidence that failure in Th3 cells, low levels of transforming growth factor β (TGF-β), and high expression of TNF-a may be involved in the pathogenesis of this disease. Patients affected by FPIAP generally present with red blood and mucus mixed with the stool, with or without diarrhea; they are generally healthy in appearance and do not report weight loss. Diagnosis is clinical, and FPIAP diagnosis is confirmed when patients respond positively to eliminating a suspected triggering food allergen after excluding other causes of gastrointestinal symptoms, such as necrotizing enterocolitis, intussusception, infectious colitis, anal fissures, and very early onset inflammatory bowel disease [11].

FPE is mainly characterized by non-bloody diarrhea, malabsorption, and failure to thrive in the first nine months of life [9]. It is triggered in formula-fed infants, but also by soybean, wheat, and egg. Diagnostic tests are not available, and diagnosis is based on clinical history, physical examination, and an oral food challenge (OFC) [9].

Mixed FAs include Eosinophilic Gastrointestinal Disorders (EGIDs), such as eosinophilic esophagitis (EoE), allergic eosinophilic gastroenteritis (AEG), and eosinophilic colitis, characterized by gastrointestinal symptoms, eosinophilic infiltration of the gastrointestinal tract, and, sometimes, peripheral eosinophilia [12]. Generally detected in the first year of life, the clinical picture of EoE includes regurgitation, vomiting, rumination, lack of appetite, burning, and pain, causing crying after feeding and sometimes immediately after starting to feed. The suspicion of EoE increases when the response to a proton pump inhibitor (PPI) is lacking. The esophageal biopsy shows a diagnostic eosinophilic infiltration (>15 eosinophils per high-power field (eos/hpf) [13].

Multiple food allergens are implicated in the onset of AEG [14], generally affecting children and adults. According to the severity of the involvement of bowel wall layers, abdominal pain, irritability, vomiting, diarrhea, weight loss, easy satiety, anemia, and hypoalbuminemia range from a mild to a severe degree. The esophageal biopsy shows eosinophilic infiltration of gastric and/or duodenal mucosa (>30 eos/hpf). Moreover, in approximately 50% of patients with AEG, peripheral eosinophilia, positive food skin prick tests (SPTs), and specific IgE antibodies can be found [15].

Eosinophilic colitis is the less common of the EGIDs. It is generally seen in adolescents affected by inflammatory bowel disease and/or celiac disease and allergy to cow's milk protein, soya, or peanuts. Currently, there is no consensus on the diagnosis of eosinophilic colitis; however, the detection of >50 eos/hpf in the ascending colon, >42 eos/hpf in the transverse and descending colon, and >32 eos/hpf in the rectosigmoid colon are considered suggestive of eosinophilic colitis [16].

The main clinical characteristics of FAs are summarized in Table 1.

Regardless of their pathogenesis and clinical pictures, and due to the lack of definitive treatment, FAs represent a significant burden on affected children and their families, due to dietary restrictions, diet adherence, fear of accidental reactions, and the self-management of anaphylactic reactions. To date, no specific treatment for FAs is available, so their therapeutic management is limited to dietary avoidance. However, allergen-specific therapy (immunotherapy) is showing encouraging results. In parallel, several therapeutic strategies are also emerging that restore immune tolerance against the offending food epitopes (Figure 1). In this regard, treatment with monoclonal antibodies has recently provided encouraging results as a therapeutic weapon modifying the disease course.

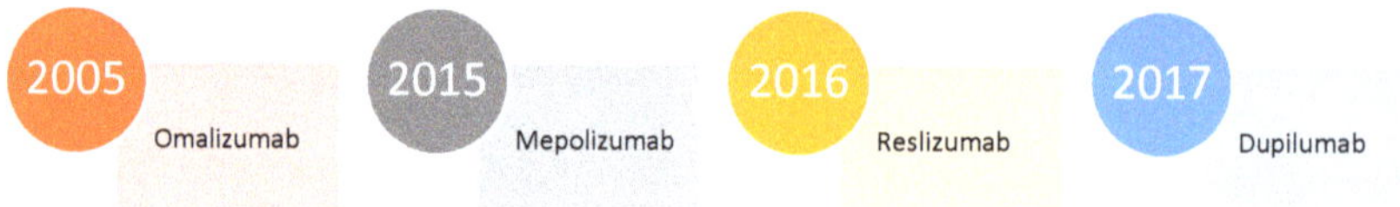

Figure 1. Developmental timing of monoclonal antibodies used for treating allergic disorders.

2. Monoclonal Antibodies in FA

2.1. Omalizumab

The critical mediator involved in FA is IgE, making it a promising therapeutic target. As a prototype of an anti-IgE, omalizumab is a humanized IgG1 monoclonal antibody that acts through various mechanisms. Firstly, it binds to free IgEs, blocking them from binding with specific high-affinity receptors (FcεRI) expressed on dendritic and mast cells. Furthermore, it decreases receptor expression on these cells, thus interfering upstream with the inflammatory cascade. It also leads to a reduction of IgE synthesis by B-cells. At present, anti-IgE therapy is one of the mainstay treatments for severe asthma, severe chronic urticaria, and severe chronic rhinosinusitis with nasal polyps (CRSwNP) [17,18]. Omalizumab is administered as a subcutaneous injection; the dosage, time interval, and frequency are based on a nomogram derived from baseline total serum IgE levels and body weight (kilograms). The following section is focused on the available literature on anti-IgE therapy for the management of FA, where anti-IgE is still off label [19].

Eighty-four individuals, aged between 12 and 60 years old, affected by peanut allergy, were enrolled in a double-blind, placebo-controlled, randomized clinical trial (RCT) to test the anti-IgE monoclonal antibody TNX-901. Patients were randomized into four groups, and three different doses of TNX-901 (150 mg, 300 mg, 450 mg) or a placebo were administered for four monthly doses. Two to four weeks after the end of treatment, the subjects underwent an OFC, which showed a significant increase of threshold dose for peanut, compared to screening at enrollment, only in those receiving 450 mg of TNX-901. Nevertheless, 25% of the patients did not develop tolerance to peanuts, evidencing variable responses among them [20].

Regarding children, an RCT including patients aged 6–13 years old raised safety concerns due to the reactions to pre-omalizumab challenges and, therefore, was stopped early [21].

An open-label study enrolling 14 individuals aged between 18 and 50 years with a history of peanut allergy evaluated the effectiveness of a six-month treatment with anti-IgE.

The median threshold tolerated dose for peanut significantly increased from an 80 mg baseline to 6500 mg after treatment. However, the study had some limitations due to the small sample size (n = 14 adults) and the need for antihistamines and epinephrine in 10 out of 14 patients at the third food challenge, after six months of omalizumab [22].

To maximize the development of tolerance and reduce safety concerns relating to immunotherapy, a synergic effect of combined therapy with anti-IgE and FA-AIT was hypothesized. Accordingly, 13 children, with a median age of 10 years, suffering from peanut allergy, were enrolled in a double-blind, placebo-controlled food challenge trial. Children underwent a course of omalizumab combined with rapid oral food allergy desensitization. Omalizumab was administered during the 12 weeks before and during oral food desensitization, until a maintenance oral dose of peanut (8000 mg) was reached. Following the peanut challenge, 92% of patients tolerated an 8000 mg dose of peanut flour, and 39% reported moderate to severe adverse reactions [23].

Another food challenge trial was conducted on 11 children with cow's milk allergy. After nine weeks of omalizumab pretreatment, 9 out of 11 subjects completed an OFC and received omalizumab until week 16. Few reactions were reported (1.6% of cow's milk doses administered), and most were mild [24].

Combined treatment with oral immunotherapy (OIT) and omalizumab has also been investigated in the setting of multiple FAs, in a phase I clinical trial enrolling 25 children (median age: 7 years) who were treated with OIT, up to five allergens, and omalizumab. The omalizumab was started eight weeks before the OIT. The safety outcome was satisfactory: reactions followed only 5.3% of administered doses, and 94% of these were mild. Only one child showed a severe reaction, and was treated with epinephrine [25].

The previously mentioned results were consistent with a subsequent phase II RCT of 48 patients aged 4–15 years with multiple FAs. Sixteen weeks of omalizumab treatment was significantly associated with a higher percentage of tolerance to up to 2 gr of at least two foods at 36 weeks, compared to a placebo (83% vs. 33%, $p = 0.004$), in a double-blind placebo-controlled food challenge (DBPCFC). Furthermore, omalizumab, compared with a placebo, significantly increased the tolerated dose, reduced the time taken to achieve a maintenance dose, and reduced the median rate of adverse reactions (27% vs. 68%), with no severe adverse events reported [26].

Contrary to previous studies, an RCT involving 57 patients aged 7–32 years with severe cow's milk allergy did not significantly improve the success rate of OIT in those treated with omalizumab vs. a placebo over 28 months. However, omalizumab allowed patients to achieve a maintenance dose with fewer OIT doses and improved the safety of the OIT. Indeed, the incidence of adverse reactions was significantly lower (2.1% vs. 16.1% of doses, $p = 0.0005$) and those that did occur were less severe in the omalizumab group (2 vs. 18 doses requiring epinephrine) [27].

These findings were consistent with a study of 14 children aged between 4 months and 11 years affected by egg and cow's milk allergies. The OIT was tolerated by all patients only if pretreatment and concomitant treatment with omalizumab took place. Nevertheless, a question arises about when omalizumab should be stopped. Indeed, six patients developed grade 3–4 anaphylactic symptoms after suspending omalizumab, suggesting the need for longer maintenance therapy with an anti-IgE [28].

As regards the underlying mechanism of omalizumab-induced desensitization, Bedoret et al. suggested that milk-specific CD4-T cells might be involved in the development of anergy. It has been suggested that a combination of omalizumab and oral desensitization with higher doses of milk is associated with an early reduction in the proliferation of T-CD4 milk-specific cells, through the development of anergy [29]. The underlying mechanism might be mediated by a reduction in antigen presentation induced by omalizumab [30]. Further, long-term desensitization was found to be associated with an increase in IFN/IL-4 ratio and IgG4, showing a shift in immune response, whose mechanism is still unclear. IgG4 could act by inhibiting IgE [29].

In conclusion, these data support the role of omalizumab as a viable therapeutic option in patients with FA through raising the threshold tolerance dose, thus reducing the risk of severe adverse reactions in the case of accidental ingestion [23–29,31]. Long-term follow-up studies are probably needed to strengthen these data. Indeed, only one study showed that, one year after the suspension of omalizumab, some patients relapsed and their specific IgE significantly reduced, although IgE levels could not be associated with the response to therapy or relapse [28].

As yet, omalizumab has not been approved as a treatment for FA, and the optimal dosage has not been determined. Basophil allergen threshold sensitivity has been suggested as a monitoring marker of response to omalizumab in patients with severe peanut allergy, and it might be helpful in individualizing therapy [32]. Currently, several ongoing trials investigate the role of omalizumab as a monotherapy or in combination with OIT. The clinical development program for omalizumab as a monotherapy or an adjunctive treatment in FA is summarized in Tables 2 and 3.

Table 2. Randomized Clinical Trials for omalizumab as monotherapy or as adjunctive treatment in food allergy.

	Number Clinical Trial	Status	Phase	Estimated Enrollment (No. Patients)	Patients' Age (Years)	Primary Outcome	Drugs	Drug Dosage	Results
1	NCT02879006 [33]	Ongoing	2	34	$\geq$6 and $\leq$40 *	Sustained unresponsiveness	Chinese herbal medication, placebo, omalizumab, multi-OIT*	Not applicable	Not yet reported
2	NCT02643862 [34]	Concluded	2	48	$\geq$4 and $\leq$55	Desensitization assessed by proportion of FA * individuals who tolerate a DBPCFC * up to 2000 mg protein for each of 2 allergens at week 36	Omalizumab, placebo	Not applicable	Not yet reported
3	NCT03181009 [35]	Ongoing	2	60	$\geq$2 and $\leq$25	Change in allergen-specific serum IgG4 * and IgE *	Omalizumab, food flour allergens	Omalizumab: subjects $\geq$4 years receive 150 mg. Subjects $\leq$4 years receive 75 mg Food flour allergens: 300 to 1200 mg	Not yet reported
4	NCT02626611 [36]	Concluded	2	70	$\geq$4 and $\leq$55	No. individuals tolerating an OFC to 2000 mg for at least 2 allergens at week 36	Omalizumab, food flour buildup	Omalizumab: not applicable Food flour buildup: up to 2000mg	Not yet re-ported
5	NCT01510626 [37]	Concluded	1	35	$\geq$4 and $\leq$55	No. adverse events in the treatment group	Omalizumab food protein	Not applicable	Not yet reported
6	NCT00949078 [38]	Concluded	2	51	$\geq$18 and $\leq$50	1. No. patients who experienced a decrease in Pn-BHR * AUC * of > 80% compared with baseline values before week 8 2. Percentage change in peanut-specific IgE from baseline to after Pn-BHR response 3. Percentage change in peanut-specific IgE after pn-BHR response 4. Total IgE after pn-BHR response 5. Dose of peanut protein inducing allergic symptoms at OFC*1 6. Dose of peanut protein inducing allergic symptoms at OFC2 7. Dose of peanut protein inducing allergic symptoms at OFC3 8. Omalizumab received before OFC2 9. No. doses of omalizumab received before OFC2	Omalizumab, food allergen	Not applicable	Not yet reported

Table 2. *Cont.*

	Number Clinical Trial	Status	Phase	Estimated Enrollment (No. Patients)	Patients' Age (Years)	Primary Outcome	Drugs	Drug Dosage	Results
7	NCT01781637 [39]	Not yet started	1, 2	36	≥7 and ≤25	Tolerance of 2000 mg 6 weeks after last dose of omalizumab/placebo	Omalizumab, placebo	Not applicable	Not yet reported
8	NCT03881696 [40]	Not yet started	3	225	≥2 and ≤55	No. participants by stage 1 treatment group, omalizumab versus placebo, who successfully consumed ≥600 mg of peanut protein without dose-limiting symptoms during the DBPCFC conducted at the end of treatment stage 1	Omalizumab, placebo, multi-allergen OIT	Omalizumab: 75 to 150 mg	Not yet reported
9	NCT02402231 [41]	Not yet started	2	23	≥12 and ≤22	Peanut challenge	Omalizumab, immunotherapy	Not applicable	Not yet reported
10	NCT01157117 [42]	Concluded	2	77	≥7 and ≤35	Percentage of subjects in omalizumab group vs. placebo group developing clinical tolerance to milk	Omalizumab, milk powder	Omalizumab: not applicable, milk powder: up to 3.84 g	Omalizumab vs. milk powder: $p = 0.42$
11	NCT00968110 [43]	Concluded	1	10	≥4 and ≤18	To assess the safety of omalizumab in young children, and the safety of oral desensitization in patients pretreated with omalizumab	Omalizumab	Not applicable	Not yet reported
12	NCT00086606 [44]	Concluded	2	150	≥6 and ≤75	Not applicable	Omalizumab	Not applicable	Not yet reported
13	NCT00932282 [45]	Concluded	1, 2	13	≥12	Percentage of subjects who pass the 20gm peanut flour (~50% peanut protein) OFC 2–4 weeks after discontinuing peanut OIT therapy	Peanut OIT, omalizumab	Peanut OIT: 0.2 mg of peanut flour to 8000 mg omalizumab: not applicable	Not yet reported
14	NCT00382148 [46]	Concluded	2	10	≥6 and ≤75	Serious adverse events	Omalizumab	Not applicable	Not yet reported
15	NCT01290913 [47]	Concluded	1, 2	13	≥7 and ≤25	No. participants that tolerated rapid oral peanut desensitization to a dose of 500 mg peanut flour	Omalizumab	Not applicable	Not yet reported
16	NCT04045301 [48]	Ongoing	2	90	≥2 and ≤75	To evaluate the efficacy of Omalizumab at reducing time-to-maintenance during a symptom-driven multi-food OIT protocol	Omalizumab, placebo, multi-food OIT	Omalizumab 16 mg/kg, omalizumab 8 mg/kg	Not yet reported
17	NCT04037176 [49]	Ongoing	4	100	≥6 and ≤18	Change in challenge threshold after 3 months of treatment in patients treated with omalizumab vs. placebo	Omalizumab, placebo	Not applicable	Not yet reported
18	NCT01040598 [50]	Concluded	1	19	12 to 76	To assess markers that will predict responders to Omalizumab	Omalizumab	Not applicable	Not applicable

Table 2. *Cont.*

	Number Clinical Trial	Status	Phase	Estimated Enrollment (No. Patients)	Patients' Age (Years)	Primary Outcome	Drugs	Drug Dosage	Results
19	NCT00123630 [51]	Concluded	2	30	12 to 60	Change in eosinophil numbers per high power field proximally and distally between baseline and post-treatment and between both groups	Omalizumab, placebo	Omalizumab or placebo: 150 to 375 mg SC * every 2 or 4 weeks.	Not applicable
20	NCT00084097 [52]	Concluded	2	30	≥12 and ≤70	To evaluate safety of omalizumab and its efficacy in reducing peripheral blood absolute eosinophil count pre- and post-administration of omalizumab	Omalizumab	maximum dose of 375 mg every 2 weeks	Not yet reported
21	NCT03964051 [53]	Ongoing	4	10	≥18 and ≤70	Change in food challenge threshold	Omalizumab	300 mg every 2 weeks for 12 weeks	Not yet reported

* N.: number; pts: patients; OIT: oral immunotherapy; FA: food allergy; DBPCFC: double-blind placebo-controlled food challenge; IgG: Immunoglobulin-G; IgE: Immunoglobulin-E; Pn-BHR: peanut allergen induced basophil histamine release; AUC: Area under curve; OFC: oral food challenge; SC: subcutaneously.

Table 3. Clinical development program for dupilumab as monotherapy or as adjunctive treatment in food allergy.

	Number Clinical Trial	Status	Phase	Estimated Enrollment (No. Patients)	Patients' Age (Years)	Primary Outcome	Drugs	Drug Dosage	Results
1	NCT04462055 [54]	Ongoing	Not stated	21	≥12	To determine the effect of dupilumab on change in clinical eliciting dose (i.e., lowest dose causing an allergic reaction) in subjects with peanut, hazelnut, walnut, cow's milk, hen's egg and/or soybean allergy	Dupilumab	Not applicable	Not applicable
2	NCT04394351 [55]	Ongoing	3	90	≥1 and ≤11	Proportion of patients achieving peak esophageal intraepithelial eosinophil count ≤ 6 eos/hpf (400×)	Dupilumab, placebo	Not applicable	Not applicable

2.2. Ligelizumab

Ligelizumab, also called QGE031, is a new humanized monoclonal anti-IgE antibody. It is administered as a subcutaneous injection at a dosage of 24, 72, or 240 mg every two weeks. It was initially tested in a phase II RCT, parallel design, dose-ranging, multi-center trial enrolling adult patients (age range, 18–50 years) affected by peanut allergy [56]. However, no results have been posted, as the recruitment was stopped.

2.3. Etokimab

Etokimab, also known as ANB020 (AnaptysBio), is a monoclonal antibody directed against IL-33, a pro-inflammatory cytokine that promotes B-class switching to IgE. The terminal half-life of etokimab is approximately 372 hours, with comparable values across all doses (10–750 mg) and regardless of route (i.v. or s.c.) of administration. In a six-week placebo-controlled phase II clinical trial enrolling 15 adults (age range, 19–54 years) with FAs, the authors showed that etokimab was safe and well tolerated. A single administration of etokimab as a monotherapy was able to induce immune tolerance to peanut, as well as reduce atopy-related adverse events in the enrolled patients [57].

2.4. Dupilumab

Dupilumab is a fully human IgG4 monoclonal antibody directed against the interleukin (IL)-4 receptor alpha (IL-4Ra) subunit, blocking IL-4- and IL-13-mediated pathways. By binding to IL-4Ra, a subunit also shared with the IL-13 receptor (IL-13R), dupilumab blocks the Th2-mediated inflammatory cascade [58,59]. Currently, dupilumab is approved in Europe for treating adolescents aged over 12 years, affected by severe asthma with an eosinophilic phenotype, or with oral corticosteroid-dependent asthma. The drug is available in prefilled syringes and is administered subcutaneously, once at a dose of 400 mg, then at 200 mg every two weeks; or once at 600 mg, then 300 mg every two weeks. The latter scheme is approved for patients who have oral corticosteroid-dependent asthma or comorbid moderate-to-severe atopic dermatitis (AD), for which dupilumab is indicated. Dupilumab is also indicated as an add-on maintenance treatment in patients older than 18 years with inadequately controlled CRSwNP [59,60]. The positive results of dupilumab studies in allergic diseases such as asthma, AD, and CRSwNP suggest that this monoclonal antibody can positively affect the course of other atopic diseases, including FA. Rial et al. [60] reported the first evidence of the efficacy of dupilumab in treating FAs in a 30-year-old woman with a positive history of severe AD and allergic rhinitis without asthma. Several ongoing clinical trials are evaluating dupilumab as either a monotherapy or an adjunct to oral immunotherapy for peanut allergy. Specifically, the NCT04462055 trial [54] is a three-year observational clinical trial to evaluate the effect of dupilumab on change in clinical eliciting dose (i.e., the lowest dose causing an allergic reaction) in subjects with peanut, hazelnut, walnut, cow's milk, hen's egg, and/or soybean allergy. It was conducted in a cohort of 21 patients ($\geq$12 years) with moderate-to-severe AD. This study is still ongoing, and no preliminary results have been published. The NCT04394351 trial [55], a prospective phase II, single-center trial, is currently ongoing, and no preliminary results have been published. In this RCT, the authors aim to demonstrate the efficacy of dupilumab, compared to a placebo, in treating 110 pediatric patients, aged 6 to 21 years, with active EoE and multiple allergies. The efficacy of the dupilumab treatment will be assessed via endoscopic visual measurements of disease activity using the Eosinophilic Esophagitis-Endoscopic Reference Score (EoE-EREFS) and histologic abnormalities as measured by the EoE Histology Scoring System (EoE-HSS). Clinical trials on dupilumab's use in treating FAs are summarized in Tables 3 and 4.

Table 4. Clinical development program for Omalizumab and Dupilumab as treatment in food allergy.

	Number Clinical Trial	Status	Phase	Estimated Enrollment (No. Patients)	Patients' Age (Years)	Primary Outcome	Drugs	Drug Dosage	Results
1	NCT03679676 [61]	Not started	2	200	≥6 and ≤21	The success rates of passing a peanut food challenge	Omalizumab, placebo, dupilumab	Not applicable	Not applicable

3. Conclusions

The high prevalence of FA, its impact on quality of life, and the risk of life-threatening reactions have highlighted the need for new treatment strategies other than avoidance of the involved food allergen alone [62]. Although OIT has shown promising results, the use of monoclonal antibodies in treating FA has been suggested based on the pathogenic mechanism. Various trials have highlighted the role of monoclonal antibodies, both as monotherapies and in combination with OIT, in improving the threshold of tolerated dose of allergens. Therefore, monoclonal antibodies may emerge as a more effective, tailored, and potentially disease-modifying therapy for FA. Nevertheless, the application of monoclonal antibodies in food allergy treatment is rather novel and not many well-controlled, large-sample-size studies are available to date; therefore, updated reviews of the literature need to be carried out on a regular basis as more data are published.

Author Contributions: Conceptualization, S.L. and G.L.M.; methodology, S.M. and A.L.; software, G.F.P.; validation, S.L. and G.L.M.; formal analysis, A.G.; resources, A.G., G.P., F.P.; data curation, G.F.P., M.P.; writing—original draft preparation, A.G., G.P., F.P.; writing—review and editing, S.M., A.L., A.G.; supervision, S.L. and G.L.M. All authors have read and agreed to the published version of the manuscript.

Funding: This research received no external funding.

Data Availability Statement: Not applicable.

Conflicts of Interest: The authors declare no conflict of interest.

References

1. Boyce, J.A.; Assa'ad, A.; Burks, A.W.; Jones, S.M.; Sampson, H.A.; Wood, R.A.; Plaut, M.; Cooper, S.F.; Fenton, M.J.; Arshad, S.H.; et al. Guidelines for the diagnosis and management of food allergy in the United States: Summary of the NIAID-sponsored Expert Panel Report. *J. Allergy Clin. Immunol.* **2010**, *126*, 1105–1118. [CrossRef] [PubMed]
2. Prescott, S.L.; Pawankar, R.; Allen, K.J.; Campbell, D.E.; Sinn, J.K.; Fiocchi, A.; Ebisawa, M.; Sampson, H.A.; Beyer, K.; Lee, B.-W. A global survey of changing patterns of food allergy burden in children. *World Allergy Organ. J.* **2013**, *6*, 21. [CrossRef] [PubMed]
3. Gupta, R.S.; Springston, E.E.; Warrier, M.R.; Smith, B.; Kumar, R.; Pongracic, J.; Holl, J.L. The prevalence, severity, and distribution of childhood food allergy in the United States. *Pediatrics* **2011**, *128*, e9–e17. [CrossRef]
4. Osborne, N.J.; Koplin, J.J.; Martin, P.E.; Gurrin, L.C.; Lowe, A.J.; Matheson, M.C.; Ponsonby, A.L.; Wake, M.; Tang, M.L.; Dharmage, S.C.; et al. Prevalence of challenge-proven IgE-mediated food allergy using population-based sampling and predetermined challenge criteria in infants. *J. Allergy Clin. Immunol.* **2011**, *127*, 668–676. [CrossRef]
5. Chinthrajah, R.S.; Tupa, D.; Prince, B.T.; Block, W.M.; Rosa, J.S.; Singh, A.M.; Nadeau, K. Diagnosis of Food Allergy. *Pediatr. Clin. N. Am.* **2015**, *62*, 1393–1408. [CrossRef]
6. Cuppari, C.; Manti, S.; Salpietro, A.; Valenti, S.; Capizzi, A.; Arrigo, T.; Salpietro, C.; Leonardi, S. HMGB1 levels in children with atopic eczema/dermatitis syndrome (AEDS). *Pediatr. Allergy Immunol.* **2016**, *27*, 99–102. [CrossRef]
7. Ivković-Jureković, I. Oral allergy syndrome in children. *Int. Dent. J.* **2015**, *65*, 164–168. [CrossRef]
8. Wilson, J.M.; Schuyler, A.J.; Schroeder, N.; Platts-Mills, T.A. Galactose-α-1,3-Galactose: Atypical Food Allergen or Model IgE Hypersensitivity? *Curr. Allergy Asthma Rep.* **2017**, *17*, 8. [CrossRef]
9. Nowak-Węgrzyn, A.; Katz, Y.; Mehr, S.S.; Koletzko, S. Non-IgE-mediated gastrointestinal food allergy. *J. Allergy Clin. Immunol.* **2015**, *135*, 1114–1124. [CrossRef]
10. Manti, S.; Leonardi, S.; Salpietro, A.; Del Campo, G.; Salpietro, C.; Cuppari, C. A systematic review of food protein-induced enterocolitis syndrome from the last 40 years. *Ann. Allergy Asthma Immunol.* **2017**, *118*, 411–418. [CrossRef] [PubMed]
11. Mennini, M.; Fiocchi, A.G.; Cafarotti, A.; Montesano, M.; Mauro, A.; Villa, M.P.; Di Nardo, G. Food protein-induced allergic proctocolitis in infants: Literature review and proposal of a management protocol. *World Allergy Organ. J.* **2020**, *13*, 100471. [CrossRef] [PubMed]

12. Calvani, M.; Anania, C.; Cuomo, B.; D'Auria, E.; Decimo, F.; Indirli, G.; Marseglia, G.; Mastrorilli, V.; Sartorio, M.; Santoro, A.; et al. Non-IgE- or Mixed IgE/Non-IgE-Mediated Gastrointestinal Food Allergies in the First Years of Life: Old and New Tools for Diagnosis. *Nutrients* **2021**, *13*, 226. [CrossRef] [PubMed]
13. Hirano, I.; Furuta, G.T. Approaches and Challenges to Management of Pediatric and Adult Patients with Eosinophilic Esophagitis. *Gastroenterology* **2020**, *158*, 840–851. [CrossRef] [PubMed]
14. Abou Rached, A.; El Hajj, W. Eosinophilic gastroenteritis: Approach to diagnosis and management. *World J. GastroIntest. Pharmacol. Ther.* **2016**, *7*, 513–523. [CrossRef]
15. Pajno, G.B.; Castagnoli, R.; Muraro, A.; Alvaro-Lozano, M.; Akdis, C.A.; Akdis, M.; Arasi, S. Allergen immunotherapy for IgE-mediated food allergy: There is a measure in everything to a proper proportion of therapy. *Pediatr. Allergy Immunol.* **2019**, *30*, 415–422. [CrossRef]
16. DiTommaso, L.A.; Rosenberg, C.E.; Eby, M.D.; Tasco, A.; Collins, M.H.; Lyles, J.L.; Putnam, P.E.; Mukkada, V.; Rothenberg, M.E. Prevalence of eosinophilic colitis and the diagnoses associated with colonic eosinophilia. *J. Allergy Clin. Immunol.* **2019**, *143*, 1928–1930.e3. [CrossRef]
17. Giallongo, A.; Parisi, G.F.; Licari, A.; Pulvirenti, G.; Cuppari, C.; Salpietro, C.; Marseglia, G.L.; Leonardi, S. Novel therapeutic targets for allergic airway disease in children. *Drugs Context* **2019**, *8*, 212590. [CrossRef]
18. Licari, A.; Manti, S.; Castagnoli, R.; Parisi, G.F.; Salpietro, C.; Leonardi, S.; Marseglia, G.L. Targeted Therapy for Severe Asthma in Children and Adolescents: Current and Future Perspectives. *Paediatr. Drugs* **2019**, *21*, 215–237. [CrossRef] [PubMed]
19. Pelaia, C.; Calabrese, C.; Terracciano, R.; de Blasio, F.; Vatrella, A.; Pelaia, G. Omalizumab, the first available antibody for biological treatment of severe asthma: More than a decade of real-life effectiveness. *Ther. Adv. Respir. Dis.* **2018**, *12*. [CrossRef]
20. Leung, D.Y.; Sampson, H.A.; Yunginger, J.W.; Burks, A.W.; Schneider, L.C.; Wortel, C.H.; Davis, F.M.; Hyun, J.D.; Shanahan, W.R. Effect of anti-IgE therapy in patients with peanut allergy. *N. Engl. J. Med.* **2003**, *348*, 986–993. [CrossRef]
21. Sampson, H.A.; Leung, D.Y.; Burks, A.W.; Lack, G.; Bahna, S.L.; Jones, S.M.; Wong, D.A. A phase II, randomized, double-blind, parallel-group, placebo-controlled oral food challenge trial of Xolair (omalizumab) in peanut allergy. *J. Allergy Clin. Immunol.* **2011**, *127*, 1309–1310.e1. [CrossRef] [PubMed]
22. Savage, J.H.; Courneya, J.P.; Sterba, P.M.; Macglashan, D.W.; Saini, S.S.; Wood, R.A. Kinetics of mast cell, basophil, and oral food challenge responses in omalizumab-treated adults with peanut allergy. *J. Allergy Clin. Immunol.* **2012**, *130*, 1123–1129.e2. [CrossRef] [PubMed]
23. Schneider, L.C.; Rachid, R.; LeBovidge, J.; Blood, E.; Mittal, M.; Umetsu, D.T. A pilot study of omalizumab to facilitate rapid oral desensitization in high-risk peanut-allergic patients. *J. Allergy Clin. Immunol.* **2013**, *132*, 1368–1374. [CrossRef] [PubMed]
24. Nadeau, K.C.; Schneider, L.C.; Hoyte, L.; Borras, I.; Umetsu, D.T. Rapid oral desensitization in combination with omalizumab therapy in patients with cow's milk allergy. *J. Allergy Clin. Immunol.* **2011**, *127*, 1622–1624. [CrossRef] [PubMed]
25. Bégin, P.; Dominguez, T.; Wilson, S.P.; Bacal, L.; Mehrotra, A.; Kausch, B.; Trela, A.; Tavassoli, M.; Hoyte, E.; O'Riordan, G.; et al. Phase 1 results of safety and tolerability in a rush oral immunotherapy protocol to multiple foods using omalizumab. *Allergy Asthma Clin. Immunol.* **2014**, *10*, 7. [CrossRef]
26. Andorf, S.; Purington, N.; Block, W.M.; Long, A.J.; Tupa, D.; Brittain, E.; Spergel, A.R.; Desai, M.; Galli, S.J.; Nadeau, K.C.; et al. Anti-IgE treatment with oral immunotherapy in multifood allergic participants: A double-blind, randomised, controlled trial. *Lancet Gastroenterol. Hepatol.* **2018**, *3*, 85–94. [CrossRef]
27. Wood, R.A.; Kim, J.S.; Lindblad, R.; Nadeau, K.; Henning, A.K.; Dawson, P.; Plaut, M.; Sampson, H.A. A randomized, double-blind, placebo-controlled study of omalizumab combined with oral immunotherapy for the treatment of cow's milk allergy. *J. Allergy Clin. Immunol.* **2016**, *137*, 1103–1110.e11. [CrossRef]
28. Martorell-Calatayud, C.; Michavila-Gomez, A.; Martorell-Aragone's, A.; Molini-Menchón, N.; Cerdá-Mir, J.C.; Félix-Toledo, R.; De Las Marinas-Álvarez, M.D. Anti-IgE assisted desensitization to egg and cow's milk in patients refractory to conventional oral immunotherapy. *Pediatr. Allergy Immunol.* **2016**, *27*, 544–546. [CrossRef]
29. Bedoret, D.; Singh, A.K.; Shaw, V.; Hoyte, E.G.; Hamilton, R.; DeKruy, R.H.; Schneider, L.C.; Nadeau, K.C.; Umetsu, D.T. Changes in antigen-specific T-cell number and function during oral desensitization in cow's milk allergy enabled with omalizumab. *Mucosal. Immunol.* **2012**, *5*, 267–276. [CrossRef]
30. Holgate, S.T.; Djukanović, R.; Casale, T.; Bousquet, J. Anti-immunoglobulin E treatment with omalizumab in allergic diseases: An update on anti-inflammatory activity and clinical efficacy. *Clin. Exp. Allergy* **2005**, *35*, 408–416. [CrossRef]
31. Licari, A.; Manti, S.; Marseglia, A.; De Filippo, M.; De Sando, E.; Foiadelli, T.; Marseglia, G.L. Biologics in Children with Allergic Diseases. *Curr. Pediatr Rev.* **2020**, *16*, 140–147. [CrossRef] [PubMed]
32. Brandström, J.; Vetander, M.; Lilja, G.; Johansson, S.G.O.; Sundqvist, A.-C.; Kalm, F.; Nilsson, C.; Nopp, A. Individually dosed omalizumab: An effective treatment for severe peanut allergy. *Clin. Exp. Allergy* **2017**, *47*, 540–550. [CrossRef] [PubMed]
33. E-B-FAHF-2, Multi OIT and Xolair (Omalizumab) for Food Allergy. Available online: https://clinicaltrials.gov/ct2/show/NCT02879006 (accessed on 27 June 2021).
34. Study Using Xolair in Rush Multi Oral Immunotherapy in Multi Food Allergic Patients. Available online: https://clinicaltrials.gov/ct2/show/NCT02643862 (accessed on 27 June 2021).
35. Multi OIT to Test Immune Markers after Minimum Maintenance Dose. Available online: https://clinicaltrials.gov/ct2/show/NCT03181009 (accessed on 27 June 2021).

36. Multi Immunotherapy to Test Tolerance and Xolair. Available online: https://clinicaltrials.gov/ct2/show/NCT02626611 (accessed on 27 June 2021).
37. Omalizumab with Oral Food Immunotherapy with Food Allergies Open Label Safety Study in a Single Center. Available online: https://clinicaltrials.gov/ct2/show/NCT01510626 (accessed on 27 June 2021).
38. Omalizumab in the Treatment of Peanut Allergy. Available online: https://clinicaltrials.gov/ct2/show/NCT00949078 (accessed on 27 June 2021).
39. Peanut Reactivity Reduced by Oral Tolerance in an Anti-IgE Clinical Trial. Available online: https://clinicaltrials.gov/ct2/show/NCT01781637 (accessed on 27 June 2021).
40. Omalizumab as Monotherapy and as Adjunct Therapy to Multi-Allergen OIT in Food Allergic Participants. Available online: https://clinicaltrials.gov/ct2/show/NCT03881696 (accessed on 27 June 2021).
41. Treatment of Severe Peanut Allergy with Xolair (Omalizumab) and Oral Immunotherapy. Available online: https://clinicaltrials.gov/ct2/show/NCT02402231 (accessed on 27 June 2021).
42. OIT and Xolair® (Omalizumab) in Cow's Milk Allergy. Available online: https://clinicaltrials.gov/ct2/show/NCT01157117 (accessed on 27 June 2021).
43. Xolair Treatment for Milk Allergic Children. Available online: https://clinicaltrials.gov/ct2/show/NCT00968110 (accessed on 27 June 2021).
44. A Safety and Efficacy Study of Xolair in Peanut Allergy. Available online: https://www.clinicaltrials.gov/ct2/show/NCT00086606 (accessed on 27 June 2021).
45. Peanut Oral Immunotherapy and Anti-Immunoglobulin E (IgE) for Peanut Allergy. Available online: https://clinicaltrials.gov/ct2/show/NCT00932282 (accessed on 27 June 2021).
46. A Study of Xolair in Peanut-Allergic Subjects Previously Enrolled in Study Q2788g. Available online: https://clinicaltrials.gov/ct2/show/NCT00382148 (accessed on 27 June 2021).
47. Xolair Enhances Oral Desensitization in Peanut Allergic Patients. Available online: https://clinicaltrials.gov/ct2/show/NCT01290913 (accessed on 27 June 2021).
48. Omalizumab to Accelerate a Symptom-Driven Multi-Food OIT. Available online: https://clinicaltrials.gov/ct2/show/NCT04045301 (accessed on 27 June 2021).
49. Behandling af Boern Med Foedevareallergi Med Omalizumab (Xolair). Available online: https://clinicaltrials.gov/ct2/show/NCT04037176 (accessed on 27 June 2021).
50. Identifying Responders to Xolair (Omalizumab) Using Eosinophilic Esophagitis as a Disease Model. Available online: https://clinicaltrials.gov/ct2/show/NCT01040598 (accessed on 27 June 2021).
51. A Pilot Study of the Treatment of Eosinophilic Esophagitis with Omalizumab. Available online: https://www.clinicaltrials.gov/ct2/show/NCT00123630 (accessed on 27 June 2021).
52. Omalizumab to Treat Eosinophilic Gastroenteritis. Available online: https://clinicaltrials.gov/ct2/show/NCT00084097 (accessed on 27 June 2021).
53. Protection from Food Induced Anaphylaxis by Reducing the Serum Level of Specific IgE (Protana). Available online: https://clinicaltrials.gov/ct2/show/NCT03964051 (accessed on 27 June 2021).
54. Effectiveness of Dupilumab in Food Allergic Patients with Moderate to Severe Atopic Dermatitis. Available online: https://clinicaltrials.gov/ct2/show/NCT04462055 (accessed on 27 June 2021).
55. Study to Investigate the Efficacy and Safety of Dupilumab in Pediatric Patients with Active Eosinophilic Esophagitis (EoE). Available online: https://clinicaltrials.gov/ct2/show/NCT04394351 (accessed on 27 June 2021).
56. Efficacy and Safety of 4 Doses of QGE031 in Patients 18–50 Years of Age with Peanut Allergy. Available online: https://clinicaltrials.gov/ct2/show/NCT01451450 (accessed on 8 May 2021).
57. Chinthrajah, S.; Cao, S.; Liu, C.; Lyu, S.-C.; Sindher, S.B.; Long, A.; Sampath, V.; Petroni, D.; Londei, M.; Nadeau, K.C. Phase 2a randomized, placebo-controlled study of anti-IL-33 in peanut allergy. *JCI Insight* **2019**, *4*, e131347. [CrossRef] [PubMed]
58. Dupilumab FDA Prescribing Information. FDA, 2020. Available online: https://www.accessdata.fda.gov/drugsatfda_docs/label/2019/761055s014lbl.pdf (accessed on 8 May 2021).
59. Dupilumab, E.M. Summary of Product Characteristics. EMA, 2020. Available online: https://www.ema.europa.eu/en/documents/product-information/dupixent-epar-product-information_en.pdf (accessed on 8 May 2021).
60. Rial, M.J.; Barroso, B.; Sastre, J. Dupilumab for treatment of food allergy. *J. Allergy Clin. Immunol. Pract.* **2019**, *7*, 673–674. [CrossRef] [PubMed]
61. Clinical Study Using Biologics to Improve Multi OIT Outcomes. Available online: https://clinicaltrials.gov/ct2/show/NCT03679676 (accessed on 27 June 2021).
62. Calvani, M.; Anania, C.; Caffarelli, C.; Martelli, A.; Miraglia Del Giudice, M.; Cravidi, C.; Duse, M.; Manti, S.; Tosca, M.A.; Cardinale, F.; et al. Food allergy: An updated review on pathogenesis, diagnosis, prevention and management. *Acta Bio-Med. Atenei Parm.* **2020**, *91*, e2020012. [CrossRef]

MDPI

Review

An Updated Overview of Almond Allergens

Mário Bezerra [1,2,3], Miguel Ribeiro [1,2,3] and Gilberto Igrejas [1,2,3,*]

1 Department of Genetics and Biotechnology, University of Trás-os-Montes and Alto Douro, 5000-801 Vila Real, Portugal; mariojbezerra02@gmail.com (M.B.); jmribeiro@utad.pt (M.R.)
2 Functional Genomics and Proteomics Unity, University of Trás-os-Montes and Alto Douro, 5000-801 Vila Real, Portugal
3 LAQV-REQUIMTE, Faculty of Science and Technology, University Nova of Lisbon, Caparica, 2829-516 Lisbon, Portugal
* Correspondence: gigrejas@utad.pt; Tel.: +351-25-935-0530

Abstract: Tree nuts are considered an important food in healthy diets. However, for part of the world's population, they are one of the most common sources of food allergens causing acute allergic reactions that can become life-threatening. They are part of the Big Eight food groups which are responsible for more than 90% of food allergy cases in the United States, and within this group, almond allergies are persistent and normally severe and life-threatening. Almond is generally consumed raw, toasted or as an integral part of other foods. Its dietary consumption is generally associated with a reduced risk of cardiovascular diseases. Several almond proteins have been recognized as allergens. Six of them, namely Pru du 3, Pru du 4, Pru du 5, Pru du 6, Pru du 8 and Pru du 10, have been included in the WHO-IUIS list of allergens. Nevertheless, further studies are needed in relation to the accurate characterization of the already known almond allergens or putative ones and in relation to the IgE-binding properties of these allergens to avoid misidentifications. In this context, this work aims to critically review the almond allergy problematic and, specifically, to perform an extensive overview regarding known and novel putative almond allergens.

Keywords: food allergy; almond; almond allergens; nutrition

Citation: Bezerra, M.; Ribeiro, M.; Igrejas, G. An Updated Overview of Almond Allergens. *Nutrients* **2021**, *13*, 2578. https://doi.org/10.3390/nu13082578

Academic Editors: Sara Manti, Gian Luigi Marseglia, Salvatore Leonardi and Francisco J. Pérez-Cano

Received: 8 May 2021
Accepted: 23 July 2021
Published: 27 July 2021

1. Introduction

Food allergies are a concerning issue affecting the worldwide population, and their prevalence has been increasing for the last couple of decades [1–3]. For example, in the United States, around twenty-six million adults [4] and six million children [1] suffer from this condition. Although there is no cure to food allergies and food avoidance is considered the best strategy, vast research has been made in this area and potential therapies can be generally divided into two categories: allergen non-specific such as the use of monoclonal antibodies and allergen specific where the treatment is performed using recombined or native food antigens [5]. However, less commonly, adverse side effects can range from mild to anaphylaxis or eosinophilic esophagitis [6] and due to their unpredictable character [7], new and innovating therapies must be pursued.

For scientific research to go further, food allergy, allergic diseases and allergens must be firstly identified and characterized. For allergens, when new ones from specific species are identified, a distinctive name is given by the WHO/IUIS Allergen Nomenclature Sub-Committee alongside the additional information about it. A vast number of allergens from more than one hundred and sixty species have been identified and most of them belong to a restricted number of protein families. Among these, the (1) tryp_alpha_amyl protein family includes the higher number of known food allergens, which includes, for example, lipid transfer proteins (LTPs) and 2S albumin seed storage proteins; (2) cupin_1 protein family including the 7S vicilin seed storage proteins and the 11S legumin, and the (3) profilin family comprising profilins, are the most prominent ones [8]. In almonds, several proteins of these protein families have been already identified as allergens, namely Pru du 6 (11S globulin legumin-like protein), Pru du 4 (profilin) and Pru du 3 (nonspecific LTP) and several other proteins belonging to other protein families and/or that do not have a name attributed by the Allergen Nomenclature Sub-Committee.

Great attention has already been devoted to this topic [8–10] and here we intend to present a comprehensive and updated overview of almond allergens, namely the description of Pru du 10, the most recent almond allergen to be added to the WHO-IUIS list of allergens. We also reviewed the legal framework of the European Union and the United States concerning food allergies and labelling, and the methods currently available for the detection and quantification of almond allergens in food products. All these topics combined offer a wide, updated, and comprehensive narrative about almond allergies and allergens. With that, this review aims to provide easy access to updated information about almond allergies to researchers, clinicians, and patients to be applied in their respective manners.

Methods

The research documents analyzed in this work were extracted from the PubMed and Elsevier Scopus online databases collecting academic documents, both including keywords such as 'almond', 'almond allergy' or 'almond allergens' or other topics considered relevant. Only publications in English were included. The articles from the search were assessed according to document type, language, and inclusion in subject category. They were further analyzed, and the results were used to write this review.

2. Food Allergy

By definition, a food allergy is "an adverse food health effect arising from a specific immune response that occurs reproducibly on exposure to a given food" [11]. It is also important to clarify that the immune reaction is key, otherwise food allergies could probably be described as food intolerances, which are a non-immune response but may reproduce food allergy clinical symptoms [12].

Evidence that shows global variation of food allergies as well as changes in their prevalence associated with migration [13] are increasing the interest on the epidemiological strand of food allergies and may promote hypothesis for why food allergy is a rising issue in some parts of the world and not in others [14]. Some authors proposed various hypotheses on the increasing prevalence of food allergy in association with geographical sites; the most accepted ones were hygiene increases, which have led to less pathogen exposure, changes in the human microbiome, avoidance of certain allergens in the early stages of life causing allergen exposure reduction, obesity, diets lacking antioxidants and vitamin D deficiency [15,16].

Tree nuts are one of the Big Eight food groups among peanut, milk, shellfish, soy, wheat, egg and fish which are responsible for more than 90% of food allergy cases in the United States [8] and, in particular, the number of people sensitized to tree nuts and peanuts has been growing concerningly in Europe and the United States [17]. In this group of foods, almond and peanut allergies are persistent and normally severe and life-threatening in opposition to allergies caused by milk or eggs, which are normally mild and transient [18–20].

Tree nut allergy prevalence data is very limited and is even more limited for a specific nut species such as almonds [10]. However, it is known that tree nut allergy rates vary according to geographical regions, ethnic differences, and dietary habits [21].

2.1. Molecular Pathway of Immunoglobulin E-Mediated Food Reaction

Food allergies can arise through several immunological mechanisms that lead to a reaction to food allergens. The most common mechanism of food allergy expression is a hypersensitivity manifestation where specific Immunoglobulin E (IgE) antibodies interact with mast cells and basophils leading to a rapid physiological response [22]. Usually, food allergy symptoms appear nearly immediately, or a few minutes later after food ingestion, however in exceptional cases it could take several hours for the symptoms to manifest [23].

In people with food allergy disorders, the absorption process of allergens in the intestinal epithelium and consequent access to the bloodstream and mucosa is increased [24]. When food allergens are ingested, an interaction occurs between them and IgE and its high-affinity fragment crystallizable receptor (FCER1) on basophils in circulation, or mast cells present in mucosal tissues leading to their activation (Figure 1). FCER1 crosslinking leads to a signaling cascade where tyrosine protein kinase SYK will promote exocytosis of granules containing mediators of hypersensitivity such as histamine, chymase and tryptase [22]. This process together with the synthesis of lipid metabolites such as prostaglandins, leukotrienes and platelet-activating factor (PAF) [25] will result in physiological responses such as the activation of nociceptive nerves that promote itching and soft muscle constriction, vasodilation, higher vascular permeability and, in the most severe cases, anaphylaxis [26].

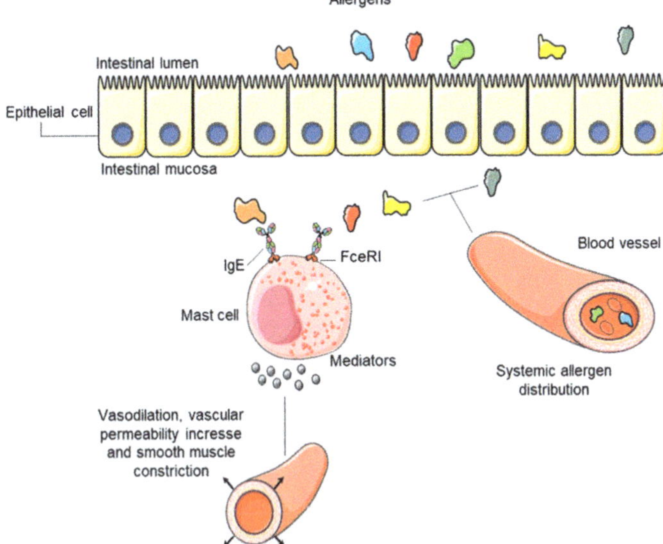

Figure 1. General mechanisms of IgE mediated response to food allergens. Interaction between food allergens and IgE and its high-affinity FC receptor (FCER1) on basophils in circulation or mast cells present in mucosal tissues leading to their activation and consequent physiological response. Adapted from Renz, Allen, Sicherer, Sampson, Lack, Beyer and Oettgen [22]. Adapted with permission from Ref. [22]. Copytright 2018. Springer Nature.

Although this is the generic mechanism after food ingestion, non-IgE mediated reactions such as the inflammatory process subjacent to eosinophilic esophagitis [27] can also occur [27,28]. The physiological response is dependent of the kind of mediators released by the mast cells and basophils but is also dependent on tissue location where these mediators would act. These two factors combined will directly influence the physiological response. [22].

2.2. Legal Framework

There are several regulatory frameworks for food allergen labeling according to countries or regions that differ significantly around the world due to the priority level that each jurisdiction applies to specific allergens. The criteria for the development of the allergen's priority list and the standards for the addition or removal of allergens from the regulations differ and they are often unclear [29].

The Regulation (EU) No. 116/2011 sets the regulation on food labelling, forbidding misleading consumers and any claims that a certain food, such as almonds, can prevent, treat, or cure human diseases cannot be made. Moreover, nutritional and allergen information must be highlighted in the list of ingredients and included in non-packed foods or any product where they are used as ingredient, with the punishment of being withdrawn from the market.

Regulation (EU) No 1169/2011 of the European Parliament and of the Council of 25 October 2011 on the provision of food information to consumers states that allergens should be indicated in the list of ingredients with a clear reference to the name of the substance or product causing allergies or intolerances and should be emphasized through a typeset that clearly distinguishes it from the rest of the list of ingredients, for example by means of the font, style, or background color. In this list of substance or product causing allergies or intolerances nuts are included, with a clear reference to almonds, hazelnuts, walnuts and others, cereals containing gluten, crustaceans, eggs, fish, peanuts, soybeans, milk, celery, mustard, sesame, lupin, mollusks, and products from each one.

In the United States, food labelling requirements are quite similar to the ones applied in the European Union, where the Food Allergen Labeling and Consumer Protection Act of 2004 states that any food source containing a major food allergen, or protein derived from them, should be printed right next to the ingredient list, and specifically have the word "contains" before it. The term "major food allergen" refers to milk, egg fish, crustacean shellfish, tree nuts (like almonds, pecans, or walnuts), wheat, peanuts, and soybeans, however any highly refined oil derived from any of the previous foods and products derived from those oils are considered exceptions.

For the appliance of the food labelling requirements, it is important to defined threshold values which correspond to the minimal concentration of a specific food allergen in a food able to trigger any reaction in a sensitized individual. However, is very difficult to establish a threshold, since they vary according to the individual/population, the allergens itself and the consequent food processing [30]. To get there, wide population tests and data are needed. For almonds, currently no thresholds are established [10], which shows a clear sign that further investigations and regulations are imperative.

3. Almond

One of the most important foods in human nutrition are tree nuts, namely due to their excellence in terms of taste as well as their versatility to be used combined with other foods and, more recently, their potential health benefits. All these characteristics mean that tree nuts are consumed all around the world in the most various of forms, according to the availability in the region and the populational habits [9,31].

The almond (*Prunus dulcis* Mill.) is a member of the *Rosaceae* family and is considered a native plant from Minor Asia [32], being one of the oldest nut trees cultivated worldwide with special relevance in the Mediterranean warm-arid countries [33,34], namely the Apulia region on southern Italy [35]. Among tree nuts, almonds present as one of the most important nuts, which is very noticeable in tree nut production data around the world (Walnut 3663, Almond 3183 and Hazelnut 864 ktons/year; [36]). Furthermore, its nutritional properties should be highlighted; high levels of mono and polyunsaturated fatty acids, phytosterols and a low glycemic index are associated with reduction of some risk factors for cardiovascular disease and diabetes [37–40]. It has also been described as having antioxidant and inflammatory activities due to its polyphenol content, including flavonoids, hepato and neuroprotective potential and, perhaps the most known, cholesterol-lowering properties [41–44]. Also, almond derived products such as their oils have demonstrated both antibacterial and antifungal capabilities [45] which makes almond a product of great interest both to the consumer and producer.

Regarding almond cultivars, European commercial cultivars such as the Spanish Marcona, Glorieta, Masbovera, Guara and Francolí cvs. and the French Ferrastar, Ferraduel and Ferragnès cvs. are the main ones produced in Europe. In the United States, the most widely produced almond variety in Nonpareil cv. represents near half of the production. On other hand, in Portugal there is a mix of traditional and local varieties such as Amendoão, Pegarinhos, Casanova and Refego cvs. [46,47]. However, in a study testing three almond varieties Nonpareil, Mission and Carmel against eight almond allergic patient's sera, no significant differences were found. New similar research must be conducted to correctly evaluate the allergic potential of each variety of interest [48].

Along with the almond nutritional value comes the agronomical properties of different cultivars. For example, Bolling et al. [49] described that the individual polyphenols synthesis was only due to the cultivar itself, however total polyphenols and antioxidant activity were significantly dependent on both genotype and environmental growing conditions. Pursuing this point of view, Summo et al. [50] performed a study aiming to determine if either the cultivar or harvest time influence the chemical composition of the fruit. From this, the team concluded that, in fact, harvest time and genotype both have a strong influence on the fruit nutritional value.

3.1. Almond Allergy

Nut allergy is associated with clinical symptoms that can range in severity from mild to life-threatening, and in this sense when a patient is diagnosed with an allergy to a certain nut it is often advised to avoid the consumption of the entire group [51,52].

Epidemiologically speaking, almond allergies have the fourth highest prevalence among the tree nuts allergies [53]. Looking at the specific cases of the United States, Korea, United Kingdom, Mexico and Sweden, almonds present the third most common tree nut to cause allergies in the United States [10], and between 9% and 15% of people pre-sensitized to tree nuts also report allergy to almonds [54]. In a study performed in a group of 134 Korean patients with previous reports of food allergies, 11.2% also reported almond allergies. Among them, 16.3% were between 19 to 29 years old, 13% in the 40–49 age group and 9.1% in the 50–59 group. Also, the same study reported that sensitivity to almonds is lower in females, with 9.8% compared to males at 13.5% [55]. In the United Kingdom, in pre-sensitized individuals, almonds represent the most common tree nut allergy, with 22% to 33% of the cases [54,56]. The higher rate of sensitization to almonds was reported in a study performed in Mexico City, reporting a 43% rate in older children with ages comprised between 6 and 17 years old [57]. A cross-sectional enquiry made in Sweden with 1042 responses from individuals between 17 and 78 years old concluded that near 32.5% of adults had food hypersensitivity and 3% were sensitive to almonds [58].

Almond allergy can cause several clinical responses. The Oral Allergy Syndrome (OAS) is a pollen-food syndrome that produces mild oral symptoms in cases of pollen sensitization triggered by nuts. Although it hardly causes anaphylaxis, it can happen in the direct confrontation of serum sIgE with PR-10 homologous [59]. Another common clinical response is allergic rhinitis, that has been associated with almond allergies in a study performed in southern Taiwan with a group of 216 individuals with ages comprised between 2 and 93 years old. Most of these people had respiratory and cutaneous symptoms, and the study reported a 36.97% prevalence of allergic rhinitis caused by almonds in the group of the non-sensitized patients. Besides allergic rhinitis, asthma has been associated with almonds with a prevalence of 7.4% in the non-sensitized nut group and 13.70% in the sensitized one. Also in Taiwan, it was reported that almonds were responsible for 42.47% of atopic dermatitis cases in a group of 33 nut sensitized individuals [60]. Other symptoms can emerge, such as gastrointestinal ones. In a group of 1024 sensitive individuals, 15% reported these, and from those, 2.7% were due to almonds [58].

Regarding strategies for prevention and therapy for an almond allergy, the main method is dietary avoidance. Individuals sensitive to almonds should take special attention looking at packages and labels to prevent the ingestion of almond or almond-based products [59]. However, there are some strategies that seem to prevent the development of almond allergies, namely the premature consumption of almonds during infancy or even during pregnancy, or lactation also showed a positive impact on its prevention [61].

Moreover, there is evidence that about 10% of tree nut allergies are outgrown by young individuals who develop tolerance due to the rise of T regulatory cells and the consequent reduction of allergen specific IgE [62]. Immunotherapy, a food allergen-specific therapy, which refers to the administration of gradual and increasing doses of an antigen over a certain time [63,64], is considered as a solid option since in the majority of cases the side effects are mild, such as itching and, if successful, immunotherapy can induce desensitization and less commonly sustained unresponsiveness, also known as tolerance [5]. Moore, Stewart and Deshazo [5] believe that tolerance induced by immunotherapy with or without the administration of monoclonal antibodies could significantly shift the allergic diseases field.

Cross reactivity between almonds and other sources of allergens is a well-known problem and there are some of these associations (summarily described in Table 1) already described.

Table 1. Almonds' most common cross reactions with other relevant sources of allergens. Green areas represent a positive association between almond allergens and other allergens of the respective sources.

<table>
<tr><th></th><th></th><th colspan="9">Possible Cross-Reaction Source</th></tr>
<tr><th>Source</th><th>Allergen</th><th>Mahleb</th><th>Peanut</th><th>Chestnut</th><th>Hazelnut</th><th>Walnut</th><th>Peach</th><th>Pollen</th><th>Profilin-Containing Plants</th><th>Maze</th></tr>
<tr><td rowspan="5">Almond</td><td>Pru du 3</td><td></td><td></td><td colspan="4">[65]</td><td></td><td></td><td></td></tr>
<tr><td>Pru du 6</td><td>[66]</td><td></td><td></td><td></td><td></td><td></td><td></td><td></td><td>[67]</td></tr>
<tr><td>Pru du 1</td><td></td><td></td><td></td><td></td><td></td><td></td><td>[59]</td><td></td><td></td></tr>
<tr><td>Pru du 4</td><td></td><td></td><td></td><td></td><td></td><td></td><td></td><td>[68]</td><td></td></tr>
<tr><td>Pru du γ-conglutin</td><td></td><td>[69]</td><td></td><td></td><td></td><td></td><td></td><td></td><td></td></tr>
</table>

Nevertheless, it is still unclear if the taxonomic proximity between tree nuts groups and peanuts is a key factor for the cross-reactivity between these two, or it comes from the high structural homology of IgE-binding epitopes [70,71]. In general, tree nut allergies are caused by non-pollen-mediated food sensitization, however, in cases such as with almonds and hazelnuts, sensitization to plane tree pollen, birch pollen or mugwort pollen may induce allergies [72,73] such as those schematically represented in Figure 2. On the other hand, tree nut allergy cross reaction is highly related to botanical family associations which, for almonds, is common regarding cross-reactivity between other members of the *Rosaceae* family [74,75]. Furthermore, within the *Rosaceae* family, a strong source of cross-reaction lies in the structural homology between allergic lipid-transfer proteins (LTP's). Specifically, in the tree nut group, almond Pru du 3, chestnut Cas s 8, hazelnut Cor a 8 and walnut Jug r 3 are the most predisposed to show cross-reactivity. Besides these, peach Pru p 3 holds higher IgE-binding affinity and a higher number of epitopes compared to other LTP's, which results in the fact that a peach is a primary sensitizer to LTP's [65] and makes it a strong cause for cross-reactivity to other plants, including nuts like almonds [76]. Other studies performed by Kewalramani et al. [77] showed extensive IgE cross-reactivity between almonds and apricot seeds, and that there may exist some cross-reactive proteins with pine nut, pecan, walnut, and sunflower seeds.

3.2. Almond Allergens

To date, ten groups of almond allergens have been identified, namely: Pru du 1, Pru du 2, Pru du 2S albumin, Pru du 3, Pru du 4, Pru du 5, Pru du 6 (amandin), Pru du γ-conglutin, Pru du 8 and Pru du 10. From these groups, only Pru du 1, Pru du 2, Pru du 2S albumin and Pru du γ-conglutin are not included in the WHO-IUIS list of allergens. Their corresponding biochemical names, biological functions, GenBank nucleotides and UniProt annotations, molecular weight, food processing effects and clinical relevance are summarized in Table 2.

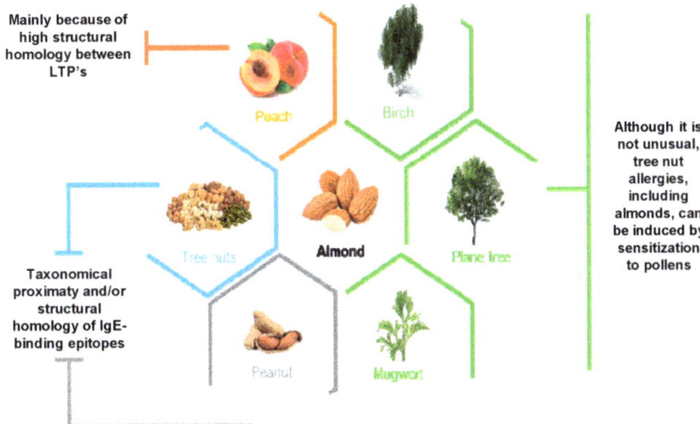

Figure 2. Most associated allergic cross-reactions with almonds. In orange, allergic cross reactions between almonds and peaches are most commonly due to high structural homology between allergic LTP's present in the Rosaceae family that both belong to; in blue and grey, it is still unclear if cross reactivity between almonds and other tree nuts groups and peanuts is a consequence of taxonomical proximity and/or high structural homology of IgE-binding epitopes; finally, in green are represented three different pollens which, although it is not usual, when sensitized to them allergies to tree nuts such as almonds could be induced.

3.2.1. WHO/IUIS Designated Almond Allergens

Pru du 6 (Amandin)

Pru du 6 or amandin is the most well and widely studied almond allergen according to its biochemical function and molecular structure [78–81]. It was first reported as an allergen in 1999 [77] but was only recognized in 2010 and added to the WHO-IUIS database.

Biochemically, amandin, also known as almond major protein (AMP), is a member of the cupin superfamily, namely the 11S seed storage globulin family [51,52]. Globulins are very abundant proteins in legumes and tree nuts, and in almonds they correspond to roughly 65% of total almond protein content [9].

As an allergen, Pru du 6 have been associated with severe allergic reactions [80]. Studies on the Pru du 6 isoforms, Pru du 6.01 and Pru du 6.02, showed that the 6.01 isoform is more broadly recognized than the 6.02 isoform. In addition, its denaturation had only slightly effects on IgE-binding intensity in sensitive subjects [82]. In fact, Pru du 6 polypeptides are highly resistant to heat treatment, which is one of the most common strategies to decrease or even eliminate the allergenic potential of foods. Due to its heat resistance, contamination of food with Pru du 6 polypeptides presents a serious threat to sensitized patients [83]. On the other hand, some experiments using in vitro models of gastrointestinal digestion suggested that this allergen is sensitive to pepsin but, interestingly, when almond flour is added to other foods, pepsin's action on Pru du 6 is a lot less effective [84]. Holden et al. [85] suggested that the reaction between Pru du 6 and α-conglutin from lupine, another 11S globulin, may be the cause of it.

Pru du 5 (60S Acidic Ribossomal Protein P2)

Pru du 5, also known as 60S acidic ribosomal protein P2, is encoded by *P. dulcis* 60S acidic ribosomal protein gene and was included in the WHO/IUIS allergen list in 2007. This name comes from the fact that this allergen is an 11 kDa protein which is a member of the 60S large subunit of the eukaryotic 80S ribosomes [8], and its biological function is related to protein biosynthesis. Pru du 5 is considered a major almond allergen due to the presence of specific IgE antibodies in 50% of sensitized patients' sera [86].

This allergen can exist as a complex with other ribosomal components/proteins or in its free state [65], with the ability to form homodimers and oligomers [72,74]. On the allergenicity front, this data is very important because oligomerization gives the allergen the capability of cross-linking IgE antibodies on mast cells and/or basophils surfaces, even if the recognition is made from a single epitope of the allergen [8].

Although being considered a major allergen and present in the WHO-IUIS allergen list, many authors believe that this classification must be supported by more studies concerning the IgE reactivity of allergic patients' sera to this allergen [9,10]. Also, studies regarding the biochemical and immunological properties of Pru du 5 in its natural state as an allergen are lacking [8], leading to the conclusion that newer and tougher studies are needed.

Pru du 3 (nsLTP)

Added to the WHO/IUIS database in 2009, Pru du 3 is a non-specific lipid transfer protein 1 (nsLTP1) belonging to the subfamily of nonspecific lipid transfer proteins (nsLTPs) [75]. This family includes proteins constituted by a hydrophobic core to ease lipid transference such as phospholipids, steroids, fatty acids, and glycolipids between membranes. Besides that, nsLTPs are also known as pathogenesis-related 14 (PR-14) proteins, a member of the prolamin superfamily [9,65], which actively participate in plant-defense mechanisms against fungal and bacterial pathogens and other environmental stresses [76].

In almonds we identified and characterized three nsLTPs [87] with identical molecular weights (9 kDa) and similar amino acid lengths: 117, 123 and 116 amino acids for Pru du 3.01, 3.02 and 3.03, respectively. In the three isoallergens, there are eight cysteine conserved residues, which allow the formation of four disulfide bonds [9].

Due to the typical accumulation of this protein family in outer epidermal layers, the peels are associated with stronger allergenicity compared with the pulps of the fruits in the *Rosaceae* family. Regarding allergenicity, this protein family is quite concerning because of its resistance to abrupt pH changes, pepsin digestion, thermal treatments, and the ability of restore folding structures and the consequent proprieties after cooling [88]. Cross-reactivity is also a major concern once the nsLTP family is characterized by a high level of conserved sequences and tridimensional structures allowing IgE recognition, which in turn results in cross-reactivity between species [76]. Furthermore, the Rosaceae fruits and seeds normally present nsLTP proteins, and with that comes a high probability of cross-reactivity between, for example, apples, peaches, cherries, apricots and almonds [89]. This latest evidence is the main reason why nsLTPs are included in the panallergens group—allergens ubiquitously spread throughout nature, showing a high level of conservation besides being from different and unrelated organisms [8].

Pru du 4 (Profilins)

Pru du 4 proteins are included in the profilin family and are encoded by the putative genes *Pru du 4.01* and *Pru du 4.02* [68] which, although present in different size fragments (1041 and 754 bp, respectively) encode two proteins with similar sequences (131 aa), molecular weights (roughly 14 kDa) and acidic properties (*p*I near 4.6) [9].

These proteins can establish high-affinity complexes with monomeric actin, leading to its polymerization into filaments. Once they are associated with actin, it is not surprising that profilin allergens are included in the panallergens group with Pru p 4.01 and Pru av 4 from peaches and sweet cherries, respectively, being the most similar and identical proteins (99 and 98%, respectively) in relation to almond profilins. In general, profilins seem to present moderate structural stability, and harsh conditions contribute to their denaturation and consequent loss of conformational structure. In almonds, Pru du 4 profilins are very difficult to detect by immunoblot screens because of their low levels and their labile character. Because almond profilins antibodies are detected in 44% of patients' sera, they are classified as minor allergens [68].

Pru du 8

Pru du 8 is one of the latest allergens included in the WHO-IUIS database. This allergen was reactive in six of eighteen sera of almond allergic patients [10,84]. Biochemically speaking, Pru du 8 is characterized by a signature repeat of a $CX_3CX_{10-12}CX_3C$ (X being any amino acid), motif which is also related to the N-terminal or the signal peptide of some vicilins [90], and it was also reported to maintain antimicrobial function of some peptides derived from macadamia vicilin [91].

The first nomenclature attempt for this allergen was based on the sequencing of two short peptides of this allergen to reveal the identity of an IgE-reacting protein several years ago. Nevertheless, the result was a misidentification of this allergen as an almond 2S albumin because of the sequence alignment of the two peptide sequences and those in other 2S albumin proteins [92]. More recently, in silico investigations and bioinformatic analyses reopened the debate, naming this allergen as Pru du vicilin (almond 7S vicilin), although some authors believe in a second misidentification [8,93]. In fact, the authors claim that this misidentification is due to the similarity between the signal peptides of vicilins of other species and Pru du 8. Besides that, it is argued that some Pru du 8 orthologs present in the NCBI database, most of them predicted by automatic genome annotations, are incorrectly named as vicilin-like proteins due to the absence of the cupin signature domains of 7S vicilins [8,90].

All this controversy shows that further studies are needed to better elucidate the actual protein family of Pru du 8.

Pru du 10

To date, this allergen was the last one to be added to the WHO-IUIS database. This allergen corresponds to mandelonitrile lyase 2 (formerly hydroxynitrile lyase 2), which is a highly effective catalytic enzyme [87]. This allergenicity was recognized after allergic response to almond ingestion where thirteen of eighteen almond allergic patients were sensitized. Also, the Pru du 10.0101 isoallergen was identified and added to the WHO-IUIS allergen information.

Besides being identified in raw almond samples, this protein was also identified in digested samples, which may indicate that this allergen is able to overcome the digestion process [89]. Still, there is a lack of information regarding this allergen which clearly shows that more studies should address this issue.

3.2.2. Allergens Not Included in the WHO/IUIS Allergen List

There are two main processes to classify a protein as a food allergen, based on immunological data such as the IgE reactivity or based on sequence similarity with proteins of other species already considered allergens. For an allergen to be included in the WHO-IUIS database, immunological data is required and because of that, some authors defend that those which cannot be supported by it should hardly be assumed as an allergen. However, bioinformatic-based investigation is very important to promote further investigation and make aware the scientific and industrial community to the dangers of food allergens.

Pru du γ-Conglutin

The IgE and serological reactivity to Pru du γ-conglutins were not associated with any clinical symptoms and because of that, they are not recognized into standard clinical nomenclature [10].

After the report and characterization of conglutins in other fruits and seeds such as lupine [94], peanut [95], soybean [96] or cashew [97], in almonds an N-terminal peptide sequence of 25 aa belonging to a IgE binding protein with a molecular weight of 45 kDa was also identified, presenting around a 40% identity rate between the mature forms of γ-conglutin from wide and narrow-leafed lupine [92]. Moreover, with a high similarity, approximately 50%, between this almond protein and 7S globulin from soybean, this allergen was considered a vicilin (7S globulins) of the cupin superfamily [8,9]. Nevertheless, some authors do not agree with this classification, stating that γ-conglutin is not a vicilin due to its biochemical properties [8]. In particular γ-conglutin presents sequence and structural similarities with xyloglucan-specific endo-beta 1,4-glucanase inhibitors, however such glucanase inhibition properties are not related to the natural γ-conglutin due to is peptidase cleavage susceptibility [98].

The same authors believe that more studies regarding immunological and biochemical properties of this protein are needed, and the confirmation of this assumption would make this protein the first food allergen from this supposed protein family.

Pru du 1-PR-10 Protein (Pathogenesis Related-10 Protein)

Pathogenesis related proteins are a common group of proteins, generally upregulated in plants to promote defense mechanisms against pathogens such as viruses, bacteria or fungi and environmental factors [8]. The PR-10 family is related to the intracellular defense processes and the response to fungal and bacterial infections. Due to its function, there are numerous isoforms which promote different IgE-binding capabilities [89]. Furthermore, PR-10 proteins are constitutively expressed in different plant parts and usually are not related to other PR proteins [99]. They are commonly seen as pollen or food allergens [100,101] and because of that they can be considered as panallergens, being responsible for cross-reaction events [76].

Although there is no immunological data to support their classification as an allergen and the high similarity and identity between almond PR-10 proteins and the peach counterparts, which are known allergens (Pru p 1), almond PR-10 proteins are assumed as an allergen and named as Pru du 1 [76].

Pru du 2 (PR-5/Thaumatin-Like Protein)

This allergen group is also known as PR-5 or thaumatin-like proteins (TLPs) and are responsible for the biological response to pathogen infection, fungal proteins, and osmotic stress. The TLP's group is known to be very resistant to proteases, heat-induced denaturation, and pH variations, possibly because of sixteen conserved cysteine residues which form eight disulfide bonds [89]. Several isoallergen genes have been identified which code for TLP, ranging in molecular weight from 23 to 27 kDa. Also, the isoallergens aminoacidic sequence length ranges from 246 aa to 330 [102].

Like PR-10 proteins, no immunological characterization of PR-5 almond proteins exists. Although, it is believed that these proteins are almond allergens due to the high sequence identity with Pru p 2, a peach allergen [103]. Moreover, due to their biochemical properties, traditional food-processing practices do not significantly influence these protein's structure and characteristics, so they could affect sensitive patients [9].

Pru du 2S Albumin

Included in the prolamin superfamily, 2S albumins are an important group of seed storage proteins involved in seed growth and in defense related mechanisms [104,105]. Besides 2S albumin, the prolamin superfamily also includes other protein groups such as the non-specific lipid transfer proteins (nsLTPs), prolamin storage proteins and α-amylase/trypsin inhibitors, which may indicate several cross-reactions [106].

2S albumins are thought to be somehow resistant to acidic pH enzyme digestion, particularly the albumins with proteolytic activity and surfactant denaturation effects. These conclusions come from the fact that is believed to this proteins cause sensitization along the intestinal tract, which could only be possible if the previous resistances were actually accurate [107].

As an allergen, the strongest data that lead to the classification of almond 2S albumins as almond allergens is the two short partial peptide sequences with high similarity with 2S albumins of other species [108] that, as discussed in Section 3.2.1, some authors believe to be a misidentification and really correspond to Pru du 8 proteins [8]. In fact, 2S albumins of other species, such as Ara h 2 (peanut 2S albumins) for example, are very potent allergens [109–111] and for this reason the assessment of whether these almond proteins are allergens or not is required and imperative.

Table 2. Almond allergens and their biological function, molecular weight, food processing effects and clinical relevance.

Allergen	Biochemical Name	WHO-IUIS	Isoallergen and Variants	GenBank Nucleotide	UniProt	Biological Function	MW (kDa)	Processing	Clinical Relevance	References
Pru du 3	non-specific Lipid Transfer Protein 1 nsLTP1	Yes (2009)	Pru du 3.0101	FJ652103	C0L0I5	Non-specific lipid transfer protein (nslTP1) and plant defense proteins against pathogens	9	Very resistant to pH, thermal and enzyme treatments	Systemic and life-threatening symptoms; cross reactivity among *Rosaceae* fruit	[112]
Pru du 4	Profilin	Yes (2006)	Pru du 4.0101 Pru du 4.0102	AY081850 AY081852	Q8GSL5 Q8GSL5	Actin-binding protein for cellular function	14	Unstable during heat processing	Mild symptoms and mainly in oral cavity	[68]
Pru du 5	60S acidic ribosomal protein P2	Yes (2007)	Pru du 5.0101	DQ836316	Q8H2B9	Protein synthesis	10	Unknown	Unknown	[86]
Pru du 6	Amandin, 11S globulin legumin-like protein	Yes (2010)	Pru du 6.0101 Pru du 6.0201	GU059260 GU059261	E3SH28 E3SH29	Major storage protein	360	Stable to dry heat but can be denatured by boiling	Severe IgE allergic reactions	[82]
Pru du 8	Antimicrobial seed storage protein	Yes (2018)	Pru du 8.0101	MH922028	A0A516F3L2	Antimicrobial and seed storage function	31	Unknown	Unknown	[90]
Pru du 10	Mandelonitrile lyase 2	Yes (2019)	Pru du 10.0101	AF412329.1	Q945K2	Highly efficient catalytical enzyme	60	Resistant to enzyme digestion	Unknown	[87,89]
Pru du γ-conglutin	Cupin superfamily	No	———	———	———	7S vicilin storage protein	45 for each subunit	Unknown	Unknown	[92]
Pru du 1	PR-10 protein	No	———	———	———	Plant pathogenic and stress response	17	Wet heat processing reduces IgE reactivity	Unknown	[99]
Pru du 2	PR-5/thaumatin-like protein	No	———	———	———	Pathogenic response	23–27	Resistant to protease, pH or heat treatment	Unknown	[113]
Pru 2S albumin	Prolamin super family	No	———	———	———	Seed storage protein	12	Stable to heat treatment	Unknown	[92]

3.3. Methods for Almond Allergens Detection

Most of the methods used for the detection of almond allergens are based in immunochemical properties, DNA techniques and, lately, in Mass Spectrometry (MS) approaches [9].

The immunochemical methods are based on the interaction between immunoglobulins and epitopes present in the target protein. For almond allergen detection, lateral flow devices (LFD), immunoblotting and especially Enzyme-Linked Immunosorbent Assay (ELISA), are very standard methods and the usual techniques for quantitative and qualitative detection of food allergens [114,115]. This comes from the fact that ELISA tests, for example, have enough sensitiveness for protein detection (in the orders of ppm), being the main advantage of the fast assessment, which is important for clinical purposes [115]. Several immunological commercial kits, such as the ones exemplified in Table 3, have been developed with the objective of delivering the most sensitive result in the shortest amount of time. As seen in the kit's characteristics, ELISA-based methods provide more sensitive results, as their limit of detection is lower than the LFD-based kits. However, the assay time is longer for the ELISA cases. Taking this into consideration, the assay type should be taken into serious consideration, according to the situation that are supposed to be used.

Table 3. Example of commercial immunological kits for almond detection and/or quantification and their main characteristics: time for results including extraction times, assay type, limit of detection (LOD), limit of quantification (LOQ) and their manufacturers.

Kit [1]	Assay Time	Assay Type	LOD (ppm)	LOQ (ppm)	Company
ELISA-based					
MonoTrace ELISA kit	40 min	Monoclonal antibody-based ELISA	0.15	1	BioFront Technologies, Tallahassee, FL, USA
SENSISpec ELISA almond	75 min	Sandwich enzyme immunoassay	0.2	0.4	Eurofins Technologies, Budapest, Hungary
RIDASCREEN FAST Mandel/Almond	50 min	Polyclonal antibody specifically for almond protein detection, sandwich ELISA	0.1	2.5	R-Biopharm AG, Madrid, Spain
AgraQuant® Plus Almond	30 min	Sandwich enzyme-linked immunosorbent assay	0.5	1	Romer Labs®, Getzersdorf, Austria
LFD-based					
AgraStrip® Almond	11 min	Lateral flow device	2	———	Romer Labs®, Getzersdorf, Austria
Reveal 3-D Almond Test	10 min	Lateral flow device	5	———	Neogen Corp., Lansing, MI, USA
Lateral Flow Almond incl. Hook Line [2]	10 min	Lateral flow device	1	———	R-Biopharm AG, Madrid, Spain

[1] Mention of commercial kits and trade names is only for exemplification purposes and the authors declare no competing financial interest.
[2] The hook line is included with the purpose of overcoming the hook effect—very high amounts of an analyte in the sample can lead to falsely lowered or negative results.

Another possible approach, instead of looking directly for the protein itself, is the DNA-based method where an amplification is performed of the gene fragment responsible for encoding the allergen by Polymerase Chain Reaction (PCR), allowing quantitative and qualitative measurement using real-time PCR or endpoint PCR assays, respectively [10]. One of the advantages of these methods is that they rely on the detection of low quantities of almond DNA even after food processing, which could promote the degradation of some allergen proteins and therefore not be detected by immunological approaches [116]. However, the presence of the gene encoding the allergens does not imply its expression and, because of that, the synergistically use of DNA-based techniques and ELISA could overcome some of the drawbacks of both techniques [117].

Proteomics play a very important role in the food allergy problematic, firstly on a fundamental investigation basis to characterize allergens and further to their application in the diagnostic routines. Namely, a variety of tests and methods must be applied to characterize allergens according to their allergenic activities, purity and folding properties. Following this line of thought, SDS-Polyacrylamide Gel Electrophoresis (SDS-PAGE) is a reliable technique to determine purity, and following 2 Dimension (2D) electrophoresis, capillary electrophoresis or High-Performance Liquid Chromatography (HPLC) are great techniques to access individual isoforms and obtain more additional information in general. Further, MS techniques are powerful tools to determine protein molecular masses, being Matrix Assisted Laser Desorption Ionization (MALDI) and ElectroSpray Ionization (ESI), as the most commonly used [118,119]. MS techniques have been the most recent methods to be explored for qualitative and quantitative purposes [120,121]. For example, the isolation and characterization of Pru du 3 allergen was conducted using MS techniques where the full sequence was obtained by Liquid Chromatography ElectroSpray Ionization Orbitrap Mass Spectrometry (LC-ESI-Orbitrap-MS) [122]. Mass spectrometry has the advantage of ELISA tests which can directly identify proteins with a high sensitivity, and therefore could provide a direct risk evaluation and, besides that, can be used for the detection of multiple allergens simultaneously [122]. MS could be the chosen technique for a standard test; however, it is a relatively recent approach which demands expensive equipment and specialized personnel. At this standpoint, further improvements are required to allow easier access and profitable use by clinical facilities [10].

Another methodology under development is based on microarrays. Namely, allergen microarrays such as the MeDALL allergen-chip have been explored for the diagnosis and monitoring of allergies. The main advantages rely on the simultaneous detection of several allergens with a minimal amount of sera in a reduced time. The development of this chip has the purpose of monitoring IgE and IgG reactivity profiles against 170 allergens in sera collected from European birth cohorts. With that information, it would be possible to make a geographical association of clinical important allergens in different populations and track the progress of food allergy itself and would allow clinical therapies to act in a prophylactic and more personalized manner [123].

It is worth mentioning the basophil activation test (BAT) as a powerful method for tree nut allergy diagnosing [124]. This is an in vitro assay based on flowcytometry protocols that, essentially, allows the evaluation of activation and/or degradation levels of basophils upon the intentional contact with the pretended food allergens [125]. However, it also has some limitations, mainly because of the level of equipment required which makes difficult the use of this technique in small medical centers; this could be overcome with the use of specialized centers and with new research to lower the costs. On other hand, results have been shown that BAT assays have very strong performances and useful results, including multi-nut sensitizations and, because of that, medical infrastructures should take this test into consideration for these kinds of diagnostics [126].

4. Conclusions

Almond production has been increasing for the last years and is currently positioned as one of the most consumed tree nuts and one of the most likely to cause mild to severe allergic reactions. Worldwide data regarding the epidemiological standpoint of almond allergies is concerningly scarce. Without this kind of information, it is hard for governmental and medical institutions to establish personalized and efficient protocols and initiatives to mitigate this problem.

On the other hand, a lot of almond proteins have been already described as potential allergens, although only a part of them have been recognized as allergenic and the authenticity of some designations have been questioned, mainly due to misidentification problems. It is expected that the development of suitable analytical methods for the efficient detection of food allergens and its characterization, for example supported by comprehensive proteomics approaches, will help in the validation of many of these proteins/allergens in the years to come.

For the near future, the develop of new techniques and the increasing usage of powerful ones like BAT should happen to take a step forward into the search for a more permanent solution. In the meantime, accurate characterization of ancient and local varieties should be made for the possible selection of hypoallergenic varieties, and breeding programs can be used for the development of varieties with hypoallergenic characteristics. Moreover, the effort of also evaluating almond-based products must be made to secure safety for the general consumer.

However, a long way is yet to be made and researchers, clinical institutions and governmental entities must work together to establish an efficient network covering all the aspects of almond allergies in order to better understand this problem and enable the development of new and more efficient preventive therapies.

Author Contributions: The authors' contributions were as follows: conceptualization, M.R.; investigation, M.B. and M.R.; data curation, M.B.; writing—original draft preparation, M.B.; writing—review and editing, M.B., M.R. and G.I.; project administration, G.I.; funding acquisition, G.I. All authors have read and agreed to the published version of the manuscript.

Funding: This work was funded by the R&D Project GLUTEN2TARGET-Optimizing natural low toxicity of wheat for celiac patients through a nano/microparticles detoxifying approach, reference POCI-01-0145-FEDER-029068 and PTDC/BAA-AGR/29068/2017, financed by the European Regional Development Fund (ERDF) through COMPETE 2020-Operational Program for Competitiveness and Internationalization (POCI) and by Foundation for Science and Technology (FCT).

Acknowledgments: The authors acknowledge the supported by the Associate Laboratory for Green Chemistry-LAQV, which is financed by FCT under the Partnership Agreement UIDB/50006/2020.

Conflicts of Interest: The authors declare no conflict of interest.

References

1. Gupta, R.S.; Springston, E.E.; Warrier, M.R.; Smith, B.; Kumar, R.; Pongracic, J.; Holl, J.L. The Prevalence, Severity, and Distribution of Childhood Food Allergy in the United States. *Pediatrics* **2011**, *128*, e9–e17. [CrossRef]
2. Gupta, R.S.; Warren, C.M.; Smith, B.M.; Jiang, J.; Blumenstock, J.A.; Davis, M.M.; Schleimer, R.P.; Nadeau, K.C. Prev-alence and severity of food allergies among US adults. *JAMA Netw. Open* **2019**, *2*, e185630. [CrossRef]
3. Kumfer, A.M.; Commins, S.P. Primary prevention of food allergy. *Curr. Allergy Asthma Rep.* **2019**, *19*, 7. [CrossRef]
4. Sicherer, S.H.; Sampson, H.A. Food allergy: A review and update on epidemiology, pathogenesis, diagnosis, prevention, and management. *J. Allergy Clin. Immunol.* **2018**, *141*, 41–58. [CrossRef] [PubMed]
5. Moore, L.E.; Stewart, P.H.; Deshazo, R.D. Food Allergy: What We Know Now. *Am. J. Med. Sci.* **2017**, *353*, 353–366. [CrossRef]
6. Sánchez-García, S.; Del Río, P.R.; Escudero, C.; Martínez-Gómez, M.J.; Ibáñez, M.D. Possible eosin-ophilic esophagitis induced by milk oral immunotherapy. *J. Allergy Clin. Immunol.* **2012**, *129*, 1155–1157. [CrossRef] [PubMed]
7. Varshney, P.; Steele, P.H.; Vickery, B.P.; Bird, J.A.; Thyagarajan, A.; Scurlock, A.M.; Perry, T.T.; Jones, S.M.; Burks, A.W. Adverse reactions during peanut oral immunotherapy home dosing. *J. Allergy Clin. Immunol.* **2009**, *124*, 1351–1352. [CrossRef]
8. Zhang, Y.; Jin, T. Almond allergens: Update and perspective on identification and characterization. *J. Sci. Food Agric.* **2020**, *100*, 4657–4663. [CrossRef]
9. Costa, J.; Mafra, I.; Carrapatoso, I.; Oliveira, B. Almond Allergens: Molecular Characterization, Detection, and Clinical Relevance. *J. Agric. Food Chem.* **2012**, *60*, 1337–1349. [CrossRef]
10. Mandalari, G.; Mackie, A.R. Almond Allergy: An Overview on Prevalence, Thresholds, Regulations and Allergen Detection. *Nutrients* **2018**, *10*, 1706. [CrossRef]
11. Boyce, J.A.; Assa'Ad, A.; Burks, A.W.; Jones, S.M.; Sampson, H.A.; Wood, R.A.; Plaut, M.; Cooper, S.F.; Fenton, M.J.; Arshad, S.H.; et al. Guidelines for the Diagnosis and Management of Food Allergy in the United States: Summary of the NIAID-Sponsored Expert Panel Report. *J. Am. Acad. Dermatol.* **2011**, *64*, 175–192. [CrossRef]
12. Green, P.H.; Lebwohl, B.; Greywoode, R. Celiac disease. *J. Allergy Clin. Immunol.* **2015**, *135*, 1099–1106. [CrossRef]
13. Prescott, S.; Allen, K. Food allergy: Riding the second wave of the allergy epidemic. *Pediatr. Allergy Immunol.* **2011**, *22*, 155–160. [CrossRef]

14. Panjari, M.; Koplin, J.; Dharmage, S.; Peters, R.; Gurrin, L.; Sawyer, S.; McWilliam, V.; Eckert, J.; Vicendese, D.; Erbas, B. Nut allergy prevalence and differences between Asian-born children and Australian-born children of A sian descent: A state-wide survey of children at primary school entry in Victoria, Australia. *Clin. Exp. Allergy* **2016**, *46*, 602–609. [CrossRef]
15. Du Toit, G.; Tsakok, T.; Lack, S.; Lack, G. Prevention of food allergy. *J. Allergy Clin. Immunol.* **2016**, *137*, 998–1010. [CrossRef]
16. Sicherer, S.H.; Allen, K.; Lack, G.; Taylor, S.L.; Donovan, S.M.; Oria, M. Critical Issues in Food Allergy: A National Academies Consensus Report. *Pediatrics* **2017**, *140*, e20170194. [CrossRef]
17. Alasalvar, C.; Shahidi, F. *Tree Nuts: Composition, Phytochemicals, and Health Effects*; Informa UK Limited: London, UK, 2008.
18. Panel, N.-S.E. Guidelines for the diagnosis and management of food allergy in the United States: Report of the NIAID-sponsored expert panel. *J. Allergy Clin. Immunol.* **2010**, *126*, S1–S58.
19. Sampson, H.A.; Mendelson, L.; Rosen, J.P. Fatal and Near-Fatal Anaphylactic Reactions to Food in Children and Adolescents. *N. Engl. J. Med.* **1992**, *327*, 380–384. [CrossRef] [PubMed]
20. Yunginger, J.W.; Sweeney, K.G.; Sturner, W.Q.; Giannandrea, L.A.; Teigland, J.D.; Bray, M.; Benson, P.A.; York, J.A.; Biedrzycki, L.; Squillace, D.L.; et al. Fatal Food-Induced Anaphylaxis. *JAMA* **1988**, *260*, 1450–1452. [CrossRef] [PubMed]
21. Luyt, D.K.; Vaughan, D.; Oyewole, E.; Stiefel, G. Ethnic differences in prevalence of cashew nut, pistachio nut and almond allergy. *Pediatr. Allergy Immunol.* **2016**, *27*, 651–654. [CrossRef]
22. Renz, H.; Allen, K.J.; Sicherer, S.H.; Sampson, H.A.; Lack, G.; Beyer, K.; Oettgen, H.C. Food allergy. *Nat. Rev. Dis. Primers* **2018**, *4*, 1–20. [CrossRef] [PubMed]
23. Burton, O.T.; Darling, A.R.; Zhou, J.S.; Noval-Rivas, M.; Jones, T.G.; Gurish, M.F.; Chatila, T.A.; Oettgen, H.C. Direct effects of IL-4 on mast cells drive their intestinal expansion and increase susceptibility to anaphylaxis in a murine model of food allergy. *Mucosal Immunol.* **2013**, *6*, 740–750. [CrossRef]
24. Perrier, C.; Corthésy, B. Gut permeability and food allergies. *Clin. Exp. Allergy* **2010**, *41*, 20–28. [CrossRef]
25. Vadas, P.; Gold, M.; Perelman, B.; Liss, G.M.; Lack, G.; Blyth, T.; Simons, F.E.R.; Simons, K.J.; Cass, D.; Yeung, J. Platelet-Activating Factor, PAF Acetylhydrolase, and Severe Anaphylaxis. *N. Engl. J. Med.* **2008**, *358*, 28–35. [CrossRef] [PubMed]
26. Williams, K.W.; Sharma, H.P. Anaphylaxis and Urticaria. *Immunol. Allergy Clin. N. Am.* **2015**, *35*, 199–219. [CrossRef]
27. Mishra, A.; Schlotman, J.; Wang, M.; Rothenberg, M.E. Critical role for adaptive T cell immunity in experimental eosinophilic esophagitis in mice. *J. Leukoc. Biol.* **2006**, *81*, 916–924. [CrossRef]
28. Clayton, F.; Fang, J.C.; Gleich, G.J.; Lucendo, A.J.; Olalla, J.M.; Vinson, L.A.; Lowichik, A.; Chen, X.; Emerson, L.; Cox, K.; et al. Eosinophilic Esophagitis in Adults Is Associated With IgG4 and Not Mediated by IgE. *Gastroenterology* **2014**, *147*, 602–609. [CrossRef]
29. Gendel, S.M. Comparison of international food allergen labeling regulations. *Regul. Toxicol. Pharmacol.* **2012**, *63*, 279–285. [CrossRef] [PubMed]
30. Derr, L.E. When food is poison: The history, consequences, and limitations of the Food Allergen Labeling and Consumer Protection Act of 2004. *Food Drug Law J.* **2006**, *61*, 65–165.
31. Rehm, C.D.; Drewnowski, A. Replacing American snacks with tree nuts increases consumption of key nutrients among US children and adults: Results of an NHANES modeling study. *Nutr. J.* **2017**, *16*, 17. [CrossRef]
32. Bottone, A.; Montoro, P.; Masullo, M.; Pizza, C.; Piacente, S. Metabolomics and antioxidant activity of the leaves of Prunus dulcis Mill. (Italian cvs. Toritto and Avola). *J. Pharm. Biomed. Anal.* **2018**, *158*, 54–65. [CrossRef] [PubMed]
33. Nanos, G.D.; Kazantzis, I.; Kefalas, P.; Petrakis, C.; Stavroulakis, G.G. Irrigation and harvest time affect almond kernel quality and composition. *Sci. Hortic.* **2002**, *96*, 249–256. [CrossRef]
34. Piscopo, A.; Romeo, F.; Petrovicova, B.; Poiana, M. Effect of the harvest time on kernel quality of several almond varieties (Prunus dulcis (Mill.) D.A. Webb). *Sci. Hortic.* **2010**, *125*, 41–46. [CrossRef]
35. de Giorgio, D.; Leo, L.; Zacheo, G.; Lamascese, N. Evaluation of 52 almond (Prunus amygdalusBatsch) cultivars from the Apulia region in Southern Italy. *J. Hortic. Sci. Biotechnol.* **2007**, *82*, 541–546. [CrossRef]
36. FAOSTAT. Food and Agriculture Organization of the United Nations. Available online: http://www.fao.org/faostat/en/#home (accessed on 10 October 2020).
37. Griel, A.E.; Kris-Etherton, P.M. Tree nuts and the lipid profile: A review of clinical studies. *Br. J. Nutr.* **2006**, *96*, S68–S78. [CrossRef]
38. Jenkins, D.J.A.; Hu, F.B.; Tapsell, L.C.; Josse, A.; Kendall, C.W.C. Possible Benefit of Nuts in Type 2 Diabetes. *J. Nutr.* **2008**, *138*, 1752S–1756S. [CrossRef]
39. Mandalari, G.; Tomaino, A.; Arcoraci, T.; Martorana, M.; Turco, V.L.; Cacciola, F.; Rich, G.; Bisignano, G.; Saija, A.; Dugo, P.; et al. Characterization of polyphenols, lipids and dietary fibre from almond skins (*Amygdalus communis* L.). *J. Food Compos. Anal.* **2010**, *23*, 166–174. [CrossRef]
40. Richardson, D.P.; Astrup, A.; Cocaul, A.; Ellis, P. The nutritional and health benefits of almonds: A healthy food choice. *Food Sci. Technol. Bull. Funct. Foods* **2009**, *6*, 41–50. [CrossRef]
41. E Berryman, C.; Preston, A.G.; Karmally, W.; Deckelbaum, R.J.; Kris-Etherton, P. Effects of almond consumption on the reduction of LDL-cholesterol: A discussion of potential mechanisms and future research directions. *Nutr. Rev.* **2011**, *69*, 171–185. [CrossRef] [PubMed]
42. Kozłowska, A.; Szostak-Wegierek, D. Flavonoids–food sources and health benefits. *Roczniki Państwowego Zakładu Higieny* **2014**, *65*, 65.

43. Mandalari, G.; Genovese, T.; Bisignano, C.; Mazzon, E.; Wickham, M.; Di Paola, R.; Bisignano, G.; Cuzzocrea, S. Neuroprotective effects of almond skins in experimental spinal cord injury. *Clin. Nutr.* **2011**, *30*, 221–233. [CrossRef]
44. Seo, K.; Lee, D.; Kang, H.; Kim, H.; Kim, Y.; Baek, N.; Lee, D. Hepatoprotective and neuroprotective tocopherol analogues isolated from the peels of Citrus unshiuMarcovich. *Nat. Prod. Res.* **2014**, *29*, 571–573. [CrossRef] [PubMed]
45. Tian, H.; Zhang, H.; Zhan, P.; Tian, F. Composition and antioxidant and antimicrobial activities of white apricot almond (*Amygdalus communis* L.) oil. *Eur. J. Lipid Sci. Technol.* **2011**, *113*, 1138–1144. [CrossRef]
46. Cabrita, L.; Apostolova, E.; Neves, A.; Marreiros, A.; Leitão, J. Genetic diversity assessment of the almond (Prunus dulcis (Mill.) D.A. Webb) traditional germplasm of Algarve, Portugal, using molecular markers. *Plant Genet. Resour.* **2014**, *12*, S164–S167. [CrossRef]
47. Monastra, F.; Raparelli, E. Inventory of almond research, germplasm and references. In *REUR Technical Series*; FAO: Rome, Italy, 1997.
48. Bargman, T.R.J.; Rupnow, J.O.H.; Taylor, S.T.L. IgE-binding proteins in almonds (*Prunus amygdalus*); identi-fication by immunoblotting with sera from almond-allergic adults. *J. Food Sci.* **1992**, *57*, 717–720. [CrossRef]
49. Bolling, B.W.; Dolnikowski, G.; Blumberg, J.B.; Chen, C.-Y.O. Polyphenol content and antioxidant activity of California almonds depend on cultivar and harvest year. *Food Chem.* **2010**, *122*, 819–825. [CrossRef] [PubMed]
50. Summo, C.; Palasciano, M.; De Angelis, D.; Paradiso, V.M.; Caponio, F.; Pasqualone, A. Evaluation of the chemical and nu-tritional characteristics of almonds (Prunus dulcis (Mill). DA Webb) as influenced by harvest time and cultivar. *J. Sci. Food Agric.* **2018**, *98*, 5647–5655. [CrossRef]
51. Ewan, P.W. Clinical study of peanut and nut allergy in 62 consecutive patients: New features and associations. *BMJ* **1996**, *312*, 1074–1078. [CrossRef] [PubMed]
52. Sicherer, S.H.; Muñoz-Furlong, A.; A Sampson, H. Prevalence of peanut and tree nut allergy in the United States determined by means of a random digit dial telephone survey: A 5-year follow-up study. *J. Allergy Clin. Immunol.* **2003**, *112*, 1203–1207. [CrossRef]
53. Geiselhart, S.; Hoffmann-Sommergruber, K.; Bublin, M. Tree nut allergens. *Mol. Immunol.* **2018**, *100*, 71–81. [CrossRef]
54. Weinberger, T.; Sicherer, S. Current perspectives on tree nut allergy: A review. *J. Asthma Allergy* **2018**, *11*, 41. [CrossRef]
55. Kim, S.R.; Park, H.J.; Park, K.H.; Lee, J.-H.; Park, J.-W. IgE Sensitization Patterns to Commonly Consumed Foods Determined by Skin Prick Test in Korean Adults. *J. Korean Med. Sci.* **2016**, *31*, 1197–1201. [CrossRef]
56. McWilliam, V.; Koplin, J.; Lodge, C.; Tang, M.; Dharmage, S.; Allen, K.J. The Prevalence of Tree Nut Allergy: A Systematic Review. *Curr. Allergy Asthma Rep.* **2015**, *15*, 1–13. [CrossRef] [PubMed]
57. Segura, L.R.; Pérez, E.F.; Nowak-Wegrzyn, A.; Siepmann, T.; Larenas-Linnemann, D. Food allergen sensitization pat-terns in a large allergic population in Mexico. *Allergol. Immunopathol.* **2020**, *48*, 553–559. [CrossRef] [PubMed]
58. Rentzos, G.; Johanson, L.; Goksör, E.; Telemo, E.; Lundbäck, B.; Ekerljung, L. Prevalence of food hypersensitivity in relation to IgE sensitisation to common food allergens among the general adult population in West Sweden. *Clin. Transl. Allergy* **2019**, *9*, 22. [CrossRef] [PubMed]
59. Stiefel, G.; Anagnostou, K.; Boyle, R.; Brathwaite, N.; Ewan, P.; Fox, A.T.; Huber, P.; Luyt, D.; Till, S.J.; Venter, C.; et al. BSACI guideline for the diagnosis and management of peanut and tree nut allergy. *Clin. Exp. Allergy* **2017**, *47*, 719–739. [CrossRef] [PubMed]
60. Cheng, C.-W.; Lin, Y.-C.; Nong, B.-R.; Liu, P.-Y.; Huang, Y.-F.; Lu, L.-Y.; Lee, H.-S. Nut sensitization profile in Southern Taiwan. *Immunol. Infect.* **2020**, *53*, 791–796. [CrossRef]
61. Ben Kayale, L.; Ling, J.; Henderson, E.; Carter, N. The influence of cultural attitudes to nut exposure on reported nut allergy: A pilot cross sectional study. *PLoS ONE* **2020**, *15*, e0234846. [CrossRef]
62. Byrne, A.M.; Malka-Rais, J.; Burks, A.W.; Fleischer, D.M. How do we know when peanut and tree nut allergy have resolved, and how do we keep it resolved? *Clin. Exp. Allergy* **2010**, *40*, 1303–1311. [CrossRef]
63. Albin, S.; Nowak-Węgrzyn, A.J.I.; Clinics, A. Potential treatments for food allergy. *Immunol. Allergy Clin.* **2015**, *35*, 77–100. [CrossRef]
64. Kulis, M.; Vickery, B.; Burks, A.W. Pioneering immunotherapy for food allergy: Clinical outcomes and modulation of the immune response. *Immunol. Res.* **2011**, *49*, 216–226. [CrossRef]
65. Egger, M.; Hauser, M.; Mari, A.; Ferreira, F.; Gadermaier, G. The Role of Lipid Transfer Proteins in Allergic Diseases. *Curr. Allergy Asthma Rep.* **2010**, *10*, 326–335. [CrossRef]
66. Noble, K.A.; Liu, C.; Sathe, S.K.; Roux, K.H. A Cherry Seed-Derived Spice, Mahleb, is Recognized by Anti-Almond Antibodies Including Almond-Allergic Patient IgE. *J. Food Sci.* **2017**, *82*, 1786–1791. [CrossRef]
67. Lee, S.-H.; Benmoussa, M.; Sathe, S.K.; Roux, K.H.; Teuber, S.S.; Hamaker, B.R. A 50 kDa Maize γ-Zein Has Marked Cross-Reactivity with the Almond Major Protein. *J. Agric. Food Chem.* **2005**, *53*, 7965–7970. [CrossRef]
68. Tawde, P.; Venkatesh, Y.P.; Wang, F.; Teuber, S.S.; Sathe, S.K.; Roux, K.H. Cloning and characterization of profilin (Pru du 4), a cross-reactive almond (Prunus dulcis) allergen. *J. Allergy Clin. Immunol.* **2006**, *118*, 915–922. [CrossRef]
69. de Leon, M.; Drew, A.; Glaspole, I.; Suphioglu, C.; O'Hehir, R.; Rolland, J. IgE cross-reactivity between the major peanut allergen Ara h 2 and tree nut allergens. *Mol. Immunol.* **2007**, *44*, 463–471. [CrossRef] [PubMed]
70. Maleki, S.J.; Teuber, S.S.; Cheng, H.; Chen, D.; Comstock, S.S.; Ruan, S.; Schein, C.H. Computationally predicted IgE epitopes of walnut allergens contribute to cross-reactivity with peanuts. *Allergy* **2011**, *66*, 1522–1529. [CrossRef]

71. Wallowitz, M.; Teuber, S.; Beyer, K.; Sampson, H.; Roux, K.; Sathe, S.; Wang, F.; Robotham, J. Cross-reactivity of walnut, cashew, and hazelnut legumin proteins in tree nut allergic patients. *J. Allergy Clin. Immunol.* **2004**, *113*, S156. [CrossRef]
72. Flinterman, A.E.; Hoekstra, M.O.; Meijer, Y.; Van Ree, R.; Akkerdaas, J.H.; Bruijnzeel-Koomen, C.A.; Knulst, A.C.; Pasmans, S.G. Clinical reactivity to hazelnut in children: Association with sensitization to birch pollen or nuts? *J. Allergy Clin. Immunol.* **2006**, *118*, 1186–1189. [CrossRef]
73. Vieths, S.; Scheurer, S.; BALLMER-WEBER, B. Current understanding of cross-reactivity of food allergens and pollen. *Ann. N. Y. Acad. Sci.* **2002**, *964*, 47–68. [CrossRef]
74. Hasegawa, M.; Inomata, N.; Yamazaki, H.; Morita, A.; Kirino, M.; Ikezawa, Z. Clinical Features of Four Cases with Cashew Nut Allergy and Cross-Reactivity between Cashew Nut and Pistachio. *Allergol. Int.* **2009**, *58*, 209–215. [CrossRef]
75. Noorbakhsh, R.; Mortazavi, S.A.; Sankian, M.; Shahidi, F.; Tehrani, M.; Azad, F.J.; Behmanesh, F.; Varasteh, A. Pistachio Allergy-Prevalence and In vitro Cross-Reactivity with Other Nuts. *Allergol. Int.* **2011**, *60*, 425–432. [CrossRef]
76. Asero, R. 7 Lipid Transfer Protein Cross-reactivity Assessed In Vivo and In Vitro in the Office: Pros and Cons. *J. Investig. Allergol. Clin. Immunol.* **2011**, *21*, 129. [PubMed]
77. KewalRamani, A.; Maleki, S.; Cheng, H.; Teuber, S. Cross-Reactivity Among Almond, Peanut and Other Tree Nuts in Almond Allergic Patients. *J. Allergy Clin. Immunol.* **2006**, *117*, S32. [CrossRef]
78. Albillos, S.M.; Jin, T.; Howard, A.; Zhang, Y.; Kothary, M.H.; Fu, T.-J. Purification, Crystallization and Preliminary X-ray Characterization of Prunin-1, a Major Component of the Almond (Prunus dulcis) Allergen Amandin. *J. Agric. Food Chem.* **2008**, *56*, 5352–5358. [CrossRef]
79. Albillos, S.M.; Menhart, N.; Fu, T.-J. Structural Stability of Amandin, a Major Allergen from Almond (Prunus dulcis), and Its Acidic and Basic Polypeptides. *J. Agric. Food Chem.* **2009**, *57*, 4698–4705. [CrossRef]
80. Roux, K.H.; Teuber, S.S.; Robotham, J.M.; Sathe, S.K. Detection and Stability of the Major Almond Allergen in Foods. *J. Agric. Food Chem.* **2001**, *49*, 2131–2136. [CrossRef] [PubMed]
81. Sathe, S.K.; Wolf, W.J.; Roux, K.H.; Teuber, S.S.; Venkatachalam, M.; Sze-Tao, K.W.C. Biochemical Characterization of Amandin, the Major Storage Protein in Almond (*Prunus dulcis* L.). *J. Agric. Food Chem.* **2002**, *50*, 4333–4341. [CrossRef]
82. Willison, L.N.; Tripathi, P.; Sharma, G.; Teuber, S.S.; Sathe, S.K.; Roux, K.H. Cloning, Expression and Patient IgE Reactivity of Recombinant Pru du 6, an 11S Globulin from Almond. *Int. Arch. Allergy Immunol.* **2011**, *156*, 267–281. [CrossRef]
83. Venkatachalam, M.; Teuber, S.S.; Roux, K.H.; Sathe, S.K. Effects of Roasting, Blanching, Autoclaving, and Microwave Heating on Antigenicity of Almond (*Prunus dulcis* L.) Proteins. *J. Agric. Food Chem.* **2002**, *50*, 3544–3548. [CrossRef]
84. Mandalari, G.; Rigby, N.M.; Bisignano, C.; Curto, R.B.L.; Mulholland, F.; Su, M.; Venkatachalam, M.; Robotham, J.M.; Willison, L.N.; Lapsley, K.; et al. Effect of food matrix and processing on release of almond protein during simulated digestion. *LWT* **2014**, *59*, 439–447. [CrossRef]
85. Holden, L.; Sletten, G.B.; Lindvik, H.; Fæste, C.K.; Dooper, M.M. Characterization of IgE binding to lupin, peanut and al-mond with sera from lupin-allergic patients. *Int. Arch. Allergy Immunol.* **2008**, *146*, 267–276. [CrossRef]
86. Abou Alhasani, M.; Roux, K.H. cDNA Cloning, expression and characterization of an allergenic 60s ribosomal protein of almond (*Prunus dulcis*). *Iran. J. Allergy Asthma Immunol.* **2009**, *8*, 77–84.
87. Yao, L.; Li, H.; Yang, J.; Li, C.; Shen, Y. Purification and characterization of a hydroxynitrile lyase from Amygdalus pedunculata Pall. *Int. J. Biol. Macromol.* **2018**, *118*, 189–194. [CrossRef]
88. Mills, E.C.; Sancho, A.; Moreno, J.; Kostyra, H. The effects of food processing on allergens. In *Managing Allergens in Food*; Elsevier: Amsterdam, The Netherlands, 2007; pp. 117–133.
89. De Angelis, E.; Bavaro, S.L.; Forte, G.; Pilolli, R.; Monaci, L. Heat and Pressure Treatments on Almond Protein Stability and Change in Immunoreactivity after Simulated Human Digestion. *Nutrients* **2018**, *10*, 1679. [CrossRef] [PubMed]
90. Che, H.; Zhang, Y.; Jiang, S.; Jin, T.; Lyu, S.-C.; Nadeau, K.C.; McHugh, T. Almond (*Prunus dulcis*) Allergen Pru du 8, the First Member of a New Family of Food Allergens. *J. Agric. Food Chem.* **2019**, *67*, 8626–8631. [CrossRef] [PubMed]
91. Marcus, J.P.; Green, J.L.; Goulter, K.C.; Manners, J.M. A family of antimicrobial peptides is produced by processing of a 7S globulin protein in Macadamia integrifolia kernels. *Plant J.* **1999**, *19*, 699–710. [CrossRef] [PubMed]
92. Poltronieri, P.; Cappello, M.; Dohmae, N.; Conti, A.; Fortunato, D.; Pastorello, E.; Ortolani, C.; Zacheo, G. Identification and characterisation of the IgE-binding proteins 2S albumin and conglutin γ in almond (*Prunus dulcis*) seeds. *Int. Arch. Allergy Immunol.* **2002**, *128*, 97–104. [CrossRef] [PubMed]
93. Garino, C.; De Paolis, A.; Coïsson, J.D.; Arlorio, M. Pru du 2S albumin or Pru du vicilin? *Comput. Biol. Chem.* **2015**, *56*, 30–32. [CrossRef]
94. Kolivas, S.; Gayler, K.R. Structure of the cDNA coding for conglutin γ, a sulphur-rich protein from Lupinus angusti-folius. *Plant Mol. Biol.* **1993**, *21*, 397–401. [CrossRef]
95. Burks, A.W.; Williams, L.W.; Helm, R.M.; Connaughton, C.; Cockrell, G.; O'Brien, T. Identification of a major peanut allergen, Ara h I, in patients with atopic dermatitis and positive peanut challenges. *J. Allergy Clin. Immunol.* **1991**, *88*, 172–179. [CrossRef]
96. Burks, A.W., Jr.; Brooks, J.R.; Sampson, H.A. Allergenicity of major component proteins of soybean determined by enzyme-linked immunosorbent assay (ELISA) and immunoblotting in children with atopic dermatitis and positive soy challenges. *J. Allergy Clin. Immunol.* **1988**, *81*, 1135–1142. [CrossRef]

97. Wang, F.; Robotham, J.M.; Teuber, S.S.; Tawde, P.; Sathe, S.K.; Roux, K.H. Ana o 1, a cashew (Ana-cardium occidental) allergen of the vicilin seed storage protein family. *J. Allergy Clin. Immunol.* **2002**, *110*, 160–166. [CrossRef]
98. Scarafoni, A.; Consonni, A.; Pessina, S.; Balzaretti, S.; Capraro, J.; Galanti, E.; Duranti, M. Structural basis of the lack of endoglucanase inhibitory activity of Lupinus albus γ-conglutin. *Plant Physiol. Biochem.* **2016**, *99*, 79–85. [CrossRef]
99. Fernandes, H.; Michalska, K.; Sikorski, M.; Jaskolski, M. Structural and functional aspects of PR-10 proteins. *FEBS J.* **2013**, *280*, 1169–1199. [CrossRef]
100. Mittag, D.; Akkerdaas, J.; Ballmer-Weber, B.K.; Vogel, L.; Wensing, M.; Becker, W.-M.; Koppelman, S.J.; Knulst, A.C.; Helbling, A.; Hefle, S.L.; et al. Ara h 8, a Bet v 1–homologous allergen from peanut, is a major allergen in patients with combined birch pollen and peanut allergy. *J. Allergy Clin. Immunol.* **2004**, *114*, 1410–1417. [CrossRef]
101. Scala, E.; Alessandri, C.; Palazzo, P.; Pomponi, D.; Liso, M.; Bernardi, M.L.; Ferrara, R.; Zennaro, D.; Santoro, M.; Rasi, C. IgE recognition patterns of profilin, PR-10, and tropomyosin panallergens tested in 3,113 allergic patients by allergen microar-ray-based technology. *PLoS ONE* **2011**, *6*, e24912. [CrossRef]
102. Chen, L.; Zhang, S.; Illa, E.; Song, L.; Wu, S.; Howad, W.; Arús, P.; Van de Weg, E.; Chen, K.; Gao, Z. Genomic characterization of putative allergen genes in peach/almond and their synteny with apple. *BMC Genom.* **2008**, *9*, 543. [CrossRef]
103. Palacin, A.; Tordesillas, L.; Gamboa, P.; Sanchez-Monge, R.; Cuesta-Herranz, J.; Sanz, M.; Barber, D.; Salcedo, G.; Díaz-Perales, A. Characterization of peach thaumatin-like proteins and their identification as major peach allergens. *Clin. Exp. Allergy* **2010**, *40*, 1422–1430. [CrossRef]
104. Roux, K.H.; Teuber, S.S.; Sathe, S.K. Tree Nut Allergens. *Int. Arch. Allergy Immunol.* **2003**, *131*, 234–244. [CrossRef] [PubMed]
105. Shewry, P.R.; Napier, J.A.; Tatham, A.S. Seed storage proteins: Structures and biosynthesis. *Plant Cell* **1995**, *7*, 945. [PubMed]
106. Shewry, P.R.; Halford, N.G. Cereal seed storage proteins: Structures, properties and role in grain utilization. *J. Exp. Bot.* **2002**, *53*, 947–958. [CrossRef] [PubMed]
107. Moreno, F.J.; Clemente, A. 2S Albumin Storage Proteins: What Makes them Food Allergens? *Open Biochem. J.* **2008**, *2*, 16–28. [CrossRef] [PubMed]
108. Clemente, A.; Chambers, S.J.; Lodi, F.; Nicoletti, C.; Brett, G.M. Use of the indirect competitive ELISA for the detection of Brazil nut in food products. *Food Control.* **2004**, *15*, 65–69. [CrossRef]
109. Koppelman, S.J.; Wensing, M.; Ertmann, M.; Knulst, A.C.; Knol, E. Relevance of Ara h1, Ara h2 and Ara h3 in peanut-allergic patients, as determined by immunoglobulin E Western blotting, basophil-histamine release and intracutaneous testing: Ara h2 is the most important peanut allergen. *Clin. Exp. Allergy* **2004**, *34*, 583–590. [CrossRef]
110. Nicolaou, N.; Poorafshar, M.; Murray, C.; Simpson, A.; Winell, H.; Kerry, G.; Härlin, A.; Woodcock, A.; Ahlstedt, S.; Custovic, A. Allergy or tolerance in children sensitized to peanut: Prevalence and differentiation using component-resolved diagnostics. *J. Allergy Clin. Immunol.* **2010**, *125*, 191–197.e113. [CrossRef]
111. Palmer, G.W.; Dibbern, D.A., Jr.; Burks, A.W.; Bannon, G.A.; Bock, S.A.; Porterfield, H.S.; McDermott, R.A.; Dreskin, S.C. Comparative potency of Ara h 1 and Ara h 2 in immunochemical and functional assays of allergenicity. *Clin. Immunol.* **2005**, *115*, 302–312. [CrossRef]
112. Buhler, S.; Tedeschi, T.; Faccini, A.; Garino, C.; Arlorio, M.; Dossena, A.; Sforza, S. Isolation and full characterisation of a potentially allergenic lipid transfer protein (LTP) in almond. *Food Addit. Contam. Part A* **2015**, *32*, 648–656.
113. Liu, J.-J.; Sturrock, R.; Ekramoddoullah, A.K.M. The superfamily of thaumatin-like proteins: Its origin, evolution, and expression towards biological function. *Plant Cell Rep.* **2010**, *29*, 419–436. [CrossRef]
114. Jackson, L.S.; Al-Taher, F.M.; Moorman, M.; DeVRIES, J.W.; Tippett, R.; Swanson, K.M.J.; Fu, T.-J.; Salter, R.; Dunaif, G.; Estes, S.; et al. Cleaning and Other Control and Validation Strategies To Prevent Allergen Cross-Contact in Food-Processing Operations. *J. Food Prot.* **2008**, *71*, 445–458. [CrossRef]
115. Wang, X.; Young, O.; Karl, D. Evaluation of Cleaning Procedures for Allergen Control in a Food Industry Environment. *J. Food Sci.* **2010**, *75*, T149–T155. [CrossRef]
116. Pafundo, S.; Gulli, M.; Marmiroli, N. SYBR®GreenER™ Real-Time PCR to detect almond in traces in processed food. *Food Chem.* **2009**, *116*, 811–815. [CrossRef]
117. Van Hengel, A.J. Food allergen detection methods and the challenge to protect food-allergic consumers. *Anal. Bioanal. Chem.* **2007**, *389*, 111–118. [CrossRef]
118. Hoffmann-Sommergruber, K. Proteomics and its impact on food allergy diagnosis. *EuPA Open Proteom.* **2016**, *12*, 10–12. [CrossRef]
119. Piras, C.; Roncada, P.; Rodrigues, P.; Bonizzi, L.; Soggiu, A. Proteomics in food: Quality, safety, microbes, and allergens. *Proteomics* **2015**, *16*, 799–815. [CrossRef] [PubMed]
120. Johnson, P.E.; Baumgartner, S.; Aldick, T.; Bessant, C.; Giosafatto, V.; Heick, J.; Mamone, G.; O'connor, G.; Poms, R.; Pop-ping, B. Current perspectives and recommendations for the development of mass spectrometry methods for the determination of allergens in foods. *J. AOAC Int.* **2011**, *94*, 1026–1033. [CrossRef] [PubMed]
121. Monaci, L.; De Angelis, E.; Montemurro, N.; Pilolli, R. Comprehensive overview and recent advances in proteomics MS based methods for food allergens analysis. *TrAC Trends Anal. Chem.* **2018**, *106*, 21–36. [CrossRef]
122. Bignardi, C.; Elviri, L.; Penna, A.; Careri, M.; Mangia, A. Particle-packed column versus silica-based monolithic column for liquid chromatography–electrospray-linear ion trap-tandem mass spectrometry multiallergen trace analysis in foods. *J. Chromatogr. A* **2010**, *1217*, 7579–7585. [CrossRef] [PubMed]

123. Lupinek, C.; Wollmann, E.; Baar, A.; Banerjee, S.; Breiteneder, H.; Broecker, B.M.; Bublin, M.; Curin, M.; Flicker, S.; Garmatiuk, T.; et al. Advances in allergen-microarray technology for diagnosis and monitoring of allergy: The MeDALL allergen-chip. *Methods* **2014**, *66*, 106–119. [CrossRef]
124. Santos, A.F.; Lack, G. Basophil activation test: Food challenge in a test tube or specialist research tool? *Clin. Transl. Allergy* **2016**, *6*, 1–9. [CrossRef]

125. Ebo, D.; Bridts, C.H.; Hagendorens, M.; Aerts, N.E.; De Clerck, L.S.; Stevens, W.J. Basophil activation test by flow cytometry: Present and future applications in allergology. *Cytom. Part B Clin. Cytom.* **2008**, *74*, 201–210. [CrossRef] [PubMed]
126. Duan, L.; Celik, A.; Hoang, J.A.; Schmidthaler, K.; So, D.; Yin, X.; Ditlof, C.M.; Ponce, M.; Upton, J.E.; Lee, J.; et al. Basophil activation test shows high accuracy in the diagnosis of peanut and tree nut allergy: The Markers of Nut Allergy Study. *Allergy* **2021**, *76*, 1800–1812. [CrossRef] [PubMed]

MDPI
St. Alban-Anlage 66
4052 Basel
Switzerland
Tel. +41 61 683 77 34
Fax +41 61 302 89 18
www.mdpi.com

Nutrients Editorial Office
E-mail: nutrients@mdpi.com
www.mdpi.com/journal/nutrients

CPSIA information can be obtained
at www.ICGtesting.com
Printed in the USA
BVHW020844071122
651333BV00007B/212